The Millstone Industry

ALSO BY
CHARLES D. HOCKENSMITH

The Millstone Quarries of Powell County, Kentucky (McFarland, 2009)

The Millstone Industry

A Summary of Research on Quarries and Producers in the United States, Europe and Elsewhere

CHARLES D. HOCKENSMITH

with a foreword by Alain Belmont

McFarland & Company, Inc., Publishers
Jefferson, North Carolina, and London

LIBRARY OF CONGRESS CATALOGUING-IN-PUBLICATION DATA

Hockensmith, Charles D.
The millstone industry : a summary of research on quarries and producers in the United States, Europe and elsewhere / Charles D. Hockensmith ; foreword by Alain Belmont.
p. cm.
Includes bibliographical references and index.

ISBN 978-0-7864-3860-0
softcover : 50# alkaline paper ♾

1. Quarries and quarrying. 2. Millstones. 3. Quarries and quarrying — United States. 4. Millstones — United States. I. Title.
TN277.H56 2009
621.2'10973 — dc22 2008048122

British Library cataloguing data are available

Manufactured in the United States of America

McFarland & Company, Inc., Publishers
Box 611, Jefferson, North Carolina 28640
www.mcfarlandpub.com

Table of Contents

Acknowledgments vii
Foreword by Alain Belmont 1
Preface 3
Introduction 5

1. Early American Millstone Documents, 1628–1829 9

2. Millstone Quarrying in the United States 19
Conglomerate Quarries • Granite Quarries • Flint and Buhrstone Quarries • Quartzite Quarries • Gneiss Quarries • Sandstone Quarries • Dolomite Quarries • Sienite Quarries • Unspecified Materials • Boulders Used for Millstones

3. Millstone Makers and Urban Factories 88

4. The Rise and Fall of the American Millstone Industry: Producers, Annual Values, and Decline 100
Millstone Producers • Millstone Values • Millstone Industry Decline

5. Foreign Millstones Imported to America 119
England • France • Germany • All Millstones Imported into the United States

6. The Millstone Quarrying Industry Outside the United States 124
Albania • Austria • Canada • Czech Republic • France • Germany • Great Britain • Greece • Hungary • Italy • Luxembourg • Norway • Slovenia • Spain • Sweden • Switzerland • Turkey • Studies on Millstones and Querns of Ancient Rome • Miscellaneous Millstone Quarries and Studies

7. Artificial Millstones 199

8. Tools Used in Making and Sharpening Millstones 202
Tools Used by Millstone Cutters • Tools Used to Sharpen Millstones

9. Working Conditions and Hazards in the Millstone Industry 207
Tool Related Injuries • Silicosis • Other Injuries

Conclusion 212
Glossary 215
Bibliography 219
Index 255

Acknowledgments

Since I began my research on the millstone industry in 1987, many people have contributed in a variety of ways. Without their assistance, this study would have been far less comprehensive. These numerous scholarly debts are acknowledged below. This section also provides insight into how this research evolved through many inquires, conferences, and communications with scholars in other countries. Because of the length of time that has passed since the beginning of this research, these acknowledgments are organized roughly chronologically.

First and foremost, the author wishes to express his appreciation to the Kentucky Heritage Council whose financial support made this study possible. The author was given staff time to conduct archival research and to write, in connection with his research on the Kentucky millstone quarries, in his capacity as a staff archaeologist. Since its creation in 1966, the Kentucky Heritage Council has taken the lead in preserving and protecting Kentucky's cultural resources. To accomplish its legislative mandate, the Heritage Council maintains three program areas: Site Development, Site Identification, and Site Protection and Archaeology (under which this work was accomplished). The Site Protection and Archaeology Program staff works with federal and state agencies, local governments, and individuals to ensure that their projects are in compliance with Section 106 of the National Historic Preservation Act of 1966. This act ensures that potential impacts to significant cultural resources are adequately addressed prior to the implementation of federally funded or licensed projects. They also are responsible for administering the Heritage Council's archaeological programs, organizing archaeological conferences, editing and publishing volumes of selected conference papers, and disseminating educational materials. On occasion, the Site Protection and Archaeology Program staff undertakes field and research projects, such as emergency data recovery at threatened sites. The agency can be contacted at Kentucky Heritage Council, 300 Washington Street, Frankfort, Kentucky 40601, or by phone at (502) 564–7005.

Past and present staff members of the Kentucky Heritage Council provided assistance with this study. David L. Morgan, former Director of the Kentucky Heritage Council, and Thomas N. Sanders, former Site Protection Program Manager, provided the encouragement and the resources necessary to undertake and complete this project. In 2005, David Pollack encouraged the completion of the manuscript. Donna M. Neary, former Executive Director of the Kentucky Heritage Council, graciously co-signed the book contracts and encouraged the author. I am very grateful to David, Tom, David, Donna, and the Kentucky Heritage Council for their faithful support of this project through the years.

Several colleagues in Kentucky provided assistance to the author. Ms. Pam Lyons, Kentucky Department of Libraries and Archives, was extremely helpful in obtaining mill and millstone publications through interlibrary loan. Ms. Gayle Alvis, Kentucky Department of Libraries and Archives, made the author aware of a great website for searching libraries and assisted in obtaining information. Mr. Donald B. Ball, retired archaeologist with the Louisville District, Corps of Engineers, shared links to new websites that yielded a wealth of information on American millstone quarries. Don also brought several publications on querns to my attention.

Many people provided information on millstones and millstone quarrying in the early stages of this research. These included Dr. Fred E. Coy, Jr. (Louisville, Kentucky), the late Tom C. Fuller (Louisville, Kentucky), James L. Murphy (Ohio State University, Columbus), Michael J. LaForest (former *Old Mill News* editor, Knoxville, Tennessee), David Rotenizer (Archaeological Society of Virginia, Blacksburg), Dale Collins (Pembroke, Virginia), Claude V. Jackson (Tidewater Atlantic Research, Washington, North Carolina), Steve Rogers (Tennessee Historical Commission, Nashville), Tom Kelleher (Old Sturbridge Village, Sturbridge, Massachusetts), John Rice Irwin (Museum of Appalachia, Norris, Tennessee), Richard Boisvert (New Hampshire Division of Historical Resources, Concord), Calvert McIlhany (Bristol, Virginia), Robert G. Schmidt (Arlington, Virginia), and Phil Gettel (Floyd, Virginia).

In an effort to locate missed or unpublished information on millstone quarries in other states, I sent letters to all State Historic Preservation Officers (SHPOs) in the United States during January 1991. Twenty-eight states responded to the information request. I also sent letters to other persons mentioned by the SHPOs who might have information. The following state offices said that they were unaware of millstone quarries within their state boundaries: Alaska, Arizona, California, Colorado, Florida, Indiana, Iowa, Louisiana, Maine, Michigan, Minnesota, Montana, Nebraska, Nevada, New Mexico, North Carolina, Texas, Washington, and Wyoming. Information on millstone quarries or related information was received from the following individuals during 1991 and 1992: Dave Poirier (Connecticut Historical Commission, Hartford), Lowell J. Soike (historian, State Historical Society of Iowa, Des Moines), John W. McGrain (Landmarks Preservation Commission, Baltimore, Maryland), Marcia M. Miller (Maryland Historical Trust, Annapolis), Edward L. Bell (Massachusetts Historical Commission, Boston), Jack D. Elliott, Jr. (historical archaeologist, Mississippi Department of Archives and History, Jackson), Julia S. Stokes (New York State Office of Parks, Recreation and Historic Preservation, Albany), Mary Mason Shell (Oregon State Historic Preservation Office, Salem), Glenn Harrison (Linn County Historical Society, Albany, Oregon), Jerry A. Clouse (Pennsylvania Historical and Museum Commission, Harrisburg), Bruce B. Brown (Greencastle, Pennsylvania), James K. Haug (State Archaeological Research Center, South Dakota State Historical Society, Rapid City), Robert G. Schmidt (Arlington, Virginia, for South Carolina), and Robert M. Frame, III (Society for Industrial Archaeology, Saint Paul, Minnesota). A special thanks is due to John McGrain who shared copies of his fact cards on millstones that included many early sources.

The first articles that came to my attention on millstone quarrying in Europe were those provided by scholars residing in Great Britain. Dr. Kate Clark (Ironbridge Gorge Museums Trust, Telford, England) provided copies of several articles and the names and addresses of the major European millstone researchers. The late Dr. D. Gordon Tucker (University of Birmingham) provided copies of his many articles on millstone quarries in England and Scotland. He also kindly assembled copies of many other major articles on millstone quarries in Europe and sent them to the author. Without Dr. Tucker's help only a small portion of the European literature would have come to light. Mr. Owen Ward (Bath, England) provided copies of his articles on the French and Welsh millstone quarries. Also, Mr. Ward stayed in touch through the years, subsequently sent copies of his later articles, and kindly made available copies of articles on millstones by other researchers. Mr. J. Kenneth Major (Reading, England) provided copies of his articles and those of other researchers on the German millstone quarries. Dr. David P. S. Peacock (University of Southampton) provided copies of his articles on Roman millstone quarries in Italy and England. Finally, Mrs. Philomena Jackson (Society for Post-Medieval Archaeology, London) provided a copy of an article from *Post-Medieval Archaeology* and published an information request for the author in the society's newsletter.

During May 1990, the author and Fred E. Coy, Jr., had a unique opportunity to collect significant information on millstone quarrying techniques. Over a two-day period, we were able to interview the last two living millstone makers in Virginia (Mr. Robert Houston Surface and Mr. W. C. Saville) and visit the Brush Mountain Millstone Quarry near Blacksburg where these men were once employed. We are grateful to these gentlemen for sharing their tremendous knowledge concerning the manufacture of millstones. Since no detailed published accounts of millstone making in the United States have come to light, the knowledge possessed by Mr. Surface and Mr. Saville was extremely significant in understanding how conglomerate millstones were manufactured. Unfortunately, Mr. Surface passed away on May 3, 1998, before the completion of this book. Mr. Saville lived a few years longer and passed away on April 11, 2003. Although these men are now gone, much of their knowledge of millstone making lives on through their interviews (see Hockensmith and Coy 1999; Hockensmith and Price 1999).

Several other individuals made the trip to the Brush Mountain Millstone Quarry productive. Mr. Lewis Dale Collins of Pembroke, Virginia, handled all the local arrangements, served as guide to the Brush Mountain Quarry, and was a wonderful host. Dr. Fred E. Coy, Jr. (Louisville, Kentucky), videotaped the interview and quarry remains at the Brush Mountain Quarry. Also, Dr. Coy provided the transportation to Virginia and served as a very gracious traveling companion. Mr. Jimmie L. Price (son-in-law of Mr. Huston Surface) provided an excellent videotape of an interview he conducted with Mr. Surface at the quarry as well as copies of his notes and drawings of the Brush Mountain Quarry. A special thanks is due to David Rotenizer (a professional archaeologist from Virginia) who first generated our interest in the Brush Mountain Quarry and who put the author in contact with Lewis Dale Collins. During the summer of 1997, the Oral History and Educational Outreach Division of the Kentucky Historical Society provided a small grant to pay for having our 1990 interview and Mr. Jimmie L. Price's 1985 interview of the millstone makers transcribed. The assistance and advice of the KHS's Division Manager, Ms. Kim Lady Smith, is greatly appreciated. Mrs. Janet Gates of Frankfort did a wonderful job in transcribing these interviews so that the information could be used in this study. The Society for the Preservation of Old Mills published an edited version of these interviews as a book (Hockensmith 1999, editor).

In another attempt to obtain information on millstone quarries in other states, I sent out letters requesting information in 1996. These letters were directed to the geological surveys in states that were known commercial producers of millstones. All of these states provided information that supplemented what we had previously collected. A special thanks is due to Lewis S. Dean (geologist with the Geological Survey of Alabama, Tuscaloosa), William Kelly (State Geologist with the Geological Survey at the New York State Museum, Albany), Robert C. Smith, II (economic geochemist with the Bureau of Topographic and Geologic Survey, Pennsylvania Department of Conservation and Natural Resources, Harrisburg), Charles H. Gardner (Director and State Geologist, State of North Carolina Department of Environment, Health, and Natural Resources, Raleigh), and Palmer C. Sweet (Commonwealth of Virginia, Department of Mines, Minerals, and Energy, Charlottesville). B. H. Rucker (Missouri Department of Natural Resources, Jefferson City) responded to an earlier request and provided information on millstones from Iron County, Missouri. Also, a very special thanks is due to Garland R. Dever (retired geologist with the Kentucky Geological Survey, Lexington) for providing addresses of other geological surveys and assisting in a variety of ways through the years.

Museums and historical societies in New York and Pennsylvania were very helpful in obtaining information. Ms. Patricia Christian with the Ellenville Public Library and Museum in Ellenville, New York, provided information on the millstone industry in Ulster County. The late Ms. Eleanor S. Rosakranase and Amanda C. Jones of the Ulster County Historical Society in Kingston, New York, were helpful. Ms. Rosakranase graciously shared copies of *The Accordian*, which included information on the Lawrence brothers and the Ulster County millstone industry. Ms. Cynthia Marquet with The Historical Society of the Cocalico Valley in Ephrata, Pennsylvania, shared copies of photographs of millstones from the Turkey Hill millstone quarry in Lancaster County. Also, Ms. Ruth Baer Gembe, Alexander Hamilton Memorial Free Library, in Waynesboro, Pennsylvania, provided copies of reference materials on millstones and copies of photographs of a millstone collection housed at the library.

In April of 1998, the author took a week of vacation time and traveled to New York and Pennsylvania. He was accompanied by his good friend and fellow scholar Dr. Fred E. Coy, Jr. They visited some of the famous Ulster County, New York, millstone quarries near the community of Accord. During the trip, they also interviewed Vincent and Wally Lawrence, two brothers over 80 years of age, whose father and uncles were millstone makers. The Lawrence brothers shared their memories of millstone making during their youth. Unfortunately, the Lawrence brothers did not live to see their information published. Wally Lawrence died in May 2000 and Vincent Lawrence died in April 2001. We were also fortunate to meet Lewis Waruch (whose mother's family included millstone makers) who has been interested in the millstone quarries near Accord for many years. Lewis shared his knowledge freely, took us to several millstone quarries, and showed us his tremendous collection of millstone making tools. I wish to express my gratitude to these men for being willing to share their knowledge about the manufacture of millstones. Dr. Coy videotaped the interviews and scenes at the millstone quarries. The Society for the Preservation of Old Mills funded the transcription of the New York interviews. Mrs. Janet Gates of Frankfort did an excellent job in transcribing these interviews. The interviews and a detailed study based on archival sources will soon be published in book form (Hockensmith 2008b, editor).

Also during April of 1998, Hockensmith and Coy visited the Turkey Hill millstone quarry in Lancaster County, Pennsylvania. This trip provided an opportunity to obtain additional information on millstone quarrying techniques and increased our overall understanding of the industry. We were also able to visit the Lancaster Historical Society's library and check some of their resources. Plans are underway to publish the Pennsylvania data in the future.

A special thanks is owed to the Hagley Museum and Library in Wilmington, Delaware. The Hagley Library kindly provided a copy of Hehnly & Wike's 1880 broadside advertisement headlined, "The Newly Discovered Turkey Hill Stones Are the Best Mill Stones in the Country, and in Consequence we have Made a reduced Price List of Cocalico Mill Stones" in Durlach, Pennsylvania. I appreciate their willingness to allow me to use the important information in this one-page advertisement. Also, Mr. Richard Brown of Louisville, Kentucky, found and shared information on the Shacklett family who made millstones at Laurel Hill in

Fayette County, Pennsylvania. This information filled an important void in the records concerning this family.

In May of 2002, the author attended the first international millstone conference at La Ferté-sous-Jouarre, France (see Barboff, Sigaut, Griffin-Kremer, and Kremer 2003). He presented a paper entitled "The Conglomerate Millstone Industry in the Eastern United States" (Hockensmith 2002, 2003b). The paper was presented at the Colloque International entitled "Extraction, Façonnage, Commerce et Utilisation des Meules de Moulin — Une Industrie dans la Longue Durée." The approximate English translation is "Quarrying, Stone-Working, Trade and Use of Millstones, Long-Term History." The conference was organized by ethnologist Mouette Barboff and historian François Siguat. About 140 participants from 16 countries attended the conference. For an overview of the conference, see Hockensmith 2003a. One of the great opportunities of the conference was to interact with scholars interested in millstones from many countries. Dr. Cozette Griffin-Kremer assisted the author in many ways and served as a translator. Several scholars shared copies of their publications during and after the conference. Special thanks are due to German archaeologist Fritz Mangartz, German researcher Fridolin Hörter, French archaeologist Alain Belmont, and Swiss archaeologist Tim Anderson who graciously shared information. The author was also privileged to meet Mr. Owen Ward of Bath, England, a millstone expert, whom he had corresponded with for a number of years. He met French millstone expert Jacques Beauvois and visited his impressive millstone museum. After the conference, Fritz Mangartz and Alain Belmont continued to share information and respond to the author's inquiries.

During the years of 2002 and 2003, the author was privileged to be in contact with two additional millstone researchers. In late 2002 and early 2003, he had the pleasure of corresponding with Dr. Curtis Runnels with the Department of Archaeology at Boston University. Dr. Runnels has done a great deal of research on Greek millstones. He kindly provided copies of most of his millstone articles. A special thank you is extended to Dr. Runnels. Another pleasant surprise was Mr. Holger Buentke's contacting the author during 2003. Mr. Buentke, who lives in Sweden, kindly shared information on the millstone industry in the Scandinavian countries. He also shared copies of some articles and graciously translated selected terms and publication titles. Without Mr. Buentke's assistance, this publication would have contained little information on the millstone industry in Sweden, Norway, Denmark, and Finland.

Additional information became available on the Ohio and Massachusetts millstone quarries during 2004 and 2005. Dr. James L. Murphy, Ohio State University, Columbus, shared information on the Flint Ridge millstones and the Raccoon buhrstones of Ohio in April 2004. Mr. Steve Parker, Lancaster, Ohio, shared photographs of Raccoon Creek buhr millstone at McArthur, Ohio, during October of 2004. In April of 2005, Mr. Parker located a Raccoon Creek buhr millstone quarry at McArthur, Vinton County, Ohio. He graciously shared photographs, a brief report, and a sketch map of the quarry. Mr. Edward L. Bell, archaeologist with the Massachusetts Historical Commission in Boston, shared information from two archaeological reports that mention possible granite millstone quarries.

During 2005, the second international millstone conference was organized by French archaeologist and historian Alain Belmont and German archaeologist Fritz Mangartz. In September of 2005, the author attended this conference in Grenoble, France, and presented a paper entitled "The Preservation, Ownership, and Interpretation of American Millstone Quarries" (Hockensmith 2005b, 2006c). The paper was presented at the Colloque International entitled "Les Meulières: Recherche, Protection et Valorisation d'un Patrimonie Industriel

Européen (Antiquité–XXIe s.)." The approximate English translation of the conference title is "Millstone Quarries: Research, Protection and Valorization of a European Industrial Heritage (from Antiquity to the 21st Century)." The conference attracted about 100 participants from many countries (Cookson 2005:30). For another overview of the conference, see Hockensmith 2006b. Conference attendees also had the opportunity to visit millstone quarries in two areas of France (Major 2005:31).

The conference afforded the author an opportunity to interact with fellow millstone scholars from several countries (Hockensmith 2006b). Professor Alain Belmont assisted the author in many ways and served as a guide to three millstone quarries in different regions of France. Alain also graciously shared copies of bibliographies he had compiled on millstones. In addition to reconnecting with Alain, it was great to become reacquainted with millstone researchers Fritz Mangartz, Tim Anderson (Switzerland), François Siguat (France), Inja Smerdel (Slovenia), and others. I also had the privilege of meeting British archaeologists David Peacock and David Williams; British researcher J. Kenneth Major; German archaeologist Stefanie Wefers; Swedish researchers Ingemar Beiron, Holger Buentke, and Hans Gustafsson; Norwegian researcher Astrid Waage, and others scholars studying millstones. Several of these individuals have generously shared literature with the author in the past. The proceedings of the second millstone conference were published in 2006 (Belmont and Mangartz 2006).

As this book approached completion, several people provided assistance. Dr. Harry Enoch of Clark County, Kentucky, shared references for ads for Red River millstones dating to 1799 and 1818. Mr. Greg Galer, Stonehill College in Easton, Massachusetts, graciously shared information on a millstone quarry on the campus. Dr. Ruth Shaffrey, an archaeologist living in Oxford, England, shared electronic copies of some back issues of the Quern Study Group's newsletter and an index of articles for earlier issues. Mr. Martin Watts (traditional millwright and milling consultant) and Sue Watts (archaeologist with Exeter University and the Devon County Council Archaeology Service) made me aware of many millstone and quern sources in the United Kingdom. Dr. Margarita Vrettou-Souli, an archaeologist living in Milos, Greece, graciously shared an English summary of her Greek-language work on the millstone industry at Milos. Mr. David Llewelyn Davies of Wales had Owen Ward send me some pages dealing with millstones from his book on the Felin Lyn corn mill in northern Wales. Mr. Ward, of Bath, England, also made me aware of Philip Hudson's millstone research for Lancaster, England. Ms. Becky Shipp, Kentucky Heritage Council, graciously translated e-mails from Spain into English and translated my responses into Spanish. She also translated some article titles into English. Without Becky's help, I could not have communicated with the scholars in Spain. Archaeologist Irene Baug of Hyllestad, Norway, graciously read my section on Norwegian millstone quarries and offered some helpful suggestions.

I also want to express my gratitude to my wife, Susie, who has graciously accompanied me to millstone conferences and climbed steep terrain to visit remote millstone quarries.

Several organizations allowed me to use information previously published in my articles and books. Ms. Esther Middlewood, Editor of *Old Mill News,* allowed me to use information from my articles published *Old Mill News* and my books published by the Society for the Preservation of Old Mills. Dr. Kit Wesler, Murray State University, Murray, Kentucky, permitted me to use information from my articles in *Ohio Valley Historical Archaeology* and the book jointly published by the Symposium on Ohio Valley Urban and Historic Archaeology. Mr. Steve Spring, editor of *The Mill Monitor*, Sterling, Virginia, allowed me to use data from my article that appeared in *The Mill Monitor* published by The International Molinology Society of America. Dr. Fritz Mangartz of the Römisch-Germanisches Zentralmuseum,

Mainz, Germany, allowed me to use information from a paper I contributed to (Belmont and Mangartz 2006). Mr. Tony Bonson, publications editor for the International Molinology Society, Cheshire, England, permitted me to use information from bibliographies that I authored and coauthored in *International Molinology*. Mr. Bernard Cesari, manager of Éditions Ibis Press, Paris, France, graciously permitted me to use information from my paper contained in a book published by Éditions Ibis Press (Barboff, Sigaut, Griffin-Kremer, and Kremer 2003).

A number of organizations graciously allowed me to use longer quotes from their publications and sent me permission letters. Ms. Alice Cross, editor of *The Accordian*, published by the Friends of Historic Rochester, Inc., in Accord, New York, allowed me to use several quotes from articles published in *The Accordian*. Ms. Esther Middlewood, editor of *Old Mill News*, granted me permission to use information published in *Old Mill News* and Society for the Preservation of Old Mills publications. Mr. Ingemar Beiron gave permission to quote from the excellent website (*www.qvarnstensgruvan.se*) that discusses "The Millstone of Lugnås" in Lugnås, Sweden. Ms. Pamela J. Bennett, Director of the Indiana Historical Bureau in Indianapolis, granted permission to use quotes from Blackburn (1942). Dr. Fritz Mangartz of the Römisch-Germanisches Zentralmuseum, Mainz, Germany, allowed me to quote from Belmont and Mangartz (2006). Senior Editor Eric Powell with *Archaeology Magazine*, Long Island City, New York, permitted me to quote from Runnels and Murray (1983). Dr. Edward E. Erb, state geologist and division director of the Division of Mineral Resources, Charlottesville, Virginia, granted permission to quote from Campbell (1925). Dr. Louise Mirrer, president of the New-York Historical Society in New York City, allowed me to use quotes from O'Callaghan and Barck (1929). Mr. Robert Hege, III, President of Meadows Mills, Inc., allowed me to quote from his company history (McGee 2001). Dr. Margarita Vrettou-Souli of Milos, Greece, graciously gave me permission to quote from her English summary of the millstone industry at Milos. Mr. P. Patrick Leahy, the American Geological Institute, authorized me to use quotes from Gary, McAfee and Wolf (1974). Ms. Tamara G. Miller with the Historical Society of Pennsylvania allowed the use of quotes from *The Pennsylvania Magazine of History and Biography*. Mr. Tony Bonson, publications editor for The International Molinology Society (TIMS), Cheshire, England, allowed me to quote from articles published by TIMS. Dr. Thomas R. Ryan, President and CEO of the Lancaster County Historical Society, Lancaster, Pennsylvania, permitted me to use quotes from Flory (1951a, 1951b). Mr. John R. Keith, Chief, Eastern Region Publications with the U.S. Geological Survey, Reston, Virginia, granted permission to use quotes from the *Mineral Resources of the United States* and the later series the *Minerals Yearbook*. Ms. Judy Wilson, Permissions Manager, Ohio University Press, Athens, Ohio allowed me to use a quote from the Smith and Swick (1997) book *A Journey Through the West*. Ms. Michelle L. Mullenax-McKinnie, Vice President of Publishing, McClain Printing Company of Parsons, West Virginia, gave permission to quote from Maxwell (1968). Ms. Deborah C. Cox, President of The Public Archaeology Laboratory, Pawtucket, Rhode Island, granted permission to use quotes from Deaton and Cherau (2003) and Leveille and Rainey (2001). Mr. John B. Skiba, Manager of the Office of Cartography and Publications, New York State Museum, Albany, New York, gave permission for quoting from New York State Musem Bulletins and other geological publications. Archaeologist Irene Baug of Hyllestad, Norway, allowed me to use quotes from the English section of her booklet on the millstone quarries of Hyllestad (Baug 2004).

Several individuals kindly shared photographs of millstones and millstone quarries for this book. Professor Alain Belmont with the University of Grenoble, France, shared some of his photographs for France and Switzerland. Mr. Ingemar Beiron of Mariestad, Sweden, shared photographs for quarries in Lugnås, Sweden. Dr. Fritz Mangartz of the Römisch-Germanisches Zentralmuseum, Mainz, Germany, shared a 19th century photograph of the Mayen region of Germany. Mr. Joern Kling, Bonn, Germany, shared several photographs of millstone quarries at Mayen and Niedermendig, Germany. Mr. Joaquín Sánchez Navarro, Ciutadella de Menorca, Illes Balears, Spain, shared photographs of quarries from the Island of Menorca, Spain. Ms. Inja Smerdel, Slovene Ethnographic Museum, Ljubljana, shared some of her photographs of millstone quarries in Slovenia. Mrs. Astrid Waage of Hyllestad, Norway, shared several of her photographs from the quarries at Hyllestad. Ms. Pilar Pascual Mayoral and Mr. Pedro García Ruiz of Logroño, Spain, graciously shared photographs from northern Spain. Mr. Philip J. Hudson, Hudson History, North Yorks, England, shared many photographs of millstone quarries in the Lancaster area of England. Finally, Dr. Fred E. Coy, Jr., allowed me to use photographs of millstone quarries in New York, Pennsylvania, and Virginia. I am grateful to all of these scholars for sharing the wonderful photographs that have greatly enhanced this book.

The author would like to thank all those people who contributed to the production of this book through the many years of its preparation. Archaeologist and mill scholar Nancy O'Malley (Museum of Anthropology, University of Kentucky) graciously read the entire manuscript and offered many helpful suggestions. Professor Alain Belmont of the University of Grenoble, France, wrote the Foreword for this book. Dr. Cozette Griffin-Kremer of Paris, France, kindly assisted me in obtaining permission to use data from my paper published in France.

Charles D. Hockensmith
Frankfort, Kentucky
Spring 2009

Foreword by Alain Belmont

"Stones of life." "Pierres à pain." "Livets steinar." "Virtutes lapidum." English, French, Norwegian, and Latin: many languages and many words that glorify former millstones and their quarries. But books didn't discuss the importance of millstones for a long time. Some twenty years ago, scholars interested in querns and millstones were considered to be strange people, outside of reality. They were wasting their time and losing money for nothing serious. When I began to research millstones, one of my colleagues at the university kindly said to me, without intending any joke, "Alain, your career is ended." In French, career and quarry use the same word—"carrière."

However, since prehistoric times, the cutting of rock to make millstones has been one of the most important industries in the world. The largest quarry sites extended for one, two, and even twenty square kilometers, employing thousands of workers, producing up to one million millstones, selling them to foreign countries, and even to new-found lands. When you know that the price of a millstone was equal to the price of a house, you can imagine how valuable this former industry was. Some French authors of the 19th century compare millstone quarries to gold mines, and La Ferté-sous-Jouarre, capital of the very famous French quarries, to California! It required great effort and great expenditure to ensure that people were fed. Without millstones there is no bread, and thus no life.

Science has at last opened its eyes to this important matter. For the last ten to fifteen years, Ph.D. students, geologists, archaeologists, historians, anthropologists, and ethnologists employed by European universities, museums and public research offices have devoted their research time to studying querns, millstones, and their quarries. There is now no year without at least four or five excavations of ancient or medieval millstone quarry sites; no year without an international conference such as La Ferté (France), Grenoble (France), Hyllestad (Norway) and very soon Mayen, in Germany. The main themes of the past are now known and, likewise, the importance of the millstone industry for human health, subsistence, and economy is now recognized. Even the general public now applauds when visiting spectacular open-air or subterranean millstone quarries that are successfully set up for tourism.

Millstone research efforts in Europe find their echo in the United States, thanks to Charles Hockensmith. No millstone enthusiast on this side of the Atlantic Ocean can ignore the face of our American millstone-addicted cousin. Many of them have led him to their digs, climbing mountains, fighting against dense trees and shrubs, dust, rain or heat, and talking, talking for hours, about giant stone wheels.

With *The Millstone Industry: A Summary of Research on Quarries and Producers in the United States, Europe and Elsewhere*, Charles offers us the work of a lifetime. The results of his visits to the Old Continent and the opportunity for non–English speaking people to discover European research is a part of it, and I greatly thank him for being our speaker. Above all, his book opens a large window on the American millstone industry. It is the first book especially devoted to that industry from the arrival of European pioneers to the last stonecutters, who met Charles and became his friends. This is a great inventory, and perhaps close to an exhaustive list, of all quarries opened for centuries from America's East Coast to the Midwest. Half of a continent, laboriously cataloged by a single man.... Crazy work, strong as a flint stone, and rich as a quarry.

"*Mensio nulla tui fuit, te mola.*" "Nobody wrote on you, millstone." What the medical doctor Georges Pictorius claimed five centuries ago is now no longer true. Thanks to you, dear Charles.

Alain Belmont
Professor of Modern History
University of Grenoble II
French National Centre for Scientific Research
Grenoble, France

Preface

During the spring of 1987, I became involved in a research project to document the millstone quarries of Powell County, Kentucky. Between 1987 and 1990 I was able to document six millstone quarries exploiting a Pennsylvanian age conglomerate. To establish a context for these quarries, I began to search for information on millstone quarries located in other areas. Initially, I found very little. As the years passed, however, I discovered many new publications and made numerous contacts with scholars in the United States and several other countries. Also helpful was the blossoming of the Internet, where many excellent websites began to appear, making available literature (both old and new) that otherwise would have been difficult to locate. As data accumulated, I wrote a series of articles sharing information on the millstone industry for the eastern United States and for several individual states where more sources were available.

I also began compiling this book, a task I could accomplish only gradually, over a period of several years, as time was available. The fact that my time was limited turned out to be a blessing, since more and more literature became available as I made slow progress on the book. The result is a much more comprehensive publication that will be of greater utility to other researchers.

While this book examines the millstone industry in a worldwide context, the bulk of its information concerns the United States. European millstone and quern studies are far more numerous, but much of the literature is published in languages other than English. As for countries not mentioned in this book, one assumes that millstones were quarried there, too; but research on the millstone industry in those countries has proved impossible to track down. Some millstone literature may be published in local or regional journals in several countries. In other instances, millstone and quern quarries may be present but not yet studied by local archaeologists. The fact that certain countries are not mentioned in this book indicates not that they don't have millstone or quern quarries but that I don't know about them.

Research for this study was accomplished in four major stages. The first stage was a search through the literature of several disciplines, including archaeology, economic geology, mines and minerals, industrial archaeology, and history. The geology library at the University of Kentucky was a vital repository for useful geological publications. The second stage was a campaign of letter writing to State Historic Preservation Offices in the United States, to geological surveys in states known to be millstone producers, and to many other individuals thought to be sources of information. A lot of obscure information on American millstone quarries was obtained through this letter writing campaign. The third stage focused on the European industry and involved writing letters to industrial archaeologists in Great Britain and making contacts at two international millstone conferences. This stage produced many articles on European millstone quarries. Finally, the Internet has been a tremendous resource in recent years. Websites such as the Making of America sites at the University of Michigan (http://quod.lib.umich.edu/m/moagrp/) and Cornell University (http://moa.cit.cornell.edu/moa/), which offer digital versions of 19th century American primary sources, have made a tremendous amount of early literature available to researchers. On these sites I was able to find many sources that would otherwise have gone undetected.

This book is organized as follows. The first chapter quotes early American millstone documents dating between 1628 and 1829. The second chapter offers information on all known millstone quarries in the United States; subsections are provided for conglomerate, granite, flint/buhrstone, quartzite, gneiss, sandstone, and other quarries. Third, a brief chapter discusses millstone makers and urban factories. The fourth chapter deals with the rise and fall of the American millstone industry. The fifth chapter discusses foreign millstones imported into America. Sixth, an overview of the millstone quarrying industry in Europe and other areas of the world is presented. Subsections are provided for Albania, Austria, Britain, Canada, the Czech Republic, France, Germany, Greece, Hungary, Italy, Luxembourg, Norway, Slovenia, Spain, Sweden, Switzerland, Turkey, and Ancient Rome, as well as for miscellaneous millstone and quern studies. The seventh chapter offers information on experiments with artificial millstones as substitutes for natural stone. Tools used in making and sharpening millstones are discussed in the eighth chapter, and working conditions and health hazards of the millstone industry in the ninth. After a brief conclusion to the book, a glossary is provided to help the reader with some specialized and technical terms. Finally, the list of cited references is partially annotated and includes millstone literature covering many different countries and published in several languages.

Although language barriers and other research difficulties have somewhat restricted the coverage, this book establishes a broad foundation of millstone and quern research for future scholars. The extensive bibliography leads readers to articles and reports where they can find much additional information. Researchers fluent in French and German will have tremendous numbers of studies that they can consult. I hope that readers will develop an appreciation for mankind's ingenuity in developing a variety of ways to grind grains and other materials, and that researchers will be encouraged by this book to undertake further study of the millstone and quern industries in their individual countries and throughout the world.

Introduction

Millstones have played an important role in human technology from antiquity to modern times. Different civilizations and cultures have produced their own styles of grinding stones to process grains. The earliest rotary millstones, known as querns, were turned by human power. Later in history, larger millstones were manufactured that required animal, water, or wind power to turn them. These larger millstones required less human effort and ground greater quantities of grain.

Before proceeding further, it is necessary to define three key terms used in this study: millstone, quern, and quarry. According to *Webster's New Twentieth Century Dictionary of the English Language* (McKechnie 1978:1143), the term millstone refers to "either of a pair of large, flat, round stones used for grinding grain or other substances." A quern is "a primitive hand mill for grinding grain, consisting of two stone disks, one upon the other, the upper stone of which was turned by hand" (McKechnie 1978:1478). A quarry is defined as a "place where stone or slate is excavated, as by cutting or blasting, for building purposes, etc.: it is usually open to the light, and in this respect differs from a mine" (McKechnie 1978: 1475).

This book focuses on quarries where millstones and querns were manufactured. Some of the quarries were pits or just rock outcroppings from which suitable stone was extracted. In some instances, underground mines were employed to reach suitable stone deposits. Other "quarries" consisted of scattered boulders on the surface that were worked into millstones. The book also includes a range of studies that describe and classify millstones and querns from various contexts.

Before the introduction of steel roller mills in the late 19th century, millstones were essential for the operation of grist mills. Without quality millstones, grains could not be adequately ground. Hughes (1869:91) noted that "as the *millstones* are the entire *key* which regulates the profits of the miller, we think much attention cannot not be expended more profitably, than that bestowed in keeping them in proper order." Considering the key role that millstones played in the American and European milling industries, it is very fitting that archaeologists begin to document the quarries that produced these important stones.

Grist mills required at least two millstones to grind grains. Larger mills often employed several sets (or runs) of millstones to increase their production. Typically, millstones were used in pairs, with one stone running above the other stone. The lower millstone was called the bedstone and it remained stationary. The upper millstone was known as the runner and it rotated. Grooves or furrows were cut into the grinding surface of each millstone to facilitate grinding. Different patterns of furrows were used at various points in time and for different grinding tasks. Both millstones had to be balanced. The distance between the stones was carefully regulated so that they would not touch but would still run close enough together to grind. The upper millstone was attached to the power source through iron hardware (rynd and spindle) and turned by a shaft called a damsel. A wooden housing covered the millstones and the flour or meal flowed out a spout where it was collected. It should be noted that there has been considerable variety in the styles of millstones and the grinding process since antiquity.

Millstones vary according to their design, grinding surface, raw material, and function. Millstones made from a single piece of rock are known as monolithic stones. Composite millstones are built from several small, shaped stones which are cemented together and bound with iron bands.

Millstones that operated horizontally were called face-grinders, while millstones that ran vertically on their edges were called edge-runners or crushers (Tucker 1977:1). Pairs of edge runners attached to the same axle were called chasers. Phalen (1908:610) stated that "chasers are larger than regular millstones. They are used for heavier work, as in grinding quartz, feldspar, barytes, etc., and as already mentioned, run on edge. They were made with a diameter as short as 24 inches, they are usually turned out with diameters ranging from 50 inches to 84 inches and with thicknesses as great as 22 inches."

Several types of stone were used for manufacturing millstones in the United States. These included conglomerate, fresh water quartz, granite, flint, sandstone, gneiss, quartzite, basalt, limestone, and occasionally other types of rock.

The European millstone industry is even more diverse in terms of raw materials exploited for querns and millstones. These quarries date from antiquity to the mid–20th century. Because of the great time spanned by these quarries, there is considerable variation in quarrying methods and the styles of millstones and querns produced. Raw materials used for millstones and querns included conglomerate, granite, flint, limestone, sandstone, gneiss, schist, fresh water quartz, andesite, trachyte, garnet mica schist, basalt, lava, and other rock types. As in the United States, millstones were quarried from both surface outcrops and underground mines. In terms of size, the quarries ranged from small operations where only a few millstones or querns were made to massive quarries that produced many thousands of stones sold over broad geographical areas. Of the European countries, the millstone and quern literature is most extensive for France, Germany, and Great Britain. Other countries are represented in the literature to a lesser degree.

Millstones had many uses in the United States. (In the eastern United States, querns were also used, but they are poorly documented.) Most people think that millstones were used only for grinding corn, wheat, and other grains, but in fact they had many other applications. Sass (1984:10–32, 55–56) has noted that special millstones were used for cleaning clover seeds, shelling oats, hulling buckwheat, pearling barley, processing split peas, chaffing wheat, regrinding middlings, making apple cider, and grinding phosphate rock. Another use was olive oil presses (Kardulias and Runnels 1995:110; Runnels 1981:225–226). They were also used in the chocolate industry, cork mills, dye mills, flint grinding mills, hemp mills, paint and color mills, plaster of Paris and gypsum grinding mills, and tanbark mills (Sass 1984:33–60). Webb (1935:217–218) also notes their use in flax mills, snuff mills, and gunpowder mills. Other applications included grinding bone (Parker 1896:927; Bost 2002:32), mica (Pratt 1901:793; Pratt 1902:793), charcoal (Williams 1885:712), cement (Day 1892:456; Phalen 1908:609; Pratt 1902:794; Pratt 1904a:879; Pratt 1905:1004), barytes (Pratt 1902:794; Pratt 1904a:879), drugs (Pratt 1905:1004), mustard (Pratt 1905:1004), glucose (Pratt 1905:1004), spices (Pratt 1905:1004), fertilizers (Pratt 1902:794; Pratt 1905:1004), plaster (Pratt 1905:1004), paste (Pratt 1905:1004), feldspar (Katz 1917:67), quartz (Katz 1917:67), and talc (Phalen 1908:609). A localized task for millstones was the hulling of rice on the southeast coast of America (Judd 1999, 2000).

In recent years, there has been a new interest in the study of millstone and quern quarries. Since 2002, several books have exposed millstone studies to a much broader audience. All of these books are collections of papers. The first, *Les Meuliers: Meules et Pierres Meulières*, by Agapain (2002), presented new studies along with reprints of earlier works on the French millstone industry. Building upon that foundation, Barboff, Sigaut, Griffin-Kremer, and Kremer (2003) edited *Meules à Grains* [Grain Millstones]*: Actes du Colloque International de La Ferté-sous-Jouarre 16–19 Mai 2002*, a collection of papers presented at the first international millstone conference held in La Ferté-sous-Jouarre, France. Three years later, the second international millstone conference was held in Grenoble, France, and Belmont and Mangartz (2006) edited a collection of papers titled *Mühlsteinbrüche: Erforschung, Schutz und Inwertsetzung eines Kulturerbes europäischer Industrie (Antike–21. Jahrhundert) [Millstone Quarries: Research, Protection and Valorization of an European Industrial Heritage (Antiquity–21st Century)]*. This book included papers on millstone quarries from several countries. A third millstone conference has been scheduled for the fall of 2008 in Mayen, Germany.

During the same period, four books dealing with European millstone and quern quarries were published by archaeologists. First, *H Milopetra tis Milou: Apo tin exorixi stin emporiki a diakinisi* [*The Millstone of Milos: From Mining to Commercial Circulation*] was authored by Vrettou-Souli (2002) and published in Greek. The following year, Anderson, Agustoni, Duvauchelle, Serneels, and Castella (2003) published *Des Artisans à la Campagne: Carrière de Meules, Forge et voie Gallo-Romaines à Châbles (FR.)*. This book is the study of a Roman era millstone quarry in eastern Switzerland. The third study, a two-volume book on the millstone quarries of France from the Middle Ages to the Industrial Revolution, was Belmont's (2006a) *La Pierre à Pain: Les Carrières de Meules de Moulins en France du Moyen Age à la Revolution Industrielle*. The fourth book, *Grinding and Milling: A Study of Romano-British Rotary Querns and Millstones Made from Old Red Sandstone* (Shaffrey 2006a), provided new information for England. Many additional studies have been reported in articles; combined with these books, they have greatly expanded our understanding of the quarrying and shaping of millstones and querns over a long period of human history.

Information on European millstone and quern quarries has been rapidly accumulating since Alain Belmont and Fritz Mangartz established the Millstonequarries.eu website (http://meuliere.ish-lyon.cnrs.fr). This website allows researchers to complete a standardized "card" or form for each individual millstone quarry. The quarries are organized by country. The reader can click on a map of Europe or France and see listings of regions for each country. Clicking on these regions produces a listing of quarries in each region or subregion. Clicking on an individual quarry name reveals text and photographs for that quarry. Versions of the website are in French, English, and German, but only the French version is currently complete. Most of the overviews for the millstone quarries in Spain are available only in Spanish. Plans are under way to translate all the quarry information into French, English, and German. Researchers from several countries have contributed information on millstone quarries. Other resources on the website include bibliographies of millstone literature for several countries and summary information for several recent millstone books. This website is a wonderful tool for scholars interested in millstone and quern quarries.

It is hoped that this book will be a worthwhile addition to the field of millstone and quarry studies, and that it will inspire new research that will continue to expand the field.

1

Early American Millstone Documents, 1628–1829

This chapter reproduces some early documents relating to millstones in the United States. These documents included here usually don't refer to specific quarries but to millstone transactions. The accounts date between 1628 and 1829. Some of these documents refer to millstones imported into the United States while others mention stones made in the United States. They offer insight into the use of millstones during America's early years. These documents are presented chronologically.

The earliest document mentioning millstones was included in the *Chronicles of the First Planters of the Colony of Massachusetts Bay, From 1623 to 1636* (Young 1846:64):

> 17th March, 1628
>
> A warrant was made for payment of £120 to Mr. Nathaniel Wright, for so much paid by him to Mr. Jarvis Kirk, Mr. William Barkley, and Mr. Robert Charlton, for the ship.
> Also, to pay for iron and steel.
> Also, to pay for buhrs to make mill-stones, 110, 2s. apiece, bought of Edward Casson, of London, merchant tailor,

London, merchant tailor,	£1100
14 c. of plaster of Paris, 18d. per c.	110
And porterage, weighing the plaster, and casting out of the buhrs, 12d. and 23d.	30
	£1240

Millstones were sent to the Dutch colony of New Netherland located on Manhattan Island in New York during 1630. O'Callaghan (1855:431) included in his long list of items sent to the colony the following entry: "To Heindrick op de Camp, for two small millstones for a small grist-mill ... 20.05."

John Winthrop (Savage 1853:458) of Massachusetts, in a March 28, 1631 letter to his son in England, requested that a number of items be sent to him including "millstones, some two foot and some three foot over, with bracings ready cast, and rings, and mill-bills."

The *Narratives of Early Carolina, 1650–1708* (Salley 1911:19) provided the following information on stone suitable for millstones observed at New Brittaine on September 4, 1650:

> About 8 of the Clock we travelled North North-East some six miles, unto the head of Farmers Chase River, where we were forced to swimm our horses over, by reason of the great rain that fell that night, which otherwise with a little labour may be made very passable. At this place is very great Rocky stones, fit to make Mill-stones, with very rich tracts of Land, and in some places between the head of Farmers Chase River and Black Water Lake,...

The early records (1639–1660) of Southampton, New York, contained a reference on page 216 to Mill Stone Swamp (Hedges, Pelletreau, and Foster 1874:236):

The highway to Millstone is the one running north from Elias Woodruff's to James Edwards.' Millstone swamp wrs [*sic*] so called from the fact that one of the millstones for the first mill in this town was procured there. The road from Samuel Jones' corner to Millstone swamp ran originally through the hollow by Elias Woodruff's house.

The *Calendar of New York Colonial Commissions, 1680–1770* (O'Callaghan and Barck 1929) included a court action between Sergeant Huybert and Jan Carreman in New Amsterdam, New York. The first mention in 1653 was as follows (O'Callaghan and Barck 1929:70–71):

Sergeant Huybert, pltf., v/s Jan Carreman, deft. Pltf. says, he has bought of deft. through Wessel Eversen a pair of millstones, but when he came at the appointed time to fetch them away, they were refused to him according to an affidavit. He demands therefore, that deft. shall be now compelled to deliver said stones at his own expense in Gravesend, and claims to have already suffered expenses and damages through the refusal of delivering them to the amount of 64 fl. as specified. Deft. admits having sold the stones to pltf. and says, he had given written orders, that Wessel Eversen should deliver them, but had not directed to refuse the stones and therefore is not to blame. After hearing the litigants Burgomaster and Schepens decide the complaint well founded and leave it at the discretion of deft. to deliver the stones sold at Gravesend or pay the above mentioned expenses, saving his action against Wessel Eversen. Deft., Mr. Kerman, agrees to deliver said stones to pltf. at Gravesend between now and May, reserving his action v/s Wessel Eversen.

Another reference to the above case was presented at the City Hall on March 24, 1653 (O'Callaghan and Barck 1929:74):

Jan Carreman, pltf., v/s Wessel Eversen, deft., asks, for what reason deft. had refused to deliver the millstones to Sergeant Huybert, who had bought them, which refusal had caused a loss of 64 fl. This loss pltf. asks deft. to make good or else to deliver the stones at his own expense to Sergeant Huybert at Gravesend. Deft. admits having refused to give the stones, belonging to Mr. Cerman, to Sergeant Huybert by order of Jan Teunissen, who had said, he wanted to buy them from Mr. Karman. Burgomaster and Schepens decide, that deft. is bound to deliver said stones in Gravesend or pay the expenses aforesaid, saving his recourse against Jan Teunissen. This is left to his option. Deft. agrees, to fetch the stones to Gravesend by the middle of April next.

Extracts from Major Pynchon's account book in Suffield, Connecticut, provided very interesting information on a millstone agreement for 1672 and 1677 (Sheldon 1879:23):

AGREEMENT FOR MILLSTONES.

Feb 5th 1672 Agreed with John Web & Zebediah Williams to make me a p^{r} of good mill stones of good greete w^{h}out flaw to be judged by Th^{o} Bancroft to be as good as mine on Westfield Millstones: to be full 5 foote: 2 inches over: 15 inches deepe in y^{e} eye & 13 inches in y^{e} skirt: wch stones are to be dlv^{d} either at Windsor or at Stony Brooke where I please next October. If at Windsor I am to allow and pay for y^{m} 23$^{£}$, & 1 Gallon of Rum; If at Stony river then to pay but 21$^{£}$ & 2 Gallon of Rum

As witness our hands
John Webb
Z. W
John Pynchon

John Webb Cr.

	£ s. d.
Nov. 27 1677 By 1 p^{r} of Millstones dl^{vd} at my mill at Stony Brook wch according to agree^{t} is	21:00:00
By allowance wch I make toward y^{e} cartage of y^{m} I give you 20^{s}	1:00:00

TRANSPORTATION OF MILLSTONES.

Thomas Gun of Westfeild Cr.

By Carrying downe ye millstones for J^{o} Web to Southfeild 2:14:00
Jan 16th 1677 Quit & Ballanced p^{r} contra
1679 By 26 appletrees at 6^{d} apiece alowing 2 Trees 00:12:10

P^{d} pr Contra.

Thomas Bancroft (of Westfeild) Cr.

By floaring my Millstones at Stony Brook, ...£3:00:00
discounted Jany 7th 1679:

The history of Woodbury, Connecticut, contained a reference to the use of millstones in a mill prior to 1681 (Cothren 1872:936):

> The committee, however, did not report till 1677, and the road was probably not built till several years later. Meanwhile, the people must have mill privileges. They accordingly procured a set of stones, and transported them on horseback, or rather, slung them between two horses, and took the weary way of their bridle-path to Woodbury. They set their mill-shed on a little brook a short distance east of Deacon Eli Summers' house, in Middle Quarter, and though but about a bushel of grain per day could be ground at this mill, yet it was all the accommodation of the kind that the inhabitants had, till 1681. These mill-stones were of small dimensions, being not more than thirty inches in diameter. One of these is still preserved, and has been attached to the base stone of the "Fathers' Monument" in the south, or ancient burial ground, for preservation, after having done service for more than a hundred years as a door-stone to the house in Middle Quarter lately occupied by Miss Lucy Sherman.

The early history of Delaware makes reference to D'Hinoyossa selling millstones in 1662 and Sweringen selling millstones in 1663 (Hazard 1850; Vincent 1870). Hazard (1850:335) describes the 1662 incident:

> They treated it with kindness, and J. Alricks promised to pay, as the purchase was approved by his lords and masters, but in vain to himself or his successor, D'Hinoyossa, &c., who received pay. Everybody complains of his unjust and fraudulent proceedings. D'Hinoyossa sold a considerable part of the city's property to the English in Maryland, such as a pair of millstones, &c.

Concerning the same incident, Vincent (1870:395) noted:

> In a letter to Stuyvesant in June, he charges D'Hinoyossa with taking away the palisades of the fort and burning them in his brewery; also with selling to the savages the new city guns which arrived in the ship Parmeland Church; also with selling to the English in Maryland the city millstones, brought in the same ship, for one thousand pounds of tobacco, and....

In 1663, Hazard (1850:343) observed that "Sweringen departed about Christmas to Maryland, it is said to receive tobacco for the millstones and galliot, &c." Finally, in 1673, Hazard (1850:409) wrote, "Two millstones lying useless at Whorekill, formerly belonging to the city's colony, are wanted at New Amstel. The magistrates ordered to send them to Alricks."

In the town meeting in Lynn, Massachusetts, on October 11, 1692 (Lynn Historical Society 1949:13), the following decision was reached:

> Clerk John Ballard & Ensine Samuell Tarbox was Chossen to bee aturneyes for ye town to prossecute a gainst any person or persons in a corse of Law that hathe Carryed away oute of the town any Mill stone or stones withoute the towns Consent.

Six years later in the May 12, 1698, town meeting in Lynn, Massachusetts (Lynn Historical Society 1949:54), it was "Voated that Robert Burges has Liberty to gitt two Milstones on the town Common & sell them oute of the town prouided he pays thirty shillings to the town."

An early reference to millstones in Massachusetts was included in *Nantucket Lands and Land Owners* (Worth 1902:98):

> Jan. 23, 1670. Edward Starbuck, John Swaine, Nathaniel Starbuck and William Worth to make a pair of mill-stones and bring them to the mill; they to be paid in corn harvest at the rate of two shillings and sixpense a day for each.

During the August 14, 1707, town meeting in Lynn, Massachusetts (Lynn Historical Society 1956:40), the following comments were recorded:

> Voated that No Millstons should be Made in ye towne Nor sold out of towne on the penalty of forty shillings for any person that shall Make such sones that is to say that grew on ye Common Vntil the first of Decemr Next.

Stoner (1947:412) provided the following information on the use of English millstones in Franklin County, Pennsylvania, in 1735:

> The first set of water-power millstones in Franklin County was installed shortly after 1735 in a log mill where Falling Spring enters Conococheague creek. These stones were brought from England and cost Benjamin Chambers, set up in his mill, about 80 pounds Sterling or close to $400 in present currency valve. In those days it required between six months and a year, from the time the order was sent in, to bring a set of stones from the Old Country, and doubtless the money to pay for them had to be forwarded with the order. This transaction shows that the founder of Chambersburg, though not much over thirty years of age, was already a man of means and a man of affairs. It was almost a super-human task to transport a set of millstones from the seaboard through the forest of Pennsylvania to the Cumberland Valley.

The importance of French Burr millstones in producing quality flour was expressed in a letter by Isaac Low shared at the October 2, 1770, meeting of the New York Chamber of Commerce (New York Chamber of Commerce 1867:108–112). Low's letter contained the following comments (New York Chamber of Commerce 1867:110–111):

> The Disrepute which the Flour, the grand Staple of this Colony, suffers at all the markets, calls aloud on the Legislature and every Member of the Community to contribute their sincere endeavors to effect a thorough reform, and, if possible, to retrieve its lost Reputation; and as nothing can be more worthily the object of the serious attention of this Corporation than to promote so desireable an end, I beg leave to propose to their consideration some of the means which appear to me most conducive to remedy the Defects so much complained of, and so severely felt by every principal Exporter of Flour in this City.
>
> First, then, I conceive, and it is demonstrably evident, that the grand Reason why Philadelphia has so much the preference of New York Flour is because the former chiefly use French Burr Stones for Grinding their Wheat, while there is scarce a single pair of them employed by any of the Millers in this Colony. For remedy whereof, I propose that the Fund of this Chamber, and as much more, to be collected by Subscriptions from each Member of it, as shall be sufficient to purchase Ten or Twenty Pair of French Burr Stones, shall be appropriated for that purpose. That application be made to the Owners of Ships in the London Trade, either to bring over the said Stones as Ballast, or at a very moderate Freight, instead of Coals or Grind Stones; and that when the said Stones arrive here, they shall be disposed of at Prime Cost to any Miller in this Colony only who shall apply for them.
>
> The prevalence of habit is such, that it is extremely difficult to induce People to abandon old Customs, especially where doing so is likely to be attended with considerable Expence, although, as in the present instance, there is a moral certainty of Success. I conceive, therefore, it is absolutely necessary to introduce, on the most easy Terms to the Purchasers, such a

number of French Burr Stones as may appear to this Chamber adequate to the Design; not only to Evince their great utility, but to excite an Emulation in our Millers, as is already the case in Pennsylvania, to use none else; which may be easily effected by this Chamber, after a few of them are introduced, by giving a due preference to the Flour which shall be ground by them.

The *Journal of Thomas Wallcut, in 1790* (Wallcut 1879:15) contained the following entry for Ohio on February 7, 1790:

I accompanied Colonel B. [Battelle] and wife as far as the stockade, and found Mr. Tylas coming down to the point with a lantern which I improved until, about half way, the wind blew it out. Very dark and muddy. Heard of the Wolf Creek men losing their millstones in going with them up Muskingum. They seem to be peculiarly unfortunate or are very careless.

An early Dutch millstone was mentioned in the Queens area of New York City for individuals to view on a historical tour (Kelly 1913:314):

A millstone from this or another old mill may be seen imbedded in the concrete sidewalk at 437 Jackson Avenue, said to have been imported from Holland by the Brouchard family, which settled here in 1657. It was placed here by Mr. W. Elmer Paynter, whose grandfather bought the mill property.

A history of Rochester, New York (Parker 1884:54), contained the following information on millstones either quarried in New York or imported from Massachusetts:

There is a story of the old millstones and irons which must not be forgotten.

The stones were taken from a neighboring quarry, we are told, although some of our old settlers affirm that they were brought from Massachusetts on wagons, and were the gift of Phelps and Gorham or the State of Massachusetts. The irons were bought in Cohocton by Allan, and brought to the mill by Indians on pack-horses. Some say Allan drove the horses alone, walking the whole way. In 1806 these stones and mill irons were carried to a small mill on the Irondequoit by Oliver Culver, Miles Northrup, and Benjamin Blossom, and set to work again. For twenty-five years did good service, and then again the sound of their grinding was low, and they were allowed to lie neglected and forgotten on the banks of Irondequoit Creek.

The John Tipton Papers contain an interesting exchange of correspondence concerning millstones in Indiana during 1827 (Blackburn 1942:728):

SAMUEL MILROY TO TIPTON, June 12, 1827

[ISL: Tipton Papers–ALS]

DEER CREEK WABASH June 12th 1827

DR SIR I forward to Mr. H B M^c^Kain the order for the millstones and Iron by M^r^. Lindsay m^r^ McKain Just on the point of starting away had unloaded the stones and was weighin they iron M^r^. Lindsy gave him up your order which he took up and recieted &c on examination to day I find the millstones to be of the meanest quallity I have seen and believe them to be not worth putting in he Indian mill — they Iron is of an inferior a quallity — I apprise you of this and shall await your orders relative to mooveing the stones from where they now are I have appraised Mr. M^c^Kain of the quality of the stons and requested him to have them inspected by a judge &c — I believe I have never seen as meen Millstones put in a Mill M^r^ M^c^Keen doutles was unaquainted with their quality or he would not have purchased them — be so oblidgeing at to let me here from you on this subject — as speedily as posible

Respectfully your friend &c

SAM^l^. MILROY

GEN^l^. TIPTON

The next correspondence between Samuel Milroy and John Tipton was on July 14, 1827 (Blackburn 1942:749–750):

SAMUEL MILROY TO TIPTON, July 14, 1827

[ISL: Tipton Papers–ALS]

POTOWATEMY MILLS TIPECANOE July 14th 1827

DR SIR — Haveing the opertunity by the hand of M^r^. Purdy of writing you a line — and have the pleasure of informing you that I have commenced the building of the mill for the Potowatomies — and anticipate the compleation thereof if no untoward event interveans by the first of Septr. I have twelve hands hear and will have more If I find I can find buisiness to advantage for them —

I recently received two lettres from you by the hand of M^r^. M^c^Keen (as also one Hundred dollars) the dificulty relative to the mill stones — in conformity to your request is adjusted entirely to the satisfaction of all concerned I presume — and particulary so as relates to myself— as I had no other object in view than to apprise you of their quality of the stones and at the same time notify M^r^. M^c^Keen that I have done so, as I conceived he well mite think I acted uncandid towards him, if I had done otherwise, I have seen him and shall as requested by him &c put the mill stones to use and that will be the best way to test ther quallity — But be their quallity what it may no dout M^r^. Keen believed them to be good — and answering the discription which you had ordered — and the deception was doutless practiced on him he not being a Judge of the quality of millstones (as but fiew is) — the want of a remark in the lettres I addressed to him and to you to this amount relative to them, was no dout the reason m^r^. M^c^Keen conceived himself in some degree injured untill I seen him, I believe he is now compleately satisfied, that no injury was intended when I informed him that I conceived it entirely superfluous to state to Gen Tipton that this man that he had honoured with friendship could not for a moment be suspected of intending to abuse the confidence reposed in him, by that friendship, and that the transaction itself was one in which I had no interest, direct or indirect, excep as related to my responsibility for the performance of the mill when completed — This much perhaps was necessary to say to you, as doutless M^r^. M^c^Keens feelings, when he seen you, was excited from a missconception of the motives of my interfearance, which when explained to him he was completely satified with — perhaps the same kind of missconception, may have operated on his mind relative to others, who passed their Judgment relative to the mill stones — whos only motive, was a wish to serve Gen Tipton — without the most distant wish to injure any individuel, either in his estemation or otherwise —

No news hear of course — Should like to hear from you frequently while hear the builders of the house near shop &c permit me to say, I feele under obligations, to Gen Tipton, that cannot be forgotten —

Respectfully Dr Sir I am Yours &c

SAM^l^. MILROY

GEN JOHN TIPTON

The final letter correspondence between Samuel Milroy and John Tipton was on September 6, 1827, (Blackburn 1942:784–750):

SAMUEL MILROY TO TIPTON, September 6, 1827

[ISL: Tipton Papers–ALS]

MOUTH OF EEL INDIANA Sep^tr^. 6th 1827

DR SIR — I have received two lettres by the hand of Mjr Hanah from you — as also the full compensation for building the Potowatomy mill the due bill will be convenient to be discharged if left or paid in the hand of M^r^. M^c^Keen — as he has an order for the Millstones Irons &c of $200 which I have accepted — I have paid 12 dollars for the transportation of iron & Smith tools, to Tiptonville — and there is due you the transportation of the flower I received which you can adjust — And I wish you to take the trouble of paying to our friend

M[r]. Joseph Holeman $19½ on account of His brotherinlaw M[r]. John Polk the money so paid will be an acomedation to them both and M[r]. Holmans Receip will be as good to me as cash and credited on the due bill you have sent me accordingly — The flower will be delivered hear (to be subject to your order) when ever you may expect to need it and aprise me thereof—

I owe yo every obligation for the prompt manner in which I have receivd the Mill compensation — and I trust you will find I have used every exertion to Compleat the mill in the shortest time, and in the best manner — The mill pond Mjr Hanah will tell you will be some time in filling it being extensive and the country dry in consequence of the excesive draught — in [it] must therefo be expected that the evaporation will nearly if not entirely equal the influx of watter untill there is an increase thereof by reign &c I have no dout of the pond filling and the mills doing well — The Indians however perhaps are not able to comprehend the reason that the mill does not go imediately — it mite be well to have it explained to them —...

SAML MILROY

GEN TIPTON

In Franklin County, Pennsylvania, millstones were manufactured (M'Cauley, Suesserott, and Kennedy 1878:122) at Chambersburg:

The manufacture of mill-stones was established in Chambersburg about the year 1792, by James Falkner, Jr., and was extensively conducted for many years. The stones were brought here in the rough, upon wagons, were shaped up and put together, and large numbers sold in the county, and to other points further west, to those having need for them.

In 1820 George Walker and George Roups carried on a "burr mill-stone manufactory" on the Baltimore turnpike, about two miles east of Chambersburg.

Andrew Cleary also manufactured mill-stones in Chambersburg as late as 1829, he being the last person who carried on the business in the county. His shop was on West Market street. None of these avocations are now carried on in our county that I know of.

The documents quoted in this chapter deal with several states and nearly 200 years of Euro-American occupation. Topics addressed include millstone quarries and millstone makers, the stone types used for millstones, the value of millstones, millstone sizes and quality, modes of shipping millstones, millstone lawsuits, and miscellaneous topics. Examining the documents chronologically with an eye to various topics allow for some general observations.

From these documents and some other early sources can be drawn some information about the quarrying and manufacture of millstones between 1628 and 1829. In the Massachusetts Bay Colony, millstones were made as early as 1628 (Young 1846:64). By 1650, stone suitable for millstones was discovered at New Brittaine (probably present day North Carolina) (Salley 1911:19). Sometime between 1639 and 1660, millstones were extracted from the Millstone Swamp near Southampton, New York (Hedges, Pelletreau, and Foster 1874:236). In 1670, reference was made to a pair of millstones being made at Nantucket, Massachusetts (Worth 1902:98). Millstones were also being made at Suffield, Connecticut, in 1672 (Sheldon 1879:23). An inexhaustible quarry of millstones was reported in 1732 as being about 10 miles from the Hudson River and at the beginning of the "Appalachian Hills" in New York (Smith 1972:216). Five years later, granite millstones were being made at New London, Connecticut (Caulkins 1852:402). Millstone or burr stone millstone factories were operated at Chambersburg, Pennsylvania, during 1792, 1820, and 1829 (M'Cauley, Suesserot, and Kennedy 1878:122).

During the early 1800s more information becomes available about millstones and millstone quarries. A gneiss millstone quarry was operated by James Parker in Jefferson County, New York, at Antwerp in 1805 or 1806 (Haddock 1895:434; Oakes 1905:1125–1126). Flint millstones were made as early as 1805 and 1807 on Raccoon Creek in Vinton County, Ohio, by

Henry Castle, Abraham Neisby, and others (Garber 1970:78; Mather 1838:33). Granite millstones were quarried in Louisiana as early as 1810 (Cuming 1810:308). Buhr millstones were quarried on Clinch Mountain in northeast Tennessee during 1818 (Clifford 1926:276) and north of Knoxville, Tennessee, in 1823 (Haywood 1823:14). Also, buhr millstones were manufactured by Alfred Munson in 1823 at Utica in Oneida County, York (Bagg 1892:594). Further, buhrstone millstones were produced in Illinois by 1823 (Beck 1823:194). Flint millstones were made on Flint Ridge in Licking and Muskingum counties, Ohio, in 1827 by Joshua Evans & Co. (Garber 1970:81). In Indiana, millstones were obtained from an unknown source for the Potowatemy Mills in 1827 (Blackburn 1942:749–750). Boulders were used for millstones in Oakland County, Michigan, in 1820 (Seeley 1912:287) and at the Van Cleve mills in Montgomery County, Indiana, in 1827 (Beckwith and Kennedy 1881:328).

The documents indicated that several types of stones were used for millstones. Granite millstones were quarried at New London, Connecticut (Caulkins 1852), and Point Pleasant, Louisiana (Cuming 1810), and those millstones from Massachusetts were probably granite as well (Lynn Historical Society 1949, 1956). Flint millstones were quarried in Franklin and Woodford counties, Kentucky (The Argus of Western America 1821a, 1821b; The Palladium 1800) and Licking, Muskingum, and Vinton counties, Ohio (Garber 1970; Mather 1838). In Ulster County, New York (Spafford 1813:110), as well as Powell and Rockcastle counties, Kentucky (Shane n.d.; The Argus of Western America 1812, 1813), conglomerate was quarried for millstones. Gneiss was quarried for millstones at Antwerp in Jefferson County, New York (Haddock 1895:434). Buhr stone was used in Illinois in 1823 (Beck 1823:194). Boulders were used for millstones in Montgomery County, Indiana, during 1827 (Beckwith and Kennedy 1881:328) and Oakland County, Michigan, during 1819–1820 (Seeley 1912:287). Unfortunately, we don't know the types of stone utilized for most of the early millstones.

Limited information pertaining to the value of early millstones was encountered. In 1628 at the Massachusetts Bay Colony, millstones were valued at £110, 2 shillings each (Young 1846:64). At Nantucket in Massachusetts, a 1670 transaction referred to a pair of millstones that were made at a rate of two shillings and six pence a day per millstone (Worth 1902:98). It is not known how long it took the workmen to make each millstone. In Suffield, Connecticut, a pair of millstones (5 feet, 2 inches in diameter) were valued at £23 plus one gallon of rum in 1672 (Sheldon 1879:23). Millstones from Colen, Germany, sold for £80 a pair in New York in 1732 (Smith 1972:216). In Franklin County, Pennsylvania, English millstones were sold for £80 during 1735 (Stoner 1947:412). By 1766, local millstones in New York were valued at £19 (Flick 1927:296). Four years later in 1770, conglomerate Esopus millstones in New York were selling for £19.5 per pair (Flick 1931:664). In 1773, a pair of New York millstones (4 feet, 8 inches in diameter) were bringing £18 (Flick 1933:939–940). Millstones were purchased in Old Chester, New Hampshire, in 1780 for £150 per pair (Chase 1869:228).

Some information is available for the sizes and prices of flint millstones in Kentucky and Ohio. In Ohio, Joshua Evans & Co. at Flint Ridge published an ad in 1827 that mentioned "they have and intend keeping constantly on hand, Stones of various sizes." Mills (1921:98) reported that the Flint Ridge quarries produced "small buhr-stones during the early settlement of the country." At the Raccoon Creek quarries between 1814 and 1820, Mather (1838a:33) noted that a pair of 4 foot diameter millstones sold for $350 while a pair of 7 foot diameter millstones sold for $500. In central Kentucky, flint millstones were manufactured in a variety of sizes in 1800 and 1821. John Tanner had millstones on hand from 4 feet 2 inches to 4 feet 8 inches (The Palladium 1800). The mentioned sizes and corresponding prices were 2.5 feet ($40), 2 feet 9 inches ($50), 3 feet ($60), 3 feet 3 inches ($50–$75), 3.5 feet ($85), 3 feet 9 inches

($100), 4 feet ($100-$125), 4.5 feet ($150), and 5 feet ($150–$180) (The Argus of Western America 1821a, 1821b).

In addition to cash payments, barter was sometimes used. There were two references to millstones being traded for whiskey. In Vinton County, Ohio, during 1805 and 1806, millstones were reportedly traded for whiskey (Howe 1888:427). During 1812, Samuel Taylor of Rockcastle County, Kentucky, offered to take a small part of the price of a millstone in whiskey (The Argus of Western America 1812).

Only in a few instances were the sizes of early millstones recorded. English millstones, two feet and over three feet in diameter, were scheduled for shipment to Massachusetts in 1631 (Savage 1853:458). At Suffield, Connecticut, millstones were listed as 5 feet 2 inches in diameter and 15 inches thick in 1672 (Sheldon 1879:23). Millstones 30 inches in diameter were mentioned in Woodbury, Connecticut, in 1681 (Cothren 1872:936). In New York, millstones were obtained locally that were 4 feet 8 inches in diameter and 13 inches thick in 1773 (Flick 1933:939–940). Flint millstones from Woodford County, Kentucky, were available from 4 feet 2 inches to 4 feet 8 inches in 1800 (The Palladium 1800). In Rockcastle County, Kentucky, during 1813, conglomerate millstones were available in a "large and general assortment" (The Argus of Western America 1813). The following year, in 1814, flint millstones were available in sizes from 4.5 to 7 feet in Ohio (Mather 1838:3). Flint millstones were also available from Miller, Railsback, and Miller in 3, 4, and 5 foot diameters in Franklin County, Kentucky during 1821 (The Argus of Western America 1821a). The same year a competing millstone quarry (Jeremiah Buckley's quarry) in Franklin County, Kentucky, offered millstones in the following sizes: 2.5 feet, 2 feet 9 inches, 3 feet, 3 feet 3 inches, 3.5 feet, 3 feet 9 inches, 4 feet, 4.5 feet, and 5 feet (The Argus of Western America 1821b). Finally, in 1827, Flint Ridge millstones were available in "various sizes" in Ohio (Garber 1970:81).

References to quality include a 1672 purchase agreement that specified the millstones were to be without flaw (Sheldon 1879:23). Millstones quarried by Charles Colyer of Rockcastle County, Kentucky, were warranted to be of "superior quality" and if they were not, the purchaser's money would either be returned or another pair of millstones would be provided (The Argus of Western America 1813). A millstone contract between Charles Colyer of Rockcastle County and Sidney Clay of Bourbon County dating to 1824 has survived in the Special Collections of The Filson Historical Society (Colyer 1824). Likewise, flint millstones produced in Franklin County, Kentucky, in 1821 were to be replaced with millstones of the same size by Miller, Railsback, and Miller if there were problems (The Argus of Western America 1821a). Quality was mentioned in Indiana where the purchased millstones in 1827 were a disappointment since they were the "meanest quality" that the mill builder had ever seen (Blackburn 1942:728).

Several sources of transportation were mentioned in the documents. The use of ships for transatlantic shipment of millstones from England to America is documented in a letter referring to Massachusetts during 1631 (Savage 1853:458). In Suffield, Connecticut, during 1672, there is a reference to millstones being shipped by river (Sheldon 1879:23). Another early reference mentions millstones being transported between two horses to Woodbury, Connecticut, in 1681 (Cothren 1872:936). German millstones were shipped to New York before 1732 (Smith 1972:216). In 1735, Stoner (1947:412) mentioned that millstones were transported from the seacoast to Franklin County, Pennsylvania, indicating long distance land transportation. Two years later, reference is made to shipping millstones from New London, Connecticut, to the West Indies in schooners (Anonymous n.d.:3–4). Finally, sloops (boats) and sleds were both utilized for the transportation of millstones in New York during 1773 (Flick 1933:939–940).

A few legal cases relating to millstones were encountered for Lynn, Massachusetts, and New York. In 1692, the city of Lynn passed a law that prevented anyone from carrying a millstone out of town without the city's consent (Lynn Historical Society 1949:13). Persons carrying millstones from Lynn would be prosecuted. Six years later, Lynn passed a law in 1698 that gave Robert Burges permission to remove two millstones from town and to sell them provided that the city received 30 shillings (Lynn Historical Society 1949:54). Lynn passed another law in 1707 that no millstones from the common area could be made in town or sold out of town (Lynn Historical Society 1956:40). In New Amsterdam, New York, a lawsuit was filed in 1653 by Sergeant Huybert against Jan Carreman for failure to deliver a set of millstones that had been ordered (O'Callagahan and Barck 1929:70–71). The court ordered that the millstones be delivered by a specified time (O'Callagahan and Barck 1929:74).

Other references to millstones include comments about someone selling millstones that belonged to a city and an accident on a river where millstones were lost in transit. In Delaware, a man named D'Hinoyossa sold the city's millstones to the English in Maryland in 1662 (Hazard 1850:335; Vincent 1870:395). Thirteen years later in 1673, the city's two former millstones were found at Whorekill and ordered to be sent back (Hazard 1850:409). An early document referring to Ohio mentioned an incident occurring in 1790 where the Wolf Creek men lost the millstones that they were taking up the Muskingum River (Walcut 1879:15).

2

Millstone Quarrying in the United States

The following discussion is intended as an overview of millstone quarrying in the United States. This discussion is not comprehensive but is restricted to information discovered during twenty years of archival research and material received in response to published information requests. No major studies of the American millstone quarrying industry have been previously undertaken. Consequently, this overview has greatly relied on information extracted from the geological reports, milling literature, directories, history books, and other sources. Because of the very limited information contained in the majority of these sources, the descriptions are usually restricted to the quarry locations and material types. In a few cases, more detailed accounts permitted the discussions to go beyond basic descriptions. Since so little is known about the American millstone industry, it was felt that all available information encountered on millstone quarries and millstones should be presented.

During the Colonial period, American mills were primarily using millstones from France, England, and Germany (Sass 1984:vii). After the Revolutionary War and the War of 1812, the number of millstones from Great Britain sharply decreased due to practical and economic reasons (Sass 1984:vii). To meet the need for millstones, millers experimented with a wide variety of stones. Consequently, American millstones were manufactured from several types of stones (Hockensmith 1993b, 1993c, 2004a, 2006c, 2006d, 2008c, 2008d; Hockensmith, Meadows, and Meadows 2008; Howell 1997). Conglomerate (Hockensmith 2002, 2003b, 2005b, 2006d) was quarried in Alabama (Hockensmith 2005a), Arkansas, Connecticut, Kentucky (Hockensmith 1988, 1990a, 1990b, 1993a, 1994a, 1999b, 2000, 2003c, 2004c, 2006c, 2008c, 2008a, editor; Hockensmith and Meadows 1996, 1997a, 1997b, 2006, 2007; Meadows 2002), New York (Hockensmith 2008b, editor, 2008c; Hockensmith and Coy 2008a, 2008b), North Carolina (Hockensmith 2004e), Pennsylvania (Hockensmith 2008b), Tennessee (Ball and Hockensmith 2005; Hockensmith 2004d), Virginia (Hockensmith 1990c, 1999a, 1999, editor; Hockensmith and Coy 1999; Hockensmith and Price 1999), Vermont, and West Virginia. Granite millstones (Hockensmith 2007b) were made in Alabama (Hockensmith 2005a), Arkansas, Connecticut, Louisiana, Maine, Maryland, Massachusetts, Minnesota, Missouri (Hockensmith 2004f, 2006a), New Hampshire, North Carolina (Hockensmith 2004e), Rhode Island, South Carolina, Tennessee (Ball and Hockensmith 2005; Hockensmith 2004d), and Virginia. Flint and buhrstone (Hockensmith 2006c) were quarried for millstones in several states including Alabama (Hockensmith 2005a), Arkansas, California, Georgia (Hockensmith 2004b), Illinois, Indiana, Massachusetts, Missouri (Hockensmith 2004f, 2006a), North Carolina (Hockensmith 2004e), Ohio (Hockensmith 2003d, 2007a, 2008b), South

Carolina, Tennessee, and Virginia. Quartzite was made into millstones in Alabama, Connecticut, and North Carolina (Hockensmith 2004e). Gneiss millstones were produced in Alabama (Hockensmith 2005a), Massachusetts, New York, North Carolina (Hockensmith 2004e), and Tennessee (Ball and Hockensmith 2005; Hockensmith 2004d). Sandstone (Hockensmith 2006f) was exploited for millstones in Alabama (Hockensmith 2005a), Arkansas, Mississippi, Missouri (Hockensmith 2004f, 2006a), Ohio, and Texas (Hockensmith 2006f). Limestone was also used on occasion for millstones (Hockensmith 2006g). Materials rarely used included dolomite in Tennessee (Ball and Hockensmith 2005; Hockensmith 2004d) and Sienite in Massachusetts. Granite boulders were sometimes used in areas lacking suitable bedrock. Finally, a number of quarries were mentioned in accounts that did not specify the raw material type. A general study of the American millstone industry was recently published (Ball and Hockensmith 2007a, 2007b, 2007c, 2007e).

This chapter on American millstone quarries is organized by stone types: conglomerate, granite, flint and buhrstone, quartzite, gneiss, sandstone, dolomite, and sienite. There is also a section on millstones of unspecified materials, and a brief look at the use of boulders for millstone manufacture.

Conglomerate Quarries

Conglomerate millstone quarries operated in several states. The most important quarries were located in Kentucky, New York, Pennsylvania, and Virginia. Conglomerate can be defined (Gary, McAfee, and Wolf 1974:149) as a "coarse-grained, clastic sedimentary rock composed of rounded (to subangular) fragments larger than 2 mm in diameter (granules, pebbles, cobbles, boulders) set in a fine-grained matrix of sand, silt, or any of the common natural cementing materials (such as calcium carbonate, iron oxide, silica, or hardened clay); the consolidated equivalent of gravel both in size range and in the essential roundness and sorting of its constituent particles. The rock or mineral fragments may be of varied composition and range widely in size, and are usually rounded and smoothed from transportation by water or from wave action."

Bowles (1939:71) in his book *The Stone Industries* informed us that: "American buhrstone is a quartz conglomerate occurring on the eastern slope of the Appalachian Mountains, notably in New York, Pennsylvania, and Virginia. The New York variety, known as "esopus" stone, occurs in a strip about 10 miles long extending southward from High Falls in Ulster County. The Pennsylvania variety, known as 'cocalico' stone, occurs in Lancaster County. In Virginia similar rock, known as 'Brush Mountain' stone is found near Blacksburg, Montgomery County."

Lesser known quarries were present in Alabama, Arkansas, Connecticut, North Carolina, Tennessee, and Vermont.

Alabama

During 1848, Michael Tuomey made a report to the Trustees of the University of Alabama on the geology of the Cahawba Valley and portions of Jefferson and Blount counties (Dean 1995). Tuomey (Dean 1995:17) described a millstone grit in Blount County, Alabama, as follows:

> East of the Village Springs, a fine opportunity is presented of observing the lower members of the coal measures in a gorge cut through the Sand Mountain by the Warrior [River]. The escarpment at this place is 200 feet in perpendicular height, composed of sandstones and

> grit. This ridge furnishes numerous localities where millstones of excellent quality can be procured. In selecting the rock it is a good test to break the imbedded pebbles with a hammer — if the cement is sufficiently hard to hold the pebbles thus fractured, it may be considered good; but if on the contrary the pebbles drop out when struck, the rock should be rejected.

Elsewhere in Alabama, "Millstone grit" was quarried near Huntsville in Madison County and a ferruginous conglomerate was quarried in Marion County (Schrader, Stone, and Sanford 1917:14). Dean (1996) spoke of "sandstone and quartz pebble conglomerates of Pennsylvania age Pottsville Formation which is found throughout north Alabama. This rock was quarried at many outcrops for millstones in the 1800s." Further, Dean (1996) noted that "one of the most historic millstone sites is Millstone Mtn. in Winston County, Alabama which is shown the Lynn, Ala., U.S. Geological Survey 7.5 minute topographic map.... This area is in the Bankhead National Forest five miles east of Natural Bridge." Additional information on Millstone Mountain, published by McCalley (1886:78), said that millstone rock "forms a prominent ledge around the ridge. It is very hard and compact, siliceous rock, resembling a good deal in looks the *knox chert* of North Alabama and the *burhstone* of South Alabama. In places it is nothing more than a hard, very fine grain sandstone.... Much of this rock has been made into millstones, and hence the name *Millstone Mountain* has been given to the mountain; it doubtless makes excellent millstones, especially for the grinding of small grain. The ledge on the outcrop is about ... 1 ft. 6 in."

Smith (1924) noted that for 1921 that "Alabama's production of abrasives is at present limited to a small number of millstones (chasers) of Pennsylvanian age, quarried and made at Dutton, Jackson County." Concerning chasers, Smith (1924) provided the following information in a footnote:

> Chasers are larger than regular millstones. They are used for heavier work such as grinding quartz, feldspar, barytes, etc., and they run on edge. Though they are made with a diameter as short as 24 inches, they are usually turned out with diameters ranging from 50 to 80 inches, and are as much as 22 inches in thickness. These chasers are run on pans paved with roughly cubical blocks of the conglomerate, with edges about a foot in length. In grinding quartz in such pans the chasers are used in the preliminary crushing; then rough blocks, usually three in number, are either attached to or carried along by lateral arms, which in turn are joined to a vertical revolving shaft. By circular movement of the blocks the material placed in the pan is ground to powder.

For 1922, Smith (1924) listed one millstone producer: "M. Morris, R. D. 1, Dutton, Ala; quarry in Jackson County." Ladoo (1925:9) noted that millstones had been made near Dutton in Jackson County from Pennsylvanian age sandstones.

Jones (1926:56, 59) in his *Index to the Mineral Resources of Alabama*, stated that "conglomerates of several formations of the State [Alabama], especially of the Weisner Quartzite, the Coal Measures and the Lafayette of South Alabama are capable of yielding good millstones. In the early days both the Weisner and the so-called Millstone Grit of the Coal Measures had quite a well-established reputation for the product. At present, the only production reported from the State comes from a quarry in Jackson County, where a limited number of millstones are taken annually from the Coal Measures."

Arkansas

In Independence County, Arkansas, Owen (1858:217) provided the following description of a conglomerate used for millstones "The conglomerate or millstone grit ... was not seen on the north side of White river, but makes its appearance in the southern part of the county, near Rocky Point post-office, where it contains embedded pebbles. This rock has been quarried, and is held in good repute for millstones. Though not more than fifty or sixty feet in thickness at the above locality, on the south side of Salido creek it increases with its associated shales, to four hundred and eighty feet."

Kentucky

Conglomerate millstone quarries were known to be present within Letcher, Madison, Marshall, Powell, Rockcastle, and Whitley counties, Kentucky. There are also some general references to conglomerate millstones being quarried in Kentucky. The following discussion provides available information on these quarries. A summary of archaeological remains is also provided for Powell County.

A millstone quarry was located near Gordon, Kentucky, in Letcher County. This quarry utilized large conglomerate boulders that had rolled off Pine Mountain (Halcomb 1992). The Letcher County quarry began operation in the early 1800s and supplied local grist mills in this rugged area of southeast Kentucky (Halcomb 1992). During May 1992, Mr. Halcomb discovered that the quarry was flooded by a lake created by a recently constructed dam (Halcomb 1992).

An early reference to a millstone quarry is McMurtrie's (1819:28) statement "Between the head waters of Silver and Station Camp creeks, it runs into a beautiful breecia or pudding stone composed of primitive pebbles chiefly of quartz, a black schorl, that seldom weigh more than an ounce, and a cement formed of lime and sand: it is extremely valuable for millstones, numbers of which are annually manufactured from it, bearing a character for durability, seldom surpassed by any of the imported ones."

Figure 1. *Millstone # 1 at the Upper McGuire Millstone Quarry, Powell County, Kentucky. Photograph taken on April 1, 1987 (facing west), by Charles D. Hockensmith, Kentucky Heritage Council, Frankfort.*

The above description probably refers to Millstone Ridge which is located east of Silver Creek in southern Madison County. The geological map indicates that Millstone Ridge contains conglomerate deposits but we have not yet located the millstone quarry. Conglomerate millstones have been observed in yards on the ridge system and conglomerate appears to be a good quality.

The first reference to a millstone quarry in far western Kentucky was in Loughridge's (1888) geological report on the Jackson Purchase region. Loughridge (1888:274) stated that the conglomerate deposits were "sometimes quarried and used for mill-stones, and seem to answer for the purpose very well, but are gotten out and dressed with difficulty. A noted locality is known as Millstone Hill, on the north side of Jonathan creek, and about one and a half miles east of the Fair Dealing and Aurora road. The rock is a conglomerate of white and dark quartz or flint pebbles, and occurs in ledges two or three feet thick."

Loughridge's (1888) comments were repeated almost verbatim by Lemon (1894) in his book on Marshall County. The author made an attempt to locate this quarry during February of 1992. Unfortunately, limited field survey of the vicinity and conversations with several older residents failed to yield any information about the quarry. The quarry may still exist but the location is so vague that it could fall within several thousand acres of forested land.

Several sources mentioned the millstone quarries of present day Powell County. Earlier in history, this area was located in both Clark and Montgomery counties. Since all these sources refer to the Red River millstone quarries, they are included under Powell County. The earliest reference to a Kentucky millstone quarry is in an interview with William Risk of Kidville in eastern Clark County. He stated "I came up the next Fall, August 1793. The road was cut through at the same time, out to the stone quarry. Sq Road. Mill stones had been cut out in the Knobs, before I came out here" (John Shane, Draper MSS. 11CC 86, Johnson 1965). This reference indicates that millstones were being quarried in present day Powell County prior to 1793. Another early reference to the Powell County quarries was during Benjamin Logan's 1796 campaign for governor. Talbert (1962:284) explained that Logan "was traveling to the town of Winchester to address a gathering. He came upon one of his own wagons which he had sent to the Knobs region near Mount Sterling to obtain millstones. The wagon had been broken down or was mired down."

Top: **Figure 2.** *Millstone # 3 at the Upper McGuire Millstone Quarry, Powell County, Kentucky. Photograph taken on April 1, 1987 (facing east).* Bottom: **Figure 3.** *Millstone # 13 at the Lower McGuire Millstone Quarry, Powell County, Kentucky. Photograph taken on March 17, 1988. Both photographs by Charles D. Hockensmith, Kentucky Heritage Council, Frankfort.*

Two ads have come to light for Red River millstones. The June 13, 1799, edition of the *Kentucky Gazette* contained an ad for the sale of five pairs of Red River millstones at Cleveland's Landing:

Left: **Figure 4.** *Photograph of Millstone # 15 at the Lower McGuire Millstone Quarry, Powell County, Kentucky. Photograph taken on April 12, 1990 (facing north).* Right: **Figure 5.** *Millstone # 14 at the Baker Millstone Quarry, Powell County, Kentucky. Note the parent boulder still attached to the millstone base. Photograph taken on March 30, 1988 (facing west). Both photographs by Charles D. Hockensmith, Kentucky Heritage Council, Frankfort.*

RED-RIVER MILL-STONES

FOR SALE, at Cleveland's landing, five pair of Red-River Mill-Stones of the best quality from that quarry, of the following size, viz.—4 feet,—3 feet 10 inches,—3 feet 8 inches,—3 feet 6 inches,—3 feet, in diameter.—Cash or good horses will be taken in payment.

Likewise, Lease of 200 acres of Land, lying on the West fork of Howard's creek, two miles from the stone Meeting house, for three years, (including the present,) together with the growing crop, consisting of 30 acres of corn, about 6 of tobacco, &c. Also, stills well fixed on distilling, with all conveniences appurtenant thereto. Apply to the subscriber, on the premises.

William Gordon

June 10, 1799

Cleveland's Landing was located on the Kentucky River where the current I-75 bridge crosses in southern Fayette County (Larry Meadows, personal communication 2006).

A second ad for Red River millstones was published in the April 1, 1818, issue of the *Kentucky Reporter*:

MILL STONES.

(OF A SUPERIOR QUALITY,)

WILL in future be made at the RED RIVER QUARRY and sold by SPENCER ADAMS & JAMES DANIEL. If any of our Mill Stones should not prove good, we bind ourselves to furnish more at the quarry until they do prove good. The prices of Stones at the quarry is as follows, to witz—For five feet stones 200 dollars per pair; for four feet 150 dollars; for three feet 80 dollars; all other sizes in the same proportion as above. The prices of a runner will be two thirds the price of a pair. All persons wishing to purchase will apply to the undersigned in Winchester Ky either personally or by letter, which will be thankfully received and duly attended to.

JAS. DANIEL

March 26, 1818

Several lawsuits have been found that mention the Red River millstone quarries or millstone makers (See Hockensmith and Meadows 2006, 2007). An April 1, 1799, lawsuit between Valentine Huff and Peter DeWitt mentioned that DeWitt was a millstone cutter by trade living on Brush Creek (Clark County, Kentucky 1799). Fayette County mill owner John Higbee filed a lawsuit against Absolom Hanks of Clark County over a December 10, 1799, deal that required Hanks to deliver two millstones to Higbee from the Red River millstone quarry (Fayette County, Kentucky 1804). In 1803, James Daniel filed a lawsuit against Martin DeWitt concerning 25 grindstones to be cut by DeWitt in 1801 (Clark County, Kentucky 1803). Joseph Wilkerson sued Spencer Adams over a pair of millstones to be made at the Red River millstone quarry in 1807 (Clark County, Kentucky 1810). During 1822, the administrator of Martin Johnson's estate sued Spencer Adams over his failure to pay Johnson for making millstones at the Red River millstone quarry in 1819 (Clark County, Kentucky 1822). In 1826, Cornelius Summers filed a lawsuit against Spencer Adams concerning a pair of millstones at the Red River millstone quarry (Clark County, Kentucky 1826). Transcribed versions of these lawsuits have been included in the book *The Millstone Quarries of Powell County, Kentucky* (Hockensmith 2008d).

Owen (1861:468) in his *Fourth Report of the Geological Survey in Kentucky Made During the Years 1858 and 1859* mentioned the Powell County millstone quarries in passing: "These knobs border the southern line of the county, and occasionally, when capped with the conglomerate, attain a considerable height, as is the case with the 'Pilot Knob,' between Black and Lulbegrud Creeks, remarkable for its millstone quarries."

In the 1884 publication *Report on the Geology of Montgomery County,* Linney (1884:65) explained, "The great sandstone, often called the millstone grit, is the top rock on the highest part of Morris Mountain, Pilot, and Kash's Knobs. On the Clark Pilot, which is in Powell county, but near the Montgomery county line, it is one hundred and thirty-five feet thick, marked in nearly all its parts with rounded pebbles of white quartz.... This rock has been quarried on Pilot Knob and made into millstones, and they are highly prized under the name of hailstone grit."

Charles Colyer operated a millstone quarry in Rockcastle County, Kentucky, during the early 19th century. *The Argus of Western America*, Frankfort, carried an ad in the January 8, 1812, edition for a millstone quarry in Rockcastle County:

MILL STONES

> I HAVE rented my QUARRY on Roundstone, about three miles from Rockcastle court house, to Mr. Colycar, who is now at work in manufacturing

Left **Figure 6.** *Millstone # 7 at the Toler Millstone Quarry, Powell County, Kentucky. Photograph taken on March 27, 1987.* Right: **Figure 7.** *Millstone # 8 at the Toler Millstone Quarry, Powell County, Kentucky. Photograph taken on March 27, 1987 (facing south). Both photographs by Charles D. Hockensmith, Kentucky Heritage Council, Frankfort.*

Figure 8. *Millstone # 23 at the Ware Millstone Quarry, Powell County, Kentucky. Note the parent boulder at the base of the millstone. Photograph taken on March 23, 1989 (facing south) by Charles D. Hockensmith, Kentucky Heritage Council, Frankfort.*

MILL STONES,

Which I believe to be equal to any in America; at least they are equal to any that I have ever seen. Mr. Colyear proposes to sell on good terms, and take a small part in Whiskey.

Samuel Taylor.

January 7th, 1812.

The following year, Mr. Colyer published his own ad for the above millstone quarry in the October 9, 1813, edition of *The Argus of Western America*:

MILL STONES

THE subscriber has on hand a large and general assortment of MILL STONES, and intends keeping on hand all sizes of MILL STONES — which he will sell at his usual price — the Stone Quarry is within three miles and a half of Mount-Vernon, Rockcastle county, Kentucky.

N.B. — The MILL TONES [sic] are of a superior quality to any in this country, and are insured-and if they are not good, the money will be returned or another pair that shall be good.

CHARLES COLYER, JR.

The Filson Historical Society of Louisville, Kentucky, has an original handwritten contract between Charles Colyer and Sidney Payne Clay of Bourbon County (Colyer 1824). This document is from the Sidney Payne Clay Papers (A\C621a, folder 5) in The Filson Historical Society's Special Collections. The text of the contract and directions to Clay's house are quoted below with the permission of The Filson Historical Society:

28 Feb 1824

I Charles Colyer of Rockcastle County the state of Kentucky do hereby agree to deliver to Sidney P. Clay at his residence in Bourbon County on or before the 15 day of April next a pair of mill stone three feet four inches in diameter the runner to be eighteen inches thick through the eye and the bed stone about ten inches thick and do warrant the same to be good & sufficient. For which the said Clay does hereby bind himself his heirs &c. to pay said Colyer ninety dollars commonwealth money on the delivery of the stones in wintess [sic, witness] whereof we have hereto let our hands and seals this 28th day of February 1824.

Sidney P. Clay (seal)
Charles Colyer

attst
George S. Shirley

A separate page attached to the contract provided directions to Sidney Payne Clay's house in Bourbon County and contained a receipt that the millstones were delivered:

Directions to where Sidney P. Clay lives. Keep the road from Lexington to Paris until you pass Bryan's station about two miles until you come to the Iron works road. Take the Right hand end of the Iron works road & in about seven miles you will come to Clay house near the side of the road.

Figure 9. *Millstone # 1 at the Ewen Millstone Quarry, Powell County, Kentucky. Photograph taken on April 25, 1990 (facing west), by Charles D. Hockensmith, Kentucky Heritage Council, Frankfort.*

> April 20th 1824 Received of Sidney P. Clay Ninety dollars for a pair of mill stones of the size mentioned in within article.
>
> Test
> Will. Hearice [or Hearne] John Wilson
>
> Sidney p. Clay.s obligation $ 90

Owen (1861:482) mentioned millstones being quarried in Rockcastle County during his geological survey conducted for 1858 and 1859: "The conglomerate member, which in the southeast, is 80 feet thick, thins out towards the head of Roundstone creek. ...On Roundstone creek, six miles above its mouth, a quarry has been opened into this rock, which was formerly extensively worked for millstones.

Collins and Collins' (1874:376) *History of Kentucky* noted that parts of the eastern coalfield contained layers of millstone grit (conglomerate) which was used for millstones. In their discussion on Rockcastle County, Collins and Collins (1874:691) stated that "on Roundstone creek, 6 miles above its mouth, a quarry was formerly extensively worked for millstones." In his description of the geology of Rockcastle County, Sullivan (1891:18) wrote that the "Conglomerate formation, in the northern and central portions of this region, tops the limestone hills, and varies from 45 to 100 feet in thickness. In the southern part, the entire mountain above drainage consist of this formation, it being 150 to 250 feet thick. Much of the Conglomerate, in this region, consist of the 'Hailstone Grit,' the ledges of sandstone being very thick and the quartz pebbles large."

A millstone quarry was reported to be in the vicinity of Cumberland Falls in Whitley County, Kentucky. Reports of this quarry have been brought to the author's attention by Tom

Sussenbach (personal communication 1991) and Clarence Halcomb (1992). The cap rock at Cumberland Falls is a well cemented conglomerate and it is suspected that this material was utilized for millstones. The exact location of the quarry is presently unknown.

During 1847, Lewis Collins (1847:158) published his *Historical Sketches of Kentucky*. In his description of the "Conglomerate Coal Series" in Kentucky, he says the "Conglomerate or pudding stone ... is composed of coarse pebbles of quartz, and fine grains of sand, rounded and cemented together by a silicious cement.... The rock is very firm, and is sometimes used for millstones to grind Indian corn."

Arnow (1983:392) in her *Seedtime on the Cumberland* noted, "Grindstones were often cut from the harder sandstones, while a particularly hard, but not too coarse, conglomerate found above or between layers of sandstone could be shaped into millstones."

Archaeological Investigations

Six millstones quarries located in northwest Powell County, Kentucky, have been thoroughly documented by archaeological investigations. Powell County is

Top: **Figure 10.** *Millstones # 11 at the Pilot Knob Millstone Quarry, Powell County, Kentucky. Note the leveling cross on the upper surface. Photograph taken on April 13, 1988 (facing west).* Bottom: **Figure 11.** *Pit on Ridge B at the Ewen Millstone Quarry, Powell County, Kentucky. Boulder # 40 is in this cluster of boulders. Photograph taken on April 25, 1990 (facing northwest). Both photographs by Charles D. Hockensmith, Kentucky Heritage Council, Frankfort.*

Above: **Figure 12.** *Boulders # 1–5 at the Ware Millstone Quarry, Powell County, Kentucky. Boulder # 1 is to the right and Boulder # 2 in the center. Photograph taken on March 16, 1989 (facing east).* Left: **Figure 13.** *Boulder # 10 at the Ware Millstone Quarry, Powell County, Kentucky. Note the closely placed drill holes used to split the original boulder apart. Photograph taken on March 17, 1989 (facing east). Both photographs by Charles D. Hockensmith, Kentucky Heritage Council, Frankfort.*

situated along the east central boundary of the Knobs Region, a semicircular band of high knobs or hills that surround the Bluegrass Region. Three quarries are located at Rotten Point, two at Kit Point, and one at Pilot Knob. The millstones manufactured at the Powell County quarries are all monolithic (one piece) and are made from a Pennsylvanian age conglomerate. It outcrops near the crest of two knobs and also occurs in boulder form on the steep slopes and narrow ridges forming the knob sides. Two of the quarries exploited in situ conglomerate deposits while the four remaining quarries utilized scattered boulders.

A variety of remains were documented at the six Powell County millstone quarries (see Hockensmith 2008d). The quarries range in size from about 1 ha to 14 ha (2.5 to 35 acres). Quarry byproducts are found on ridge tops, slopes, and within streambeds. The raw material used for the millstones is a conglomeratic sandstone which ranges from a light gray to light tan and contains numerous rounded quartz pebbles (mostly white with some yellow, brown, and pink). Most of the quartz pebbles are less than 2 cm in diameter, but occasionally pebbles up to 5 cm in diameter are present. Chert, sandstone, and fossil inclusions are also present in some of the conglomerate. The archaeological evidence associated with these quarries include millstones in various stages of completion, boulders with drill holes, shaping debris, quarry excavations, tool marks on boulders and millstones, and historic artifacts such as iron tool fragments.

Measurements were recorded for 131 millstones documented at the six quarries. As a whole, the millstones range in size from 50 cm to 1.6 meters (20 to 64 inches) in diameter, with most specimens clustering between 80 cm (32 inches) and 1.15 meters (46 inches). These millstones range in thickness from 15 to 70 cm (6 to 28 inches), but most cluster between 25 and 45 cm (10 and 18 inches). Twenty-four of the millstones contained central holes or eyes. The frequency of millstones per quarry are as follows: McGuire Quarry (38 millstones), Baker Quarry (28 millstones), Toler Quarry (15 millstones), Ware Quarry (33 millstones), Ewen Quarry (5 millstones), and Pilot Knob Quarry (12 millstones). Examples of millstones from these quarries include those from the Upper McGuire Quarry (Figures 1–2), Lower McGuire Quarry (Figures 3-4), Baker Quarry (Figure 5), Toler Quarry (Figures 6–7), Ware Quarry (Figure 8), Ewen Quarry (Figure 9), and the Pilot Knob Quarry (Figure 10).

Millstones were encountered in various degrees of completeness at the six millstone quarries. Documented stages include nearly complete millstones, advanced performs, oval/rounded preforms and rectangular preforms. The oval/rounded preforms are present at all the quarries and are the most common category. Rectangular preforms were only observed at four quarries, McGuire, Baker, Ware, and Ewen. Advanced stage preforms were encountered at all six quarries but were most common at the Toler and Ware quarries. Nearly complete millstones were found at all the quarries except the Ware and Pilot Knob quarries.

Leveling crosses were present on millstones at three quarries and a fourth quarry may have one example on a boulder. The leveling crosses were placed into four categories. The first is the complete cross that still retains two perpendicular troughs across the millstone forming a "+". The second category is a T-shaped cross where half of the cross (portions of both troughs) was removed during the leveling process. Basically, the leveled area forms the top of the "T" while the remaining segment of trough forms the vertical part of the "T." The third category is the L-shaped cross where one-fourth of the millstone has been leveled, leaving an L-shaped portion of the troughs. The fourth category is the I-shaped cross which is the initial stage of leveling. At this stage, only one trough has been cut across a millstone. They appear to represent one method of leveling the tops of millstones. First, a level trough was cut across one axis of a millstone. Logic would suggest that the depth of the cross was slightly lower than the lowest surface on the preform. Next, a second trough was cut across the millstone at a perpendicular angle. It is assumed that a simple level with a bubble was used to ensure that the troughs were level across both axes. Four quadrants were created when the cross was cut. The cross troughs provided level sighting lines across both axes of the millstone so that the worker could use these as reference points. Each quadrant was removed individually until the entire surface of the millstone was leveled. A paint staff was probably used to identify the high spots, which were pounded down with a bush hammer.

Twenty-four of the millstones contained central holes or eyes. The eyes ranged from 17 to 23.5 cm (6.8 to 9.4 inches) in diameter with most eyes clustering between 16 and 20 cm (6.8 and 8 inches) in diameter. All of the eyes were round. However, one exception is a millstone in the collection of the Red River Historical Society's museum. This millstone was found in the general vicinity of the Pilot Knob Millstone Quarry and has a square eye (22 × 22 cm).

Historical accounts suggest that eyes were first cut round and then squared later if so desired. Since the millstones left at the quarries were rejected, final details such as squaring the eyes for bedstones would not have been done in most instances. Some of the eyes were cut all the way through the millstones while some eyes were cut part way through.

The millstone maker's hard labor was not always fruitful. At any time during the manufacturing process, a critical mistake could be made or a flaw uncovered inside the rock that ruined the millstone. Sometimes a careless blow or a small seam in the rock would result in the edges of the millstone being severely damaged. In other instances, the stone would break in an irregular manner when being separated from the parent rock. At other times the millstone would survive until the eye was cut and then the rock would crack or split. Millstones were rejected for several major reasons. Often two or three flaws were obvious on the same millstone. These flaws included irregular breaks, undercutting, edge damage, surface depressions, cracks, inclusions, splitting apart, and unknown causes. Overall, the millstones at the six quarries were most commonly rejected because of irregular breaks during shaping and the undercutting of the sides.

During the investigations, 229 boulders with drill holes were documented. Holes were drilled in the boulders so that the stone could be split into desirable sizes and shapes. In terms of size, the boulders ranged between 30 cm and 2.9 meters (1 to 9.6 feet) in length with most clustering between 38 cm and 1.6 meters (1.3 to 5.3 feet). Boulder width ranged between 20 cm and 2.4 meters (8 inches to 7.9 feet) with most clustering between 30 cm and 1.1 meters (1 and 3.6 feet) in width. Boulder thickness ranged between 15 cm and 1.75 meters (6 inches to 5.8 feet) with most clustering between 25 and 80 cm (10 to 32 inches) in thickness. The boulders usually contained between one and six drill holes with most boulders containing between two and four drill holes. Some very large boulders contained between 14 and 19 drill holes each. No attempt was made to document those boulders lacking drill holes or tool marks. Examples of boulders from these quarries include those from the Ewen Quarry (Figure 11) and the Ware Quarry (Figures 12–13).

Several observations were made on drill holes. Drill holes were recorded in the following diameters: 2.5 cm (1 inch), 3 cm (1 3/16 inches), 3.5 cm (1 3/8 inches), 4 cm (1 9/16), 4.5 cm (1 3/4 inches), and 5 cm (2 inches). The most common sizes were 3.5 cm, 3 cm and 4 cm while the 2.5, 4.5, and 5 cm holes were very rare. Drill holes varied in depth from 3 to 25 cm (1.25 to ca. 10 inches) with most clustering in depth between 9 and 15 cm (3.5 to 6 inches). The holes were spaced between 7 and 53 cm (2.75 to ca. 21 inches) apart with most ranging between 18 and 31 cm (7.25 to 12.25 inches) apart. Fracture types produced when the stones were split along the drill holes were straight, irregular, concave, and convex (in order of frequency).

Fragments of conglomerate associated with the shaping of millstones are present in varying quantities across the six quarries. In some cases the shaping debris was apparently discarded where it fell. In areas where extensive millstone shaping occurred, large quantities of conglomerate fragments were generated. These fragments were either piled into low mounds or disposed of by throwing them down slope. The greatest quantities of fragments occur near the knob crests where conglomerate outcrops were exploited. Quarry activities on slopes, ridge tops, and in streambeds generated smaller quantities of debris since only scattered individual boulders were being shaped.

Excavations undertaken to expose conglomerate deposits include shallow oval pits, linear pits, and benches. The shallow oval pits (n = 69) appear to be associated with the uncovering of individual boulders or clusters of boulders. These pits range in size from one to six meters (ca. 39 inches to 19.75 feet) in diameter with most ranging between two and three meters in diameter. Their depth ranged from 30 cm to 1.5 meters (1 to 4.95 feet) but most pits were between 50 cm and 1 meter (20 to 40 inches) in depth. Sometimes portions of quarried boulders with drill holes remain in the bottoms of these pits. Occasionally, partially quarried boulders showing a sequence of slabs being removed are still present. One extremely large oval pit (17 × 18 meters and 1.5 meters deep) was documented near a knob crest where in situ bedrock was being exposed.

The linear pits and bench appear to be associated with the removal of over burden to expose in situ conglomerate deposits. The linear pits (n = 8) range in length from 6 to 33 meters (19.8 to 108.9 feet), in width from 2 to 5 meters (6.6 to 16.5 feet), and in depth from 30 cm to 3 meters (1 to 9.9 feet). The benches (n = 9) range from 30 to 88 meters (99 to 290.4 feet) in length, 5 to 20 meters (16.5 to 66 feet) in width with highwalls ranging 2 to 9 meters (6.6 to 29.7 feet). Most linear pits and benches occur near the knob crest and adjacent slopes where conglomerate outcrops were being exposed. However, linear pits and benches are sometimes found on ridges.

Several iron artifacts were discovered by local individuals at the Baker, McGuire, Toler, and Ware millstone quarries. They are currently within the Red River Historical Society Museum collections in Clay City, Kentucky. The recognizable tool types recovered from the Powell County millstone quarries include points (n = 3), wedges (n = 10), feathers (n = 8), and miscellaneous (n = 3). The points were struck with a hammer to shape the millstones. Feathers and wedges were used in each drill hole to split the stone apart

New York

The best known and most productive conglomerate millstone quarries were located at High Falls in Ulster County, New York, at a place known as the Trapp (Howell 1985:144–146; Howell and Keller 1977:69; Ladoo and Myers 1951:9; Parker 1896:586; Reis 1910:284; Sass 1984:viii; Schrader, Stone, and Sanford 1917:225; Swisher 1940:59). These millstones were known as Esopus stones. They were manufactured from single blocks of light gray to white Shawangunk Conglomerate grit and were usually banded with iron hoops (Howell and Keller 1977:69; Sass 1984:viii). Craik (1870:293, 1882:293) described this material as "a bed of small white pebbles congealed together in a darker matrix, which completely fills the interstices, and leaves no empty cells like those in the burr." An 1875 advertisement for the Esopus Millstone Company indicated that they produced millstones for millers, paving and color mills, paint and chemical mills as well as potteries and china works (Howell and Keller 1977:71).

Millstones were quarried in Ulster County as early as 1732. William Smith's *The History of the Province of New-York from the First Discovery to the Year 1732* (Smith 1972:216) stated, "At the commencement of the range of the Apalachian Hills, about 10 miles from Hudson's River, is an inexhaustible quarry of millstones, which far exceeded those from Colen in Europe, formerly imported here, and sold at ... £80 a pair.

The papers of Sir William Johnson include the following note (with misspellings and incomplete words) from J. Hasbrouch of Kingston, dated July 1, 1766 (Flick 1927:296):

> [ ] favor of the 9th of June last I Received [ ] Mill Stones have Laid at the landing [ ]d not Oppertunity to Send them to Albany [ ] [ ] the Mill Stones Now up by Mr. Wm. W[ ] to Doctor Stringer you are to pay the [ ] the Mill Stones, I hope the Mill Stones [ ] your likeing. [You] Wrote me If I Could not Appoint a per[son at Alban]y to receive the Money for the Mill Stones [ ] the price of the Mill Stones which is £ 19. and [ ]ck Swart of the City of Albany, whom I Sha[ ] it for me and his Receipt Shall be your D[ischarge] Not Ready made Another pair of Mill Stones [ ] Git them Made as Soon

Figure 14. *View of Rubble at One of the Work Areas at the Lawrence Hill Millstone Quarry in Ulster County, New York. Photograph taken on April 15, 1998, by Charles D. Hockensmith, Kentucky Heritage Council, Frankfort.*

> as possible and Send [ ] Albany for you, I am with Due regard your Sincere friend and Humble Servant
>
> J. Hasbrouch

On May 12, 1770, Samuel Stringer of Albany sent the following correspondence (with misspellings) to Sir William Johnson (Flick 1931:664):

> Last week your Mill Stones arived from Eusopus [sic], and according to my opinion are fine ones; I have measured them, & they agree very well with the dimensions you sent me. The price of them is £19..5.., and £ 3/. for the freight. Immediately on their arival I wrote to Mr. Van Eps, beging him to send down for them the next morning, as they coud be more conveniently put on a waggon out of the sloop, than after they had been landed; but he let me know that it was impossible to have them rid before it raind, the Road being so dry & sandy. I have this day sent to him again, leting him know you were in great want of them.
>
> I hope before his, you have experienced some good Effects, as to your Health, from your Sackendaga Amusements.
>
> My Comp^t. To M^r. Daily if you please.
> I am S^r.
> Y^r. Most Obed^t. Hum^l. Servt.
> SAM^L. STRINGER

A third millstone related entry (with misspellings and incomplete words) in the papers of Sir William Johnson included a December 9, 1773, note from J. Hasbrouch of Kingston, to Sir William Johnson in the County of Tryon (Flick 1933:939–940):

> I have your favor of [ ] and at your Request I have Provided you a pair of Mill Stones, Delivered at our Landing, I Could not Git Any oppertunity To Send Them To you, To albany The Season of the Year, Was So far Spent that there Was no Sloops, to Carry them up, To Albany, So you Will Be oblidged to Send for them this Winter, To fetch Them With Sleds;—I hope the Stones Will Please you, They Be of the Best Sort, they be 4 feet 7 or 8 Inches Diameter ab[t]. 13 Inches Through the Eye, I Could Git Them no Cheaper Than £18 for the pair Delivered at the Landing....
>
> A HASBROUCK

Horatio Gates Spafford (1813:110) in his publication *A Gazetteer of the State of New-York* wrote, "The Esopus mill-stones, have as high as widely extended reputation; and are found in vast abundance in the Shawangunk mountains of this county which little if any inferior to that imported for the Burr mill-stones."

William Mather (1843:357) mentioned the millstones in his *Geology of New-York* (Printed with permission of the New York State Museum, Albany, N.Y.): "The firm coarse grits have been long quarried for mill-stones, and have been extensively used. They are known in market by the name of *Esopus millstones*. They are still quarried to a small extent; but since the French buhr stone has been brought into common use for millstones, the Esopus stones are in less demand. Many small millstones, for family use, in grinding corn among the planters of the southern States, are still manufactured and sent to market."

Figure 15. *Roughly Shaped Millstone Preform and Rubble at the Lawrence Hill Millstone Quarry in Ulster County, New York. Photograph taken on April 15, 1998, by Charles D. Hockensmith, Kentucky Heritage Council, Frankfort.*

Four years later, Joseph H. Mather's (1847:31) *Geography of the State of New York* contained the following comment: "The gray sandstones and conglomerate of the Champlain and Erie groups, furnish *grindstones* of superior quality, and from the Shawangunk grits, *millstones* have been manufactured, which compared well to the French buhrstone."

Hall (1852:36) provided the following information on the Oneida conglomerate or Shawangunk grit found in New York:

> This rock is a coarse sandstone or conglomerate, resting upon the Hudson River group, and from which there is sometimes a gradual passage. It forms the line of division between the Lower and Upper Siluran....
>
> This rock is used for millstones, and the Esopus millstones are made from it. It forms the Shawangunk Mountain, in New York, and the continuation of the same range in New Jersey. It received the name Oneida conglomerate from its occurrence in Oneida county, being the only conglomerate rock in that part of the State of New York. It there rests conspicuously on the rocks of the Hudson River; and the Medina sandstone being absent, it is suceeded by the Clinton group.

In 1870, Craik (1870:293) noted, "Formerly mill stones were made of granite, or some other flinty conglomerate rock.... These stones are scarcely ever used at the present day, except in very remote new settlements, being almost entirely superseded by the French Behr, or 'burr stones.' This displacement, however, has been gradual, and one kind of these rock stones, known as Esopus or 'Soper' (Yankee) stone, maintained its ground against the burr stone

Figure 16. *Roughly Shaped Millstone Preform at the Lawrence Hill Millstone Quarry in Ulster County, New York. Photograph taken on April 15, 1998, by Charles D. Hockensmith, Kentucky Heritage Council, Frankfort.*

innovation until very recently, and a very few of these still linger in good mills, in various parts of the country, to the present time."

The *Gazetteer and Business Directory of Ulster County, N.Y. for 1871–2* provided the following information on Esopus millstones (Child 1871:64): "At an early period the Esopus grit was largely quarried and manufactured into millstones which were said to exceed those imported from Colen, in Europe, at the cost of £80 a pair, while the Esopus stones cost less than one-fourth of that sum."

In the October 1879 issue of the *Manufacturer and Builder* (1879: 226–227), a brief article appeared:

ULSTER COUNTY MILL STONES,

> The industry of quarrying mill stones along the Shawangunk mountains, in the vicinity of Alligerville, N.Y., and through that section for a number of miles has become quite a large business. These stones in some cases are very large, often measuring in diameter from 5 to 6 feet, and weighing as much as 4 tons. After being lifted from the quarry, they are rounded by workmen, and a hole drilled through the center for the axle. They are extremely hard and flinty, so that the hardest kinds of instruments have to be used in working on them, which work is done entirely by hand with the hammer and point. The ordinary tools used in the bluestone business would be pointless after one or two clips on this hard, gritty rock. They are principally shipped from Alligerville.
>
> These stones are mostly used for grinding in grist mills, and for grinding cement. Three of these stones recently quarried about two miles from the bank of the canal, are enormous, measuring, after being rounded, 6 feet in diameter and 20 inches thick; they weigh 4 tons, and it required a great deal of labor to get them from the quarry to the dock, owing to the softness of the road bed, the wheels of the wagons cutting in to the hubs every little distance. The roads will have to be considerably improved if this business continues to increase.

Figure 17. *Nearly Complete Millstone from the Lawrence Hill Millstone Quarry in Ulster County, New York. Note crack across the stone and remaining parent rock on left side. Photograph taken on April 15, 1998, by Charles D. Hockensmith, Kentucky Heritage Council, Frankfort.*

Swift (1988:6) in his discussion of the millstone industry noted that the "1880 list of the Principal Manufacturers, Merchants and Farmers of Ulster County lists six millstone dealers in Accord, two in Alligerville and one in Kyserike. One Accord entry reads: 'Gilbert Lawrence, Farmer and dealer in Mill Stone, from 20 inch up to any size, Grit from Reddish to White.'"

Grimshaw (1882:288) wrote that "Esopus stones from Ulster County, N.Y., are good for corn or oats, making soft pleasant meal for family use. They do not keep their edge long."

Craik (1870:293–294; 1882: 293–294) noted some problems with the conglomerate Esopus millstones:

> These stones resemble in texture a bed of small white pebbles congealed together in a darker matrix, which completely fills the interstices, and leaves no empty cells like those in the burr, the sharp edges of which preform so important a part in the process of grinding. The Esopus stone has therefore no cutting edge except the dress made by the mill pick, and when this is worn out, it takes a polish as smooth as glass, and to grind with it in this condition would be like trying to saw with the back of a saw. It is also much softer than the burr stone, and for this reason the dress is much sooner worn out; but this softer nature tells in its favor in one respect, that it is far easier to dress than the burr stone, and this fact helps to reconcile millers to its use.

An article entitled "Shawangunk Mountain" in the March 17, 1894, edition of *National Geographic Magazine* contained the following reference to the Ulster County millstone industry (Darton 1894:25):

> The lakes for which the mountain is famous lie in basins of moderate depth and are all near the top of the range. They are nearly surrounded by cliffs of Shawangunk grit of greater or less height, which add greatly to their beauty. The grit is mainly a messive white or gray quartzite or conglomerate, averaging 250 to 300 feet thick. The proportion of pebbles is large but variable, many beds being fine. The pebbles and grains are quartz, and the matrix is siliceous. The conglomerate is the famous Esopus millstone, and has been largely quarried for two centuries.

Newland (1907:43–44) in his report *The Mining and Quarry Industry of New York State* for 1906 provided the following information (Printed with permission of the New York State Museum, Albany, N.Y.):

> Millstones are obtained in Ulster county. The industry is a small one, but it has been established for more than a century and still furnishes most of the millstones made in this country. The product is known as Esopus stone, Esopus being the early name for Kingston, once the principal point of shipment.
>
> The millstones are quarried from the Shawangunk grit, a light gray quartz conglomerate found along the Shawangunk mountain from near High Falls southwest toward the Pennsylvania border. The Cocalico stone obtained in Lancaster county, Pa., and the Brush mountain stone, found in Montgomery county, Va., are of similar character. In Ulster county, the grit rest upon the eroded surface of gray Hudson River shales and is overlain by red shale. It has generally been correlated with Oneida conglomerate of central New York, though recent investigations have shown that it belongs higher up in the series, namely in the Salina. Its thickness ranges from 50 to 200 feet.
>
> The grit is composed of quartz pebbles of milky color enclosed in a siliceous matrix. The pebbles are more or less rounded and vary from a fraction of an inch up to 2 inches in diameter. The texture is an important factor in determining the value and particular use of the finished millstones.
>
> The size of the stones marketed range from 15 to 90 inches. The greater demand is for the smaller and medium sizes, with diameters of 24, 30, 36, 42 and 48 inches. A pair of 30-inch millstones commonly sells for $15, while $50 may be paid for a single stone 60 inches in diameter. The largest sizes bring from $50 to $100. Besides the common type of millstones, disks are furnished which are employed in a roll type crusher known as a chaser. The pavement of such crushers is also supplied by the quarrymen, in the form of blocks. Quartz, feldspar and barytes are commonly ground in chasers.
>
> Most of the Ulster county quarries are situated along the northern edge of the Shawangunk mountain. Kyserike, St Josen, Granite and Kerhonkson are the principal centers of the industry while the distributing points include New Paltz and Kingston in addition to those named. The industry is carried on intermittently, many of the producers engaging in other occupations during a part of the year.
>
> The market for millstones has been curtailed of late years by the introduction of rolls, ball

> mills, and other improved forms of grinding machinery. The roller mill process has displaced the old type of cereal mills, particularly in grinding wheat. The small corn mills distributed throughout the southern states, however, still use millstones and furnish on of the important markets for the New York quarries. A part of the product is sold also to cement and talc manufacturers.

Phalen (1908:609) reported in his discussion of abrasive materials for the year 1907 that the "material suitable for millstone is quarried from Shawangunk grit, a quartz conglomerate found near the western base of Shawangunk Mountain in the valley of Rondout River. The material suitable for millstones is exceedingly limited, being confined in linear extent to a strip extending from High Falls on the north to Kerhonkson on the south, a distance of approximately 10 miles. Beyond these limits the texture and other properties of the rock have been found unsuitable for the highest grade of stones."

Additional updates for the Ulster County millstone industry were published in *The Mining and Quarry Industry of New York State* including reports by Newland (1907, 1908, 1909, 1911, 1921), Hartnagel (1927), Newland and Hartnagel (1932, 1936, 1939), and Hartnagel and Broughton (1951). Much of the information in these reports was repetitive but new details and production figures were added from time to time. Newland (1909:38) provided more information on where the millstones were produced (reprinted with permission of the New York State Museum, Albany, N.Y.): "The quarrying operations are carried on along the northern border of the Shawangunk mountain, in Rochester and Wawarsing townships, Ulster co., mainly along the line of the New York, Ontario and Western Railroad at Wawarsing, Kerhonkson, Accord, Kyserike, Granite, St Josen and Alligerville, while New Paltz and Kingston are shipping points."

Concerning the quarrying of the stone, Newland (1916:38–39) shared the following account (Printed with permission of the New York State Museum, Albany, N.Y.):

> The work of quarry requires only small equipment, the stone being pried out by hand, after the use perhaps of a drill and plugs and feathers. Sometimes a little powder may be employed, but care has to be exercised in its use to avoid weakening the stone. The spacing of the natural joints determines the size of the stone that may be produced, the joints occurring in two sets approximately parallel to the dip and strike of the formation. The rough blocks thus obtained are reduced to shape by the hammer and point and then undergo a final tool dressing which varies with the use to which the stone is to be put. The hole or "eye" in the center is drilled by hand.

Additional information on making New York millstones was provided by Ladoo (1925:9). Newland (1921:155) gave more information on drilling the eye of the millstone in a subsequent report (Printed with permission of the New York State Museum, Albany, N.Y.): "The last operation is to drill the center hole or eye. This is usually done by starting a hole in the center of one of the sides and drilling half way, then reversing and cutting a second hole on the opposite face to meet the first. The round eye thus formed may be squared up."

Hartnagel (1927:56) also mentioned the use of a "bull point" along with a hammer for finishing millstones.

Between 1919 and 1924, the market for New York millstones included the following (Hartnagel 1927:56, printed with permission of the New York State Museum, Albany, N.Y.): "Handmade millstones are still being produced for some corn and mustard mills in the South. Millstones are also used by manufacturers of cement and talc and for grinding quartz, feldspar, barytes and mineral paint."

In addition to other uses mentioned above, Hartnagel and Broughton (1951:64) stated that "at one time there was a market for very small stones only a few inches in diameter. These were mainly for use in hand-driven spice mills."

The millstone industry was so successful that ten companies were still producing millstones as late as 1934 (Sopko 1991). Katz (1920:217) noted, "For many years Ulster County, N.Y. led in the production of millstones and chasers (stones which run on edge or on a horizontal shaft) but in recent years the state yielded first place to Virginia in production of millstones, although still leading in output of chasers."

Michael Swift wrote an excellent discussion of millstone making in the Accord area in 1988. The following quotes are from Swift's (1988:6–7) brief article:

> Any type of stone could be made into a millstone, but a quality millstone was hard and fine-grained. Early on, stones were often imported from France, where Europe's best millstones were produced. But somebody discovered that a very good millstone could be made from Shawangunk or Esopus granite (actually not a granite but a conglomerate). With the opening of the D & H Canal in 1823, shipment by water to all parts of the country made it possible for the millstone business to boom, and Accord, Alligerville, Kyserike and Kerhonkson all produced and shipped out stones by the thousands.
>
> This conglomerate was originally laid down so that it had a grain not unlike that of wood. Like firewood it was much easier to divide along the grain than against it. The stonecutters would drill down to a natural seam, say twenty inches thick, then pack black powder into the hold [sic] and blast so that the stone would split along the seam. This yielded slabs of various sizes and generally of twenty inch thickness. The slab was cut off into rough blanks of approximately square dimensions. A circular line was drawn on the blank from a small hole drilled in the center. The stone would be broken off around the line with the drill, or with sets of iron wedges known as feathers. The feathers went into a drilled hole and then were forced apart by a "plug" that was hammered between the two feathers, forcing them apart and breaking the stone. The drill was not mechanical, but was simply a bar with a tempered cutting bit on the end. One man held the bar and turned it while two men with eight-pound sledges hammered on the top end, one after the other. With such hard stone, and such basic tools, a blacksmith was an important member of the crew, constantly sharpening, straightening and tempering the tools. The eye — a hole in the center of the stone — was cut last....
>
> The millstone rock was worked very close to the surface; the land was worthless for farming or anything else, so there were no questions of mineral rights or boundaries. Stone cutters worked where they wanted, and moved their shacks, forges and tools from one area to another at will. The best stone was found along the ridge between High Falls and Kerhonkson. The stones were hauled out by horse-drawn "stone boats" on a network of small roads, now mostly overgrown. In 1893 about 350 tons of millstones were shipped, mostly by canal, and some through the Rosendale railroad station....
>
> In their youth, Accord residents Wally and Vinnie Lawrence hauled stones from the mountains in a pickup truck, down to the loading dock in Accord where they were loaded onto trains. This was well after the biggest years of millstone production. By the time the Lawrence brothers were hauling stone, specialty products like Gulden's mustard were the biggest buyers of grinding stones....
>
> Later, the Lawrence brothers worked stone with pneumatic tools, but they remember when the old hand methods prevailed. They remember the old-time stonecutters as a rough breed — hard-working, often hard-drinking too. But they were always very kind to the young brothers. Stone was a brittle, unpredictable material, and many stones broke at some stage of shaping. Two or three days of work could be wasted when the stone cracked as the eye was drilled. Failed stones can still be found, with woods and brush grown up around them. Stonecutters sometimes hung failed stones from tree branches, a rueful comment on the vicissitudes of their trade.

Weaver (1995:1–5) interviewed the Lawrence brothers of Lawrence Hill at Accord. It appears that Swift (1988) also obtained much of his information from the Lawrence brothers. The interview deals with their lives in general, but refers briefly to their involvement in the millstone industry (Weaver 1995:2–3):

> The brothers also made millstones to grind matzo flour for a place in Gardiner. Years later they helped make a documentary about the area stone cutting industry with Gordon Duffy in conjunction with Ulster Community College (UCCC). Duffy, a director for the Phoenix Theater in New York, was Vincent's neighbor at the time. He has since moved to California. "It showed all the quarries in the area," Vincent explains, "It showed stone cutting from quarry to finish. We demonstrated the process on stage in UCCC's theater, and we brought millstones. Daniel Smiley [from Mohonk Mountain House] worked on one of the millstones. It's a shame that the film got lost in the shuffle. We have no idea what happened to it. It was really well done."

Vincent Lawrence also shared the following information about millstone makers working at Accord (Weaver 1995:2):

> "When we were in high school, there were gangs throughout the mountains cutting stone. A gang consisted of three people: a blacksmith and two stone cutters. There could be three to four gangs at a time in our mountains. They were a hardy group. An eight-pound striking hammer was used, a man would use it all day long. Oh the mountains were alive in those days." Vincent sighs....
>
> "Every gang had a shop which was waterproof so they could get out of the rain. Sometimes on a rainy weekend the men would go back up there to get away from their homes and people to play cards and drink all day! They were a tough, but good bunch," Vincent says with a smile.

In a 1993 interview with Franklin Kelder, Ebert (1993:3) stated that "Shawangunk millstones were also shipped out by Wally and Vinnie Lawrence. Lavoid Coddington and Joachim Coddington also cut millstones. The local stone is the hardest in the world with the exception of a location in France.... Many of the millstones were used by paint companies."

In an interview with Vincent Dunn, Robertiello (1994:3) noted that "The grinding wheels or grinding stones were cut up in the Minnewaska Mountains someplace. Back on top of the hill — Lawrence Hill — there were a lot of quarries." He added, "You still find remnants of grinding stones around."

The following year, Robertiello (1995:7) interviewed Otto Paul Tolski who had purchased an old farm on Lawrence Hill 62 years earlier. Mr. Tolski supplied the following information related to the millstone industry (Robertiello 1995:8–9):

> The property Otto purchased from George A. Hoornbeek and his wife Charlotte in 1933 had been owned by William H. Rose (Otto refers to him as "Hip Rose") from 1890–1924. Before that it had been a Davenport farm, but that family had moved down to the flats below the hill.
>
> When Mr. Rose owned the farm, according to Otto and his son Fred, it was a sort of headquarters for millstone workers. The workers quarried the stone on Lawrence Hill in the summertime. In the winter they worked it into millstones. The building that became Otto's garage was an eight-room, two-story bungalow used to house the men. Another building, which had been a dining place for boarders, with a kitchen on the side, became one of Otto's chicken coops.
>
> Fred remembers seeing piles of millstone chips on the property, especially under one large oak tree where the men liked to work.

Cross (1996:9) reported the following information on the shipping of millstones by rail at Accord:

> A remarkable railroad-related activity at the Accord station was the loading of millstones and other stone quarried from the ridges of Shawangunk "grit" on Lawrence Hill, Rock Hill and in the Clove. A "cut" had been excavated for a siding just north of the station. When down in the cut, the sides of two gondola cars at a time would be level with the ground to facilitate loading of the stone. A scale in the ground next to the track weighed each truckload of stone before it was offloaded into the railroad car.
>
> Millstones of several sizes from a number of small quarries of white stone were accumulated near the station until there were enough to order a gondola car to take them away for sale.

Steve Hirsch (personal communication 2005) reported that he had personally observed at least 50 small millstone quarries on the Shawangunk Ridge. These quarries are east of the communities of High Falls, Alligerville, and Accord. Since he has explored only a portion of Shawangunk Ridge, there are undoubtedly hundreds of small quarries surviving along the 10-mile segment of the ridge known to have quarries. Also, see Snyder (1981) for information on the Shawangunk Mountains and a brief discussion of the millstone industry.

During April of 1998, the author and Dr. Fred E. Coy, Jr., visited some of the Ulster County, New York, millstone quarries near the community of Accord (Figures 14–17). Vincent and Wally Lawrence, mentioned above, two brothers in their eighties, whose father and uncles were millstone makers, were interviewed. The Lawrence brothers shared their memories of millstone making during their youth (Hockensmith and Coy 2008a). Lewis Waruch, whose mother's family included millstone makers, was also interviewed (Hockensmith and Coy 2008b). The Society for the Preservation of Old Mills is publishing the New York interviews and archival overview of the Esopus millstone industry (Hockensmith 2008b, editor).

North Carolina

Millstones were also quarried from the Triassic conglomerates in Moore County, North Carolina (Ladoo 1925:9; North Carolina Land Company 1869; Schrader, Stone, and Sanford 1917:234). In his 1852 geological survey of North Carolina, Emmons (1852:156) made the following comments on millstones:

> MILLSTONES.—I am not sufficiently well informed, as to what state of perfection the millstones of Deep River may be brought. They are among the best stones for grinding corn. Whether art can make them best, or as good as the French burr stones, will be determined by those acquainted with the manufactory of them than myself. They are esteemed for corn, and this fact has given them credit and a market to almost any extent; and it will increase, provided means of cheap transportation are provided: as they can be furnished much cheaper than French burr stone, and are equally good for some purposes.

In his 1856 *Geological Report of the Midland Counties of North Carolina*, Emmons (1856) provided several comments about millstones. On page 229, Emmons (1856) gives the following narrative for conglomerate in the Deep River Coal Field:

> The lower beds are made up of pebbles of quartz strongly compacted together, without the intervention of a cement. So firm are these beds of conglomerate that they make an excellent corn stone, which, when broken from the quarry, split across the pebbles with out removing from their beds. The pebbles are derived from the adjacent and inferior Taconic slates, or the auriferous slates, with their series of imbedded minerals.
>
> The most conspicuous part of the conglomerate is quartz, which is rounded by attrition, and has often assumed a flatten or oval form....
>
> The thickness of the beds varies from six inches to two feet. These solid beds are parted by soft or marly matter, which gives an opportunity to split them horizontally from the planes upon which they were deposited.

Emmons (1856:242) discussed deposits in Chatham and Moore counties used for millstones:

> But the unchanged slates emerge in an unconformable position at numerous places in Chatham and Moore counties, among which I may mention the millstone quarries on the waters of Richland creek, the tributaries of Indian creek within one mile of Evans' bridge, and on the road leading to Salem, and again about a mile above the mouth of Line creek, which enters Deep river not far above the same bridge. An important point which exhibits the junction of the series is about one-and-a-half miles from Farmersville, on the Pittsborough road, where a deep ravine divides the lower conglomerate and red sandstone from the slates of the Taconic system.

Finally, Emmons (1856:266–267) stated in his *Geological Report of the Midland Counties of North Carolina*:

> Millstones.—Beneath the red sandstone, the conglomerate is so perfectly consolidated that it forms a valuable millstone. This is made up almost entirely of compacted quartz pebbles, which are so firmly imbedded that their fracture is often directly across the axis of the pebble, where it would be expected to break out. These pebbles are derived from the quartz veins of the Taconic system, and hence, consist of milky quartz.
>
> The beds vary in thickness from six inches to eighteen, or even two feet.
>
> The stone is adapted to the grinding of indian corn. They are said to be better cornstones than the French Burrhstones; for grinding wheat, the latter have been always preferred, as they are far less liable to heat the flour. Several quarries are open in Moore county, and from them the country is principally supplied. The conglomerate at or near the base of the upper sandstone is less consolidated, and is not so well adapted to the formation of millstones. The thickness of the beds is from forty to sixty feet; but it is a mass which thins out, and hence its thickness at several points is extremely variable. The lower sandstone, with its conglomerate, is better developed in the south-west part of Moore county than elsewhere. We find, even at the Gulf, the conglomerate ceases to be an important stratum.

In 1869, it was noted that "one of the best millstone grits in the country is found on McLennon's creek, in Moore county" (North Carolina Land Company 1869:106).

The 1875 *Report of the Geological Survey of North Carolina* (North Carolina State Geologist 1875:305) boasted: "The conglomerates of the Triassic series, which are associated with and replace the sandstones above mentioned have been long and widely used for millstones. They have been principally obtained from Moore county, on McLennan's Creek, where they are obtained of excellent quality; and have been distributed from this point over a large number of the intervening counties, to the Blue Ridge. Some of these stones have been in use for 50 years; and they are occasionally found to be nearly equal to the French buhr-stone."

The magnitude of the Moore County millstone industry was illustrated in the following quote, which appeared in the April 29, 1887, issue of *The Morning Star* published in Wilmington, North Carolina:

> "Moore County grit," from its nature, requires less picking and dressing than any other stone, not excepting the French burr. A company is now turning out one mill per day, complete with all its fixtures, ready for grinding ... shipped to Europe, Asia, South America, Australia, Mexico and nearly every State in the Union. There are now several thousand millstones stacked upon the yard seasoning and ready for use whenever they are needed. The company work about a hundred hands and average sixty millstones a month. Two 90-horse power boilers are being put in to be run by an 80 horse-power automatic Buckeye engine.

A book dealing with western North Carolina (Anonymous 1890:21) commented, "The lower conglomerate furnishes good and very hard mill-stones; the best found is in Moore county."

In his *Geology of the Deep River Coal Field North Carolina, North Carolina*, Reinemund (1955:122) provided the following information on the conglomerate and its use for millstones:

> Parts of the basal conglomerate of the Pekin formation are characterized by an abundance of white or gray quartz pebbles, a dark gray or brown sandstone matrix, and quartz cement. This rock, which has been described in detail in the discussion of Triassic sedimentary rocks, is known as the "millstone grit." It is a facies of the basal conglomerate that, locally, is in lenticular beds as much as 30 feet thick....
>
> The "millstone grit" is irregularly disturbed along the northwestern edge of the Sanford basin, and it is thickest and most extensive south of Putnam, where it has been quarried and used for millstones at many places. The largest of the quarries is on a branch of Richland Creek, a few hundred feet west of State Highway 22. This quarry was operated intermittently for almost 100 years, beginning in the early part of the last century, but it is now flooded and overgrown with dense vegetation. "Millstone grit" also crops out locally north and east of the Carolina Mine and north and west of Gulf, but has not been quarried there.

Conley (1962:28) in his *Geology and Mineral Resources of Moore County, North Carolina*, stated, "The Millstone Grit, a quartz conglomerate in the Pekin formation along the northwestern border of the Deep River Triassic basin in Moore County, was extensively quarried prior to 1900 for millstones. A number of old quarries are located east and southeast of Hallison."

Pennsylvania

Millstone quarries have been reported in several southern and eastern Pennsylvania counties. Well known millstone quarries were located in Lancaster County, Pennsylvania, near the Cocalico Township (Howell 1985:146; Katz 1920:217; Ladoo 1925:9; Parker 1895:586; Sass 1984:viii; Schrader, Stone, and Sanford 1917:264; Swisher 1940:59).

At Ephrata in Lancaster County, Pennsylvania, a millstone maker was mentioned (Zerfass 1921:25): "Wm. Konigmacher, who for many years took a prominent part in the affairs of the Seventh Dayers, was the pioneer millstone maker and dealer. He also furnished the stone for the present court house and jail at Lancaster. He died in 1881."

Flory (1951a:76) stated that former millstone maker William D. Nagle told him that mill stones were made in "the area of the South Mountains in which the 'Cocalico' stones were found in West Cocalico Township, extending into Clay Township beyond Hopeland and Clay. Another section where millstones were cut in Lancaster County was a section known as Turkey Hill, near Terre Hill." These stones were manufactured from a "pink and tan pebbly conglomerate stone" (Sass 1984:viii). Flory (1951a:77) described this stone as "a very hard pebbly conglomerate, the pebbles varying in size from cherry stones to one inch, or one and one-half inches; the 'cherry stone' size being the most preferable." The Hagley Museum and Library has a ca. 1880 broadside advertisement by Hehney & Wike on the Turkey Hill millstones (Hehney & Wike 1880). Sold under the name "Cocalico," these millstones were made from one piece of conglomerate and were banded with iron hoops (Flory 1951a:79; Sass 1984:viii). Both Flory (1951a:81) and Sass (1984:x) reproduced an 1869 advertisement for a Cocalico millstone quarry owned by S. P. A. Weidman and Benjamin Wissler which was the last and most widely known quarry (Flory 1951a:80). In addition to Cocalico Millstones, they manufactured a variety of other stone products such as hearths, curb, cope, head, paving, and building stones as well as stone troughs, door sills, and window sills (Barnes & Pearsol 1869:311; Flory 1951a:81; Sass 1984:x). Weidman and Wissler provided the following information about their millstones: "Manufacturers of Grain and Spice Cocalico Mill Stones, Just Open a New Quarry Of Superior Quality at Reduced Rates" (Barnes & Pearsol 1869:311).

Figure 18. *View of the Turkey Hill Millstone Quarry in Lancaster County, Pennsylvania. Photograph taken on April 16, 1998, by Fred E. Coy, Jr., Louisville, Kentucky.*

An excellent newspaper article entitled "Stone-Age Skill Cut Millstones from Mountain: Boulders Scattered Over 50 Acres in Northeast Lancaster County Show Stonecutters' Work" described the Turkey Hill millstone quarry (Anonymous 1962:11). The article noted that millstones had been cut as early as 1751. The last millstone makers were William Weinhold and Sam Reifsnyder. At the time of the article, William Weinhold's widow was living in Ephrata in the home of her daughter, Mrs. William K. Wise. During her childhood, Annie Wise visited her Dad at the millstone quarry (Anonymous 1962:11):

> "Father spent most of his life cutting millstones on Turkey Hill," she recalls. "He started when he was 22, and stopped in 1918. Towards the last, he was cutting mostly hammermill stones, which are smaller than the oldtime gristmill stones." The older type might measure six or seven feet in diameter, and weigh a couple of tons. She remembers the painstaking work of hand-drilling holes in the rocks, splitting them off, cutting holes in the middle, and finally dressing the stones in the geometrical pattern of grooves.

The article included photographs of quarry remains and historic photographs dating to the period when the quarry was still operating. Two historic photographs in the possession of Annie Weinhold provided insight into the quarry operation. One photograph illustrated a number of completed millstones with iron bands around them. In the second photograph, several millstones were visible along with a small structure (a possible blacksmith shop), and a small derrick. It was noted that the millstones were moved to the derrick on wooden rollers. After being loaded into wagons, teams of either two or four horses were used to transport the millstones out of the quarry (Anonymous 1962:11).

The millstone makers at Turkey Hill did not own the quarries but rented the property.

Figure 19. *Roughly Shaped Millstone Preform at the Turkey Hill Millstone Quarry in Lancaster County, Pennsylvania. Photograph taken on April 16, 1998, by Fred E. Coy, Jr., Louisville, Kentucky.*

According to Annie Wise, her father initially worked for Sam Reifsnyder and later became his partner in the business. The property was owned by a Mr. Needhawk who lived in Reading. Annie Wise thought that a Mr. Haimley ran the millstone quarry prior to Reifsnyder. Further, Harry Stauffer of Farmersville was told that the property was previously owned by a Mr. Lichty who earned $60 annually for renting the quarry to millstone makers (Anonymous 1962:11).

Harry Stauffer provided some information concerning the tools used at the Turkey Hill millstone quarry. For splitting the stone, hand drills were used to place holes in a straight line. Wedges and feathers were placed in the holes to split the stone apart. The descriptions suggest that points were hit with a 10-pound sledge hammer to shape the sides of the millstones. It was noted that discarded hammer heads could be observed at the quarry and wedges were still visible in some rocks (Anonymous 1962:11).

The newspaper article provided observations about the remains at the Turkey Hill quarry scattered over 50 acres. The quarry had reverted to forest and rejected millstones were left on the quarry floor. Apparently, the abandoned millstones represented different stages ranging from very rough specimens to nearly complete millstones. Frequently, rejected millstones were observed on a base of other rocks to produce a desirable working height. The quarry excavations included some shallow pits where suitable stone was obtained but most stone was obtained from the surface (Anonymous 1962:11).

Another possible conglomerate millstone quarry was mentioned by Harry Stauffer in the Clay Township. A nearly finished 40-inch millstone was discovered on the right bank of Middle Creek below a gap in the mountains (Anonymous 1962:11).

A conglomerate millstone quarry has been documented in southwestern Pennsylvania in Fayette County. The quarry is located on Chestnut Ridge (Laurel Hill) in the Allegheny Mountains (Michael 1983). Quarry remains extended over an area 300 by 500 feet that was covered by conglomerate boulders (Michael 1983). Ronald Michael (1983) noted, "Millstones of various sizes ranging from about 24 to 48 inches in diameter remain in stages of manufacture from initial roughed-out blanks still attached to boulders; to detached blanks, finished on one side only; to nearly completed millstones which fractured during cutting, to complete or nearly finished whole millstones."

According to Michael (1983), it was not known who operated the quarry at Laurel Hill, which he thought was abandoned during the 19th century. However, it appears that this quarry was operated by the Shacklett family according to information published on the history of Meade County, Kentucky (Ridenour 1977). The first Shacklett family members left Fayette County, Pennsylvania, in 1796 and settled in Jefferson County for two years (Ridenour 1977:27). The Shacklett family moved to Meade County, Kentucky, in 1798 (Ridenour 1977:27). After the death of John Shacklett and the settling of his estate in Fayette County, Pennsylvania, in 1810, his children came to Kentucky to join other Shacklett family members (Ridenour 1977:29). According to Ridenour (1977:29–30), "They landed at Solomon Branenburg's Landing and Ferry, which was called 'Buzzard Roost,' bringing their household goods, stock and supplies with them. They 'paid for their land with bar iron, castings and millstones, the latter quarried and dressed by themselves at their father's quarry in Pennsylvania, (Laurel Hill).'"

The above quote suggests that the Shacklett family were skilled millstone makers and brought a number of the Laurel Hill millstones to Kentucky by flatboat. Further, it appears that the millstone quarry was established sometime in the 18th century.

Laurel Hill millstones were advertised in northern Kentucky and across the Ohio River in Cincinnati, Ohio. Garber (1970:77) noted that "on March 17, 1820, Thompson, Neave & Brothers advertised in the *Liberty Hall and Cincinnati Gazette* that they had Laurel Hill millstones 'for sale by the quantity.'" The author discovered a wonderful ad on Laurel Hill millstones while looking at *The Eagle* newspaper published in Mayville, Kentucky. The April 13, 1825, edition (and several other editions) carried the following ad:

LAUREL-HILL
MILLSTONES

PERSONS wishing to purchase these superior MILL-STONES, can be supplied through Messrs. January & Suthertland, of Maysville, who are my agents for the state of Kentucky. The following are the prices of said stones, exclusive of freightage:

Those of	2 ft 6 inches	diameter	$20
"	3 ' 0 "	"	30
"	3 ' 6 "	"	45
"	4 ' 0 "	"	60
"	4 ' 6 "	"	80
"	5 ' 0 "	"	100

For all of intermediate sizes, proportionable charges will be made.

SAMUEL JONES

Bridgeport, Pa. April 1825

Maxwell (1968:482, 485) provided a biographical sketch of Hugh Turner of Barbour County, West Virginia who made millstones at Laurel Hill. It is uncertain whether Turner ever made millstones in nearby (to the north) Laurel Hill in Pennsylvania or a different Laurel Hill in West Virginia:

> HUGH TURNER. This eccentric man, called "the Hermit" lived for the most of his life, and died in what is now Barbour County. The circumstances of his birth and early life are unknown. He was the first contractor to build the first court house in Randolph County, 1788; but he failed to fulfill his contract. He withdrew from the settlements and lived along Laurel Hill at several places in caves or in vaults made of building stone walls against faces of cliffs. He was a Scotchman. He eked out a living by making millstones from the rock found along Laurel Hill suited to that purpose (Pottsville Conglomerate) and sold them to millers who built the first mills in Gladie District. John G. Johnson says there are some of the unfinished millstones on Teter's Creek on Mrs. Cartherine Poling's land. Some of his old camping places are near by. John Harris who died in 1882 in his ninety-third year remembered him, and said that Turner seldom came into the settlements, and could never be induced to lie in bed, always lying on the floor before the fire. He was called a "woman hater." He had a camp on the top of Laurel Hill on the old road which led from Belington by way of Beaver Creek to Beverly. The camp was ten or twelve feet square, built of stones beside a cliff, and the ruins may still be seen. About 1860 Lewis Corley (who hung himself at the Poor Farm in this county) found a copper rule one foot long, with inches and fractions marked on it, and some mason's tools, in a crevice of a cliff. The tools were probably his. The old man was found dead by some persons going from Glady Creek to court at Beverly. He had a fire near the roadside, and it was supposed that he had fainted and fallen into the fire. He had rolled into the road. Some of his millstones were near the place in recent years. Others had been rolled down the steep eastern face of Laurel Hill (toward Leading Creek) and they acquired such force that they knocked down trees.

Stoner (1947:416–417) provided the following information on an unfinished millstone discovered in Franklin County, Pennsylvania:

> An unfinished millstone was discovered a short time ago in the South Mountain, three or four miles from Scotland in Greene Township, Franklin County. It was lying among a dense growth of trees and undergrowth alongside a little stream which courses its way down a narrow ravine known as Devil Aleck's Hollow.
>
> This stone was partly shaped. Some preliminary dressing had been done on the side and the round hole in the center had been chipped out three or four inches deep. For some reason or other it was never completed. The stonecutter after working on it for a while probably considered it was not suitable to be made into a millstone; or the portion broken off the top edge may have decided him to abandon it, or the poor man may have died before his work was finished.
>
> The location was undoubtedly a millstone quarry as there is another stone nearby which shows evidence of a little chipping on the sides but still very much in the rough. There are also several boulders in the neighborhood out of which millstones might have been fashioned, had the quarry continued in operation.

In a history of families in Bradford County, Pennsylvania, Heverly (1915:363) mentioned a place called "Millstone Run" in the townships of Monroe and Overton, "so called from the conglomerate rock abounding there from which mill-stones were cut. John Northrup and others engaged extensively in getting out mill-stones her a century ago."

Several other conglomerate millstone quarries are known in Somerset, Tioga, Berks, and Carbon counties, Pennsylvania. In Somerset County on Allegheny Mountain, millstones were quarried from a conglomerate by John Deeter between 1789 and 1814 (Brown 1990:22, 1991). Deeter recorded information about the sale of his millstones in a personal ledger. Millstones were quarried along an outcrop on Deeter's property near his mill (Brown 1991). Deeter made at least 23 sets of millstones and recorded to whom he sold them and the price (Brown

1990:22). The size of stones manufactured ranged from three feet and six inches to four feet and four inches in diameter (Brown 1990:22). In Tioga County, millstones were quarried at Niles Valley in 1836 (Clouse 1991). Geologist Thomas Berg (1986:3–6) described a small conglomerate millstone quarry in Tioga State Forest northwest of Asaph in Tioga County. Berg (1986:3) noted that James Hesselgessel quarried an unknown number of millstones from the Olean Conglomerate near the base of the Pottsville Group in 1836. Another conglomerate millstone quarry has been observed near Steve Kindig's mill in the Oley Valley of Berks County (Clouse 1991). The Strohecker family at Reading in Berks County were involved in the millstone industry (Strawhacker 2004). Family papers mention several individuals involved in making millstones during the early 19th century including Gottleib Strohecker, John Strohecker, Sr., John Strohecker, Jr., Samuel Strohecker, and Henry Ege (Strawhacker 2004). Finally, Howell (1985:146) notes a quarry that probably exploited conglomerate at Bowmanstown in Carbon County.

An interesting travel account from 1809 mentioned a conglomerate millstone quarry in southwestern Pennsylvania (Gilpin 1927). The quarry was encountered in a segment of the journey between the communities of Bedford (Bedford County) and Somerset (Somerset County) on Allegheny Mountain (Gilpin 1927:172). Since it mentioned Allegheny Mountain, it was probably was in Somerset County and may have been John Deeter's quarry. Gilpin's (1927:172–173) account provided the following information:

> We ascended the allegany [sic] which is about 1½ mile to the top, & the ascent being neither very steep or rugged we came [to] it in 35 minutes, at the top we found a large new stone house the handsomest we had seen for a long time — it is kept by Weyand — the top of the mountain consist wholly of a coarse freestone formed of small round pebbles inbeded in a hard sand — the pebbles appear of a very even size in each mass, but the different masses exhibit them from the size of a hazelnut until it becomes so small as to unite with the sand in a close compact freestone — a number of mill stones were cut, & laying on the sides of the road formed out of the coarsest stone which is hard & in compact masses, we found the species of stone on all the west side of the mountain as far as Somerset which is 14 miles from the top — the descent of the mountain is about 2 miles & very easy in point of steepness — but in many places rugged....

During April of 1998, the author and Fred Coy, Jr., visited the Turkey Hill millstone quarry in Lancaster County, Pennsylvania (Figures 18–19). Some of the quarry remains were photographed and brief notes were taken on abandoned millstones. Unfortunately, a modern subdivision had been developed on the slope below the quarry at the time of our visit. We also visited the Lancaster Historical Society's library, which produced some newspaper clippings. We hope to publish our information on the Turkey Hill quarry at a future date.

Tennessee

Conglomerate millstones were quarried in Coffee, Cannon, and Trousdale, Tennessee. In Trousdale County, Tennessee, near Hartsville, and Coffee County near Manchester, conglomerates were quarried for millstones (Killebrew 1874:260–261). Another conglomerate millstone quarry was located on Short Mountain in Cannon County, Tennessee (Butler 1971:1–2). A few miles north of Short Mountain, on Clinch Mountain in northeast Tennessee, a millstone quarry was observed by J. D. Clifford during an 1818 trip. According to Clifford (1926:276), "Clinch Mountain is composed of sandstone dipping in the same manner. Mr. M. Henry informed us that a person from New York was working a quarry of burr stone about twelve miles off the road, and that the millstone had been pronounced as good as any procured in our Atlantic cities."

A fourth Tennessee quarry exploiting conglomerate is located in Union County and three of the unfinished stones from this quarry are on display at the Museum of Appalachia at Norris, Tennessee (John Rice Irwin, personal communication 1993). Safford (1869:511) noted that "millstones for grinding corn have frequently been made from the conglomerate of the Coal Measures."

Virginia

A well known millstone quarry was located on Brush Mountain near Blacksburg in Montgomery County, Virginia (Coons and Stoddard 1929:217; Ladoo and Myers 1951:9; Ladoo 1925:9; Rotenizer 1989; Schrader, Stone, and Sanford 1917:321; Swisher 1940:59; Watson 1907:401; Worsham 1986a, 1986b). The conglomerate at this quarry is a white or gray siliceous sandstone containing rounded quartz pebbles of various sizes (Watson 1907:401). A more detailed description of the Ingles conglomerate was provided by Campbell (1925:26): "this bed, in places, consists entirely of white or gray sandstone, but generally white quartz pebbles occur in the sandstone, either scattered through it in thin and irregular layers or deposited in a thick bed which consists of well-rounded quartz pebbles, ranging up to 1½ inches in diameter, cemented by gray or white quartzose sand."

This quarry was in operation by 1838 and continued until the mid–twentieth century (Rotenizer 1989). A news item found by McGrain (1991) in the July 9, 1822, edition of the *Baltimore American* suggests that the quarry had an earlier beginning: "The *Lynchburg Press* states that there has lately been found about five miles from Montgomery Court House, Va.,

Figure 20. *View of the Brush Mountain Millstone Quarry in Montgomery County, Virginia. Photograph taken on May 30, 1990, by Fred E. Coy, Jr., Louisville, Kentucky.*

Figure 21. *View of the Brush Mountain Millstone Quarry in Montgomery County, Virginia Showing Piles of Rubble. Photograph taken on May 30, 1990, by Fred E. Coy, Jr., Louisville, Kentucky.*

a quantity of stone, which has been pronounced by those who have used them better than the French burr stones. Several mills in the neighborhood have in use solid stones from this quarry. They are sold in a finished state at one or two hundred dollars per pair, according to size."

During the eighteenth century, the quarry was operated by Michael and Jacob Price (Worsham 1986a:157). By 1850, Israel Price and Company operated the Brush Mountain Quarry. In an 1853 advertisement which appeared in the *Lynchburg Daily Virginian*, Price stated, "The undersigned are now procuring from their celebrated quarry, at Brush Mountain, Montgomery County, Mill Stones and Burrs of the very best quality. The reputation of these stones is widely known, but the railroad giving us increased facilities in filling orders, we have appointed McDaniel, Hurt & Preston and Lee & Johnson, Lynchburg, our agents for the sale of them who will receive orders and have them fulfilled at the shortest notice" (Worsham 1986a:157).

In the upper valley and southwest Virginia, "pebble-stone suited for millstones was present" (Fontaine 1869:47). Pratt (1902:793) noted, "In the vicinity of Prices Fork, Montgomery County, Va., a sandstone conglomerate occurs on Brush Mountain. Quarries have been opened in it for a distance of 3 miles. The stone occurs in various colors from white and gray to bluish, and is also of different grades of grit. These stones are known on the market as 'Brush Mountain' stone. There is practically the same variation in the sizes of the stones manufactured at all these different quarries."

Watson (1907:401) provided the following information on the Brush Mountain stone:

Figure 22. *Millstone Preform with Leveled Top and Roughed Out Sides at the Brush Mountain Millstone Quarry in Montgomery County, Virginia. Photograph taken on May 30, 1990, by Fred E. Coy, Jr., Louisville, Kentucky.*

> About 5 miles west of Blacksburg in the vicinity of Prices Fork, Montgomery county, a sandstone-conglomerate occurs in Brush Mountain, in which quarries have been opened for a distance of 3 miles. The stone is somewhat variable in color but is usually white or gray. Likewise, variation in size of pebble is shown. The rock is made up of well rounded pebbles of quartz compactly embedded in a fine siliceous sandstone matrix, the whole forming an exceedingly tough and hard mass. This stone is known on the market as "Brush Mountain" stone. Practically the same variation in the sizes of stone made at the different quarries obtains. The grindstones made from this rock are of excellent quality and they find a ready market. The age of the Brush Mountain stone is Mississippian (Lower Carboniferous) or Vespertine of Rogers.
>
> Siliceous conglomerates similar to the above are found in the crystalline area, east of the Blue Ridge, and are rather abundantly distributed over the Mountain district west of the Blue Ridge, but so far as the writer is aware the Brush Mountain quarries, in Montgomery county, are the only producing ones in the State.

A souvenir publication for Montgomery County, Virginia (Montgomery County 1907:5–6), contained similar statements on the millstone industry: "In the vicinity of Prices Fork, quarries have been opened for a distance of three miles in a sandstone conglomerate occurring on Brush Mountain, and the stone used for burhstones or millstones. The stone is admirably suited for this purpose. It bears an excellent reputation and is known on the market as the "Brush Mountain" stone. It varies in color from white and gray to bluish and is made up of different grades or sizes of grit."

Phalen (1910:613) described the Virginia millstone quarry as follows:

> The millstone industry in Virginia is confined to quarries near Price's Fork, Montgomery County, about 5 miles west of Blacksburg, the site of the Virginia Polytechnic Institute. The rock is regarded as of Mississippian (lower Carboniferous) age. The material from which the stones are quarried varies from a normal conglomerate to a fine-grained quartzitic rock. It includes pebbles, some of them as large as walnuts, though most of them are smaller. The rock has a bluish cast. Its bedding planes are very distinct, and layers only an inch thick may be observed. It is extremely hard and tough and resists erosion to a marked degree. It underlies Brush Mountain for miles, and for this reason the millstones are frequently known as Brush Mountain stones. The stone can not be quarried by blasting, and is therefore extracted by hand power, with drill and hammer, plug and feathers. Millstones and drag or rider stones are the principal products made at the Virginia quarries.

Campbell (1925:27) provided the following description of the Brush Mountain millstone quarry:

> The Ingles conglomerate is quite prominent on Brushy Mountain from Millers Cove on the east to New River on the west. It generally crops out at the crest or high up on the southern slope of the ridge. In places it is impossible to find an outcropping ledge of the conglomerate, but its presence is indicated by blocks of sandstone containing white quartz pebbles, that are more or less abundant on the surface. In other places the Ingles conglomerate is thick and massive and it forms very conspicuous ledges on the sides or at the crest of the ridge. Where it is thick and of uniform texture, it has been extensively used for millstones and many old pits in which the rock has been quarried for this purpose can be seen on Brushy Mountain. With the advent of roller mills for the grinding of grain the demand for mill-

Figure 23. *Nearly Completed Millstone at the Brush Mountain Millstone Quarry in Montgomery County, Virginia. Note that the eye is started and edge damage is visible on the left side. Photograph taken on May 30, 1990, by Fred E. Coy, Jr., Louisville, Kentucky.*

> stones has decreased, until the industry of quarrying and shipping them is almost a thing of the past. There is still, however, a small demand which is met by local farmers, working at odd times in quarrying the rock and dressing it for the market. One of the most extensive quarries in the Ingles conglomerate lies on the south slope of Brushy Mountain just north of Poverty Gap. Pl. IV represents the rock in this quarry and some of the millstones as they showed in 1914.

McGill (1936:22) provided the following information on millstones made from the Mississippian series (Price or Pocono) sandstone formation: "Stone from quarries in the conglomerate or conglomeratic sandstone member near the base of the Price formation, on Brush Mountain, a few miles west of Blacksburg, has been used for a number of years for grindstones and millstones which are well known under the trade name of 'Brush Mountain stone.'"

The last two millstone makers who worked at the Brush Mountain Quarry were discovered in the Blacksburg area in 1990. During May of 1990, the author and Fred E. Coy, Jr., traveled to Montgomery County, Virginia, and interviewed millstone makers Robert Houston Surface and W. C. Saville (Hockensmith 1990c, 1999a). This interview was recorded on video and audio tape. Portions are included in this book, and the interview is published in detail elsewhere (Hockensmith and Coy 1999, Hockensmith and Price 1999). Jimmy Price's 1985 interview with Robert Houston Surface was also published (Hockensmith and Price 1999). During 1995, the *Roanoke Times & World News* published an excellent article about Mr. Surface and Mr. Saville (Freis 1995a, 1995b). A story about Robert Houston Surface's death appeared in May of 1998 (Freis 1998). The Brush Mountain Quarry was visited and photographs were taken of the remains (Figures 20–23).

A recent history of the Meadows Mills Company (McGee 2001:15) notes, "From about 1930 to 1936 the mill stones used by the Meadows were called "Pebble Stones" and were purchased from R. E. Snider's quarry in Cambia [sic], Virginia." The reference apparently is to the community of Cambria in Montgomery County. R. E. Snider of Cambria, Virginia, was listed as a millstone producer in 1934 (Bowles and Davis 1934:901).

Calvert McIlhany (personal communication, 1992) advised the author that quartz pebble conglomerate millstone quarries were reported to him during an archaeological survey in Pulaski County, Virginia. Millstone quarrying took place on Cloyd Mountain during the 19th century. Since Pulaski County is adjacent to Montgomery County on the western border, the stone probably is very similar to that at the Brush Mountain Quarries. Campbell (1925:28) noted that the Ingles conglomerate that was exploited at Brush Mountain was also present south of the community of Pulaski, which is in Pulaski County.

In the Valley of Virginia, on Massanutten Mountain, conglomerate deposits were exploited for millstones. In a reprint of geological work for the period of 1835 to 1841, Rogers (1884:93) stated that "beds of a coarse conglomerate, of very peculiar structure, constitute an important portion of its mass, and furnish the material of mill-stones now much in use."

West Virginia

Conglomerate millstones were made in Fayette County, West Virginia (Peters and Carden 1926:119):

> The stones which did the crushing of the grain were not shaped by novices, but their making required the hand of a man who knew that business. A coarse grained, very hard rock was needed, and a pair of stones was necessary, the upper and the nether. The coarse formation known to geologists as the Pottsville conglomerate was so well suited for millstone that one of its names still is "millstone grit." The formation extends north and south along the moun-

> tain range in the eastern part of the state, and the people within reach of it generally made their millstones of that material. Many parts of the state were too far away from that supply, and they used something else. A set or pair of old-fashioned millstones weighed from 600 to 1,000 pounds. The two were about the same weight. The upper turned upon the fixed one and the grain between them was crushed.

Other States

Conglomerate millstones were also produced in Connecticut and Vermont. A "quartz-shot sandstone" was quarried from Mount Tom in Connecticut (Clark 1929:178; Howell 1985:146; Howell and Keller 1977:71; Rawson 1935:136–137; Storck and Teague 1952:102). Several years ago, Charles Howell and Owen Ward looked for the millstone quarry at Mount Tom but did not locate it during their brief visit to the area (Howell 1993:17). In Vermont, Katz (1920:217) indicated that a "quartz conglomerate rock similar to the New York Esopus stone is found near Fair Haven, Rutland County, Vt. No millstones have been made of this rock in recent years." See also Ladoo (1925:9) concerning the millstones quarried near Fair Haven, Vermont.

Granite Quarries

Granite was the second most common stone quarried for millstones in America. Gary, McAfee, and Wolf (1974:307) define granite as a "plutonic rock in which quartz constitutes 10 to 50 percent of the felsic components and in which the alkali feldspar/total feldspar ratio is generally restricted to the range of 65 to 90 percent.... The origin of granite is in dispute, with some petrologists regarding it as igneous, having crystallized from magma, and others considering it as the product of intense metamorphism of pre-existing rocks."

Millstone quarries exploiting granite were located in Arkansas, Connecticut, Massachusetts, Missouri, New Hampshire, North Carolina, Oregon, and Virginia (Hockensmith 2007b). Lesser known quarries were located in Alabama, Maine, Maryland, Minnesota, Rhode Island, South Carolina, and Tennessee.

Alabama

Little information is available on the granite millstone industry in Alabama. Hockensmith (2005) published an overview of the millstone industry in Alabama. Katz (1926:328) noted that "millstones are also made of granite, particularly in Alabama, Maine, Minnesota, New Hampshire, and North Carolina. The production recorded in Mineral Resources includes only the stones made for other than purely local use." Two years later, Katz (1928:172) stated that "millstones are also made of granite, particularly in Alabama, Maine, Minnesota, New Hampshire, and North Carolina." Dean (1996) indicated that "granites in Coosa, Tallapoosa, Chambers, and Randolph counties were widely quarried from 'flat-rock' exposures for millstones."

Arkansas

In the *Second Report of a Geological Reconnoissance of the Middle and Southern Counties of Arkansas*, Owen (1860:31) commented that in Pulaski County, "[n]oble quarries of granite could be opened, both on the north slope of the granite range in the Fourche Cove, and in the cedar glades on the waters of Hurricane and Lost Creek. At this latter locality, some very good millstones have been got out, which, though not equal to the burr-millstone, make nevertheless excellent stones for grinding corn."

Owen (1860:69–71) provided the following details:

> This granite is eminently felspathic, the felspar containing both soda and potash, though it has the white color, lustre, and cleavage of Cleavlandite. The quartz is pale gray; a few crystals of hornblende are disseminated, with occasionally small flakes of black mica. This rock might, perhaps, be called a granite-syenite, but the proporation of hornblende and mica is so small that it is better designated as a felspathic granite.
>
> On the southwest quarter of Section 34, Township 1 north, Range 12 west, granite of similar appearance and composition has been quarried for millstones; a pair of which are now in use in Wool's horse-mill. They can only be regarded, however, as a poor substitute for porous silicious burr millstone....
>
> On the southeast slope of the granite, on the southwest of Section 34, this rock has more of a porphyritic character, and contains more black mica, and is therefore more porous in its structure. Here is the locality where they have got out some millstones.

For Saline County, Arkansas, Owen (1860:108) noted that "several millstones have been quarried out of the above-named granite; the largest sell for $70 or $80." Finally, Owen (1860:408) reported that "Mr. Hastings, some years ago, quarried a set of millstones out of this granite in the Indian country, which we saw running in a small mill within the State, and obtained specimens of the rock from the fragments broken off in the fashioning."

Connecticut

In an early history of New London, Connecticut, Caulkins (1852:402) provided the following information on millstones: "In 1737, Major Buor leased the farm to Benjamin Ellard, for sixteen years, at an annual rent of £107, 16s. This lease included 'the ferry, boat, oars, rope and other utensils,' but the owner reserved to himself 'the sole priviledge of taking off millstones.' This reservation indicates that the well known granite quarry at Millstone Point was wrought at that period."

Two granite millstone quarries have been reported in southeast Connecticut. Millstones were being manufactured in New London County as early as 1737 (Anonymous n.d.a:3). The millstones manufactured at Millstone Point "were a profitable sideline, for they were not only used locally, but were shipped in schooners to the West Indies, where they were traded for slaves and Jamaica rum" (Anonymous n.d.a:3–4). This quarry which began producing granite for other uses in 1830 was the second quarry founded in the United States and eventually became the largest in Connecticut (Anonymous n.d.a:3–4). A millstone quarry was mentioned in a modern history of New London County at Waterford (Anonymous 1922:213).

A second millstone quarry has been recorded in northeastern New London County in Pachaug State Forest. This quarry is described as a 30-foot ledge containing a partially finished millstone (5 feet in diameter and 2.5 feet thick) and circular grooves suggesting that other stones were removed (Gradie 1982:2). The quarry is thought to date to the 18th and 19th centuries (Gradie 1982:2). Finally, Shepard (1837:68) noted that "millstones are occasionally hewn out of granite in various parts of the state."

Louisiana

In his *Sketches of a Tour to the Western Country...*, Cuming (1810:308) provided the following information related to millstones in the vicinity of Point Pleasant, Louisiana, on Black River, a tributary of the Mississippi River: "A quarter of a mile below Big Black, a ridge of hills called the Grand Gulph hills, terminates abruptly at the bluff on the left bank. At the base of the bluff, are a heap of loose rocks, near which is a quarry of close granite, from which some industrious eastern emigrants have cut some excellent mill and grindstones."

Massachusetts

Granite millstones were quarried in Massachusetts near the communities of Easton, Lynn, Medfield, Quincy, and Worcester. *A Gazetteer of Massachusetts* ... (Spofford 1828:234) included the following comment for Lynn, Massachusetts: "granite, suitable for millstones, abounds in this town." A small millstone quarry utilizing granite has been reported near Medfield, Massachusetts (Bell 1991). The quarry is located in Norfolk County, just southwest of Boston, in the east central portion of the state. A letter written by Robert Mannino (1988) provided the following description "...two grist mill stones ... were left by stone cutters some 150–200 years ago. It appears that there have been several other mill stones cut from the ledge outcropping, as well. Several of the splitting spikes are still embedded in the granite. The ledge covers an area of approximately 1000–2000 square feet and is in a wooded area."

A granite millstone quarry was located at Worcester, Massachusetts (Anonymous 1888b:37): "Millstone Hill, about a mile east from the Court House, is the common property of the people of Worcester, who may procure from thence stones for building, underpinning, or for door stones or steps, or other purposes. This stone was thought very well adapted for mill stones, being fine and hard grained."

The quarries at Millstone Hill were owned Webb & Batchelder (Anonymous 1888b:112).

Mr. Tom Kelleher (1993) advised the author that this quarry was the same one that was mentioned in his article on the Kingsbury Mill (Kelleher 1990:8). He also noted that a granite millstone quarry was operated during the 19th century at Quincy (Norfolk County), Massachusetts (Kelleher 1993).

An archaeological survey for the proposed Saugus Residential Development in Saugus, Massachusetts, located a granite millstone that may be associated with a quarry (Leveille and Rainey 2001:30). The report provided the following information (Leveille and Rainey 2001:30): "Generally speaking, the 89-acre project area abounds with small overhangs and crevices where an individual could conceivably seek shelter. ...There is also a high frequency of extensive vertical rock faces that could have been used for construction of lean-to shelters of larger capacity. ...The suspected cairn at the base of the shelter area has been identified as a granite millstone preform (Figure 6–2). Roughed out from exfoliated granite at the immediate outcrop, the stone has been scored for additional finish work, but was never completed."

It is suspected that this area was a small granite millstone quarry. A photograph suggests that the millstone makers had placed the millstone on small rocks so that it would be easier to work. The millstone had been roughly rounded and was rejected for some reason.

Another possible millstone quarry was located during an archaeological survey for the Worcester Vocational High School at Worcester, Massachusetts (Deaton and Cherau 2003). The authors provided archival information on the Putnam Family Quarry which was located during the survey (Deaton and Cherau 2003:44): "Samuel Putnam worked one of the older quarries at Millstone Hill, which later became the property of the Green family. The abandoned Putnam quarry was reworked by Elmer Hewitt in 1867 (Worcester Daily Spy 1867). Its exact location on Millstone Hill is not identified in the newspaper article. There were other quarries at Millstone Hill: a large quarry worked by Amos Black, a large one worked by Brigham Converse, and another held by the Green estate and leased to Webb and Batchelder (Worcester Daily Spy 1867). A kettle hole in this area reportedly contains a stone with a row of drill holes and a broken drill."

Concerning the Worcester Vocational High School project area, Deaton and Cherau (2003:45) noted "many of the ledges and outcrops within the park area as a whole, and the project impact area specifically, display round drill marks that are evidence of nineteenth-century quarrying." The following information was provided on the Putnam Family Quarry, a Derrick Stone, and a granite millstone (Deaton and Cherau 2003:46; numbers referring to illustrations in the report have been omitted):

> The "Putnam Family" Quarry area is an example of combined woodlands/historic component of the project parcel. It is located on the west side of Skyline Drive, opposite the southern proposed school entrance and the suspected blacksmith shop site. This small quarry, dominated by a steep precipice or ledge, presently contains mature oak trees, mosses and lichens, and some grass. Many of the weathered ledges still display round drill marks. At the base of the slope there is one notably ridged stone with very distinctive drill marks. There are many rocky outcrops and ledges in this area that appear to have been quarried. The South Quarry was supposedly worked in the mid-nineteenth century; the round drill marks are consistent with this date. Therefore, this quarry is most likely the South Quarry or South Ledge described in the *Worcester Daily Spy* article of 1867 announcing the construction of Joseph Chamberlain's small blacksmith shop. The kettle hole containing the quarried stone and broken drill is situated about mid-way between Skyline Drive and the former Belmont hospital buildings. The boulder is identified by a row of drill holes, which form a distinctive quarried edge or edge in the stone's surface.
>
> The former Town Quarry/Landfill is situated on the west side of Skyline Drive, surrounded by a number of Green Hill Park walking trails. It was along one of these trails that the reported Derrick Stone was located and recorded. The Derrick Stone is an approximately 75 × 75 cm, roughly square crane footing with a circular opening in the center to support the wooden pole portion of the overall derrick. ...A partially buried, quarried rounded stone was also identified about 20 m south of the derrick stone adjacent to the same walking trail. The exposed portion of the stone is a slightly rounded shape, and suggests that it was being worked for use as a mill stone before being abandoned. The haphazard placement of the stone suggests that it was relocated during earthmoving activities related to the landfill operation.

Missouri

Broadhead (1874:55), in his discussion of Missouri millstones, stated that "the granites of the South-east have formerly been used for this purpose." B. H. Rucker (1996) reported a quarry that produced granite millstones in southeast Missouri. The quarry was located in Iron County near the towns of Ironton and Graniteville. Said Rucker (1996), "These quarries have produced distinctive red granite for architectural elements, tombstones, monuments, and cobbles for the whole St. Louis waterfront. Undoubtedly, some millstones were produced as a sideline by someone or some quarry in this area of the state. Again, however, there were no specific citations found as to the making of millstones as an independent business."

New Hampshire

Millstone quarries were also located in New Hampshire (Howell 1985:146; Howell and Keller 1977:71; Sass 1984:viii). Richard Boisvert (personal communication, 1992) visited a quarry where the Shakers quarried granite and millstones from in New Hampshire. The quarrying took place on Moose Mountain near the city of Enfield in the west central part of the state. Boisvert noted that circles were cut out of granite slabs. In one area, 8 to 12 millstone extraction depressions were noted in line along a ledge. Generally, quarrying of the granite extended over 3 or 4 acres and spoil piles were present (Boisvert, personal communication 2006). The quarry dates to about 1850.

North Carolina

The most famous granite millstone quarries in America were located in North Carolina. In his *Geological Report of the Midland Counties of North Carolina*, Emmons (1856:31) provided the following comments:

> The geographical position of the Raleigh belt of granite may be defined approximately, by giving the names of the places through which its extreme outer edges pass, and connecting those places by lines; thus the western edge runs three miles west of Henderson, and one-and-a-half or two miles west of Raleigh, and from thence south-westwardly through the Buckhorn falls, on the Cape Fear river. From the latter place, it is concealed by sands, but appears to sweep around towards Rockingham and Richmond county, as it appears there, and also upon the head waters of Turkey creek. Upon Turkey creek there are fine quarries of millstone.

The 1875 *Report of the Geological Survey of North Carolina* (North Carolina State Geologist 1875:305) noted, "The coarse porphyroidal granites and gneisses which are scattered over so large a part of the State, are however the most common material for mill-stones."

In Rowan County, a quartz bearing granite was quarried for millstones (Howell 1985:146; Howell and Keller 1977:71; Ladoo 1925:9; Sass 1984:viii; Schrader, Stone, and Sanford 1917:234). Additional information is not currently available for the Rowan County millstone quarry.

A major quarry was located at Parkewood in Moore County, North Carolina (Beaman 1985:18–19). This quarry was established in the early 1800s by William Donnelly. Several people managed the quarry until its last owners went bankrupt in 1893 (Beaman 1985:18). The millstones were manufactured from a blue granite and ranged in size from 24 to 48 inches in diameter (Beaman 1985:18–19). Beaman (1985:18) provided excellent information on the history of the Parkewood millstone quarries:

> Conception of a town began before the Civil War when an early settler, William Donnelly, cut and worked beds of millstone rock....
>
> The Civil War interrupted quarrying operations, but after the war a man named Devotion Davis worked the 12-acre quarry. Davis sold it to Louis Grimm, a large blond haired man who installed boiler and engine equipment. Grimm, a northerner with vision, had come across Davis' old millstone quarry as he had explored the surrounding countryside. Grimm hurried to Baltimore, Maryland, to tell his employers, Ed and George Taylor, of his find. He convinced them of a money-making venture, and the North Carolina Millstone Company was formed November 9, 1880. At first Grimm and George ran the industrial company, a booming business that made and shipped portable corn mills built into wooden frames. The completed mills were shipped in great quantities to New England, Pennsylvania, Mississippi, and other parts of the United States from 1885 to 1887. Grimm found the best millrock was in the bed of a stream running through the valley. The best rock was a blue granite impregnated with chips of white flint.
>
> More than 75 employees worked for the company and lived in houses built by their employers. The workers bought food, clothing and other supplies from a small, general company store or commissary....

McGrain (1991) found an ad in the October 1882 edition of the *Maryland Farmer* for Moore County Grit millstones for sale (Moore County, North Carolina). The ad also mentioned "lately discovered Moore County grit of North Carolina ... vein discovered in 1879 ... purchased by J. E. Taylor of Western Maryland ... manufacturing stones of 30 and 36 inch diameter ... also tried 48-inch stones" (McGrain 1991).

Beaman (1985:18) commented on the failure of the North Carolina Millstone Company: "It should be noted that severe financial panic in this nation caused 15,000 business failures in 1893, about the same time the North Carolina Millstone Company went broke. But history remembers Ed Taylor's spendthriftedness as one reason for its failure. Another reason being that the best millstones last a long time; therefore, customers do not reorder the stones."

Today, said Beaman (1985:19), Parkewood's remains lie along Highway NC 22, some six miles from Carthage. All that is left is a big hole in the ground that once was a quarry for millstone."

Two brochures for the North Carolina Millstone Company have come to the author's attention. First, an undated brochure (ca. 1880s) describes the "Moore County Grit" stone and price lists for mills and millstones, Parkewood, North Carolina (North Carolina Millstone Co. n.d.). The original is in the University of North Carolina Library, Chapel Hill. The second brochure is "Moore County Grit, the best stone for grinding corn meal for table use in the world" (North Carolina Millstone Co. 1885). It describes the company's millstone material as "a blue-colored cement stone, filled with white flint, which when dressed has a much sharper and better cutting edge than any other stone." The original of this brochure is in the University of Delaware Library.

In addition to millstones, the company made both under and upper runner portable mills that used their stones. The company sold millstones from 14 inches to 48 inches in diameter. Table 1 provides the sizes and prices of "Moore County Grit" monolithic millstones which appear to be sold as pairs.

Sass (1984:x) provided the following information: "Meadows Mill Company, North Wilkesboro, North Carolina. Reputedly, the last company still manufacturing natural stone mills in U.S.A. Millstones large in size from 8" to 30"." A recent history of the Meadows Mills Company (McGee 2001:15–16) provided additional information: "From 1936 on up through 1970 the mill stones came from Gardner Brothers Granite Works in Faith, North Carolina (about five miles south of Salisbury, N.C.). The Gardner stones were described as 'North Carolina white flint pebble stone' for which Rowan County was famous."

Table 1. Sizes and Prices for Moore County Grit Millstones in North Carolina.

Diameter	*Quarry Dressed*	*Banded, Plastered, & Balance Iron*	*Banded, Plastered, Furrowed & Four Balance Irons*
14 inches	$25.00		
20 inches	$60.00		
30 inches	$90.00	$110.00	$125.00
36 inches	$110.00	$135.00	$150.00
42 inches	$145.00	$175.00	$195.00
48 inches	$185.00	$215.00	$240.00

Reproductions of old catalogs included in the history of the Meadows Mills Company (McGee 2001) provided additional references to millstones. Meadows Mill Company Catalog No. 23, page 5, noted that "only the best obtainable hard, sharp white flint granite burrs are used in Meadows Mills. The burrs are from the same quarry as have been furnished in Meadows Mills for over a quarter of a century, and have proven to be the best." An illustration on the same page shows monolithic millstones labeled as 12, 16, 20, 24, and 30 inches in diameter. The Williams Mills Pay catalog dated ca. 1919 (McGee 2001) included the following quote:

WHITE FLINT BUHRS

> The buhrs are the best that money can buy. They are the White Flint grit, which is the equal, if not the superior of the French Buhr. Our buhrs are hard and flinty, but tough in texture, which makes them especially adapted to grinding soft, fine meal and the finest grade of buhr flour. They require little sharpening to keep them in good grinding condition.

Virginia

An early gazetteer of Virginia (Martin and Brockenbrough 1835:218) mentioned granite millstones in the section on Louisa (Louisa County) in the eastern part of the state: "A range of granite from 5 to 8 miles wide, extends nearly across the county in much the same direction as the gold range, and in part coinciding with it; though generally the granite lies higher up the country, tapering off in the form of an ellipse, 6 or 8 miles to the S. W. of the Court House. Good millstones have been made of this rock."

A small granite millstone quarry was reported on the farm of John Hawkes near Poole Siding in Dinwiddie County, Virginia (Sass 1982). The wooded quarry site contained outcrops, boulders, and partially completed millstones (Sass 1982:6). A soft mica flecked granite was quarried for the millstones. Sass (1982:6–7) documented two millstones at the quarry and examined seven millstones moved from the quarry by a former land owner. The millstones range in diameter from 37 to 46 inches (Sass 1982:7). Making observations from the rejected millstones, Sass (1982:6–7) was able to provide some very useful comments on the manufacture of these millstones.

Other States

Brief references to granite millstone quarries were also encountered for Alabama, Connecticut, Maine, Maryland, Minnesota, New Hampshire, Rhode Island, South Carolina, and Tennessee. A granite millstone quarry was reported at Westerly, Rhode Island (Howell 1985:146; Howell and Keller 1977:71; Sass 1984:viii). Quarries were also located in Maryland (Sass 1984:viii). A small granite millstone quarry was reported in Abbeville County, South Carolina (Elliott 1987:11). Killebrew (1874:260) reported that a syenitic granite was obtained for millstones in Carter County, Tennessee. In addition, quarries in Minnesota reportedly produced a limited number of millstones (Ladoo and Myers 1951:9). Katz (1928:172) noted that "millstones are also made of granite, particularly in Alabama, Maine, Minnesota, New Hampshire, and North Carolina."

Flint and Buhrstone Quarries

Flint and buhrstone was quarried for millstones in several states (Hockensmith 2006e). Before proceeding further, these terms will be defined for the reader. Gary, McAfee, and Wolf (1974:95) defined buhrstone as a "hard, compact, cellular, siliceous rock suitable for use as a material for millstones: e.g. an open-textured, porous but tough, fine-grained sandstone, or silicified fossiliferous limestone. In some sandstones, the cement is calcareous."

Their definition for flint (Gary, McAfee, and Wolf 1974:265): "A term that has been considered as a mineral name for a massive, very hard, somewhat impure variety of chalcedony, usually black or of various shades of gray, breaking with a conchoidal fracture, and striking fire with steel. ... A term that is widely used as a syn. of chert or for the homogeneous, dark-gray or black variety of chert."

Davis (1990a:6) noted, "The United States had a French buhr crisis during the War of 1812 (1812–1815) when trade was cut off with France. A substitute for French buhr was needed."

Flint was utilized as a raw material for millstones in Alabama, Arkansas, Georgia, Indiana, Missouri, North Carolina, Ohio, South Carolina, Tennessee, and Virginia. The terms "buhr stone" or "burr stone" were often used for these millstones. Silicified shells beds were also classified as buhr stones. Both stones contain silica. American buhr stones were often compared to the famous "French buhr" millstones imported from France. This type of millstone was discussed in the May 2, 1857, edition of *Scientific American* (Scientific American 1857:268):

> BURR STONE.
>
> This is a quartz rock containing cells. It is as hard as rock crystal; and its peculiar value for grinding is owing to its hardness and cellular texture, which gives a rough surface. In the best stones the solid and cellular parts occupy about equal spaces. The "French burr stones" are obtained near Paris from the tertiary formation. To make millstones the rocks are cut in wedge-shaped panes, which are cemented and bound together with iron hoops. A cement for the purpose consist of about one part, by measure, of calcined alum ground into powder, mixed with twenty parts of plaster of Paris, by measure, made into a proper consistency with water.
>
> Good burr stone is found in Ohio, Georgia, and Arkansas. In Ohio, at Hopewell, Richland, Elk, and Clinton, the manufacture of burr stones is carried on to a considerable extent.

St John (1854:274–275) in his *Elements of Geology...* provided the following comments: "*Buhrstone*, used almost exclusively for millstones, is a cellular variety of quartz, and owes its value for this purpose to the hardness and sharpness of the inequalities of its surfaces. The finest stones have usually been imported from France where they were found in the Tertiary of the Paris basin, but stones of excellent quality are obtained in Muskingum county and other localities in Ohio, where it is associated with carboniferous sandstones. As some of the cavities contain lime, it is conjectured that removal of that substance by solution has produced the cells. It is found also in Georgia and in Arkansas."

The September 2, 1868, issue of *Scientific American* (Scientific American 1868:151) included the following information in a section entitled "Answers to Correspondents": "P. J. P., of Ohio.—The French buhr stone used for millstones is simply a variety of quartz, but it is in part composed of pure silex or flint. We have before us now a piece chipped from a rough millstone which is pure semi-transparent flint, of a yellowish, creamy color, honey-combed with holes in which were imbedded minute specimens of marine shells. A substitute for the French stone is found in the bituminous coal measures of northwestern Pennsylvania and eastern Ohio, but the French product is preferred. It is filled with the remains of minute fossil shells."

The 1872 version of *Chamber's Encyclopaedia* noted that "the most important substitute for the French B. in the United States is the B. rock of the bituminous coal measures of Northwestern Pennsylvania and Eastern Ohio" (Chambers 1872:415). Tuttle (1875:680) mentioned that "a rock very similar to the French burr millstones" had been found in southern Wisconsin. The following pages will discuss these millstones alphabetically by state.

Alabama

Flint millstones were reported from Jackson and Winston counties, Alabama. In Jackson County, Alabama, Michael Tuomey (Dean 1995:19) made the following observations during 1848: "The mountains bordering the valley on the right are composed for the greater part of beds of limestone, and siliceous strata of chert and hornestone; ...North of Sauta Creek, a ridge of hornstone occurs ten or twelve miles in length. Judging from the porous structure of this rock, I am inclined to think that a substitute for burr-millstones may be found here. It must be borne in mind that French burrs are not composed of a single piece, but often ten or twenty, cemented together."

McCalley (1886:77) mentioned a Millstone Mountain in Winston County, Alabama, which "is a ridge some three-fourths of a mile long and is much higher than the surrounding country." He also provided the following information about the stone at Millstone Mountain and its use (McCalley 1886:78): "*Millstone Rock*; forms a prominent ledge around the ridge. It is very hard and compact, siliceous rock, resembling a good deal in looks the *knox chert* of North Alabama and the *burhstone* of South Alabama.... Much of this rock has been made into millstones, and hence the name *Millstone Mountain* has been given to the mountain; it doubtless makes excellent millstones, especially for grinding of small grain. The ledge on the outcrop is about ... 1 ft. 6 in."

Dean (1996) observed that "a siliceous rock, the 'Tallahata burstone' was quarried in several south Alabama counties for use as a burrstone."

Arkansas

Buhr stone was exploited for millstones in Izard, Lawrence, and Polk counties, Arkansas. Izard County contained buhr stone suitable for millstones. Owen (1858:44–45) provided the following information about this stone:

> There are very fine buhr millstone rocks in the ridges of the North Fork not far from Ware's mill, but on the opposite side of the river, according to J. E. Ware....
>
> Lower down, on the same branch, are extensive beds of buhr millstone in "Camp creek hollow," some of which are of excellent texture and hardness for grinding corn, while other varieties are equally good for wheat. This buhrstone lies some 200 feet below the level of the ridge over which the Yellville road runs, above the Camp spring....
>
> J. E. Ware is of opinion, the best quality of buhrstones, of any required dimensions, can be obtained either in Camp creek hollow or the ridges opposite his flouring mill, on the North Fork, equal in quality to the French buhr.

A cellar quartzose rock was used for millstones in Lawrence County, Arkansas according to Owen (1858:213–214): "On Big creek, a branch of Strawberry, there is a white cellular quartzose rock found in abundance, intercalated amongst the sandstones of the section of this county, which may afford good millstones; indeed, millstones have been made out of it for some of the mills in the vicinity. A pair of stones made from this rock, may be seen running in Jone's mill on Big creek, six miles from the mouth, and have proved of excellent quality for corn."

In his report for field work conducted in 1859 and 1860, Owen (1860:96) provided the following information for Polk County, Arkansas: "There is one porous variety of silicious rock found in the vicinity of the Gap Springs, which might afford tolerable burr millstones."

Dana (1857:360) also noted that buhrstone occurs "in Arkansas, near the Cove of Wachitta."

California

An 1854 geological survey report dealing with the coastal mountains and part of the Sierra Nevada in California mentions a fossiliferous rock used for millstones. Trask (1854:16) provided the following information:

> The hills were encountered a short distance south of Laguna Seca, and followed thence southerly for eighteen miles. Small patches of fossiliferous rocks are sparingly distributed over the route, the predominating rocks being igneous and composed principally of trachytes and lavas; from Laguna Seca an almost continuous dike of the latter extends along the western base of the ridge for a distance of near twenty miles.
>
> As you approach the Rancho Cantine it becomes more cellular than farther north at any point where it was observed. In the vicinity of Gilroy's it has been used for milling purposes,

> to which it seems admirably adapted, being unequalled in hardness to the best French Buhr. I saw at Gilroy's three sets of these stones which have been in use at that place for several years in flouring wheat. These stones are capable of being split out to the diameter of four feet and the requisite thinness may be required for uses of the mill.

Also for California, Ladoo (1925:10) noted that "it was reported many years ago that on an eminence known as Little Butte, in Owens River Valley, a rock suitable for the manufacture of buhrstones was found. It is said to be hard, brecciated and very much like the French buhrstone."

Georgia

Several early sources mention that burr millstones were made in Georgia (Hockensmith 2004b). McGrain (1991) cited a February 8, 1812, report published in the *Niles Register* that mentioned suitable stone was in Georgia for making burr stones. Jones (1861:12) in another early source on Georgia noted that "here, are also found inexhaustible quarries of the finest burr-stone, which has upon trial proved equal and in some respects superior to the French burr mill stones." Five years later, Bishop (1866:149) also commented on the Georgia burr: "In the last-mentioned State, there is one of the few localities in the Union, if not the only one, that furnishes Burr millstones, identical, in composition and geological position, with the French burrs. The manufacture of these was carried on about fifty years ago near Philadelphia, by Oliver Evans, and extensively at the present time in Savannah."

The August 18, 1849, edition of *Scientific American* contained an excellent article on the Georgia Burr millstone (Scientific American 1849a:380):

> GEORGIA BURR MILL STONE
>
> The stone most commonly used for grinding wheat, is known by the name of "French Burr," because they are imported from that country. This species of stone is a porous silicious mineral, so very hard, that a pair of millstones will last quite a number of years without being worn out. The French burr stones, owing to their great price, has from time to time stimulated both the Americans and the English, to many efforts to supersede them. During the last war between France and England, when it was impossible to get burr stones; the London Society of Arts offered a premium of a gold medal, or one hundred pounds for the discovery of a quarry producing stones equal to the French Burrs.—A quarry was discovered in Wales with stones similar to the French, and answered tolerably but they were not equal to the French. A number of masses of rock were also discovered at Stirling, Scotland, and made into stones, some of which indeed gave better satisfaction than the French burr, as they were of a more even texture, but the French stone carried the bell. In our country a substitute for it has long been a desideratum. This has now been obtained. In Burke County, in the state of Georgia, a large quarry embracing an area of 17,000 acres has been discovered; and a Company named the LaFayette Burr Mill Stone Co., has been formed to work it and furnish American Millstones equal to the French Burr. The principle [*sic*] office of this Company is in Savannah. About 1,000 sets of stones have already been put up, and are now in operation, and some of them alongside of the French, where in every instance they have equaled—and a little more, the very best French Burrs. Samples of this stone have been in our office for sometime. We have contrasted them in every way with the French burrs, from which without knowing that the one came from France and the other from Georgia, no person could point out a difference. Those who have used the Georgia stone, prefer it for a more enduring fine sharpness, and in that case, it is more economical to use. From what we have seen of the Georgia stone, and heard about it from the most respectable sources, respecting its practical results, we are confident that the quarry must be of immense value.

The December 22, 1849, edition of *Scientific American* contained a brief note concerning the Georgia Burr millstone (Scientific American 1849b:106): "A load of French Burr Stones having arrived at Savannah Geo., a correspondent writing to the *Republican*, states that they are far inferior to the Georgia Burrs — that he had examined and compared the two made into stones, and found the Georgia, Lafayette Co., the best in every point of view."

The next reference to the Georgia Burr millstone was contained within a general story that appeared in the February 2, 1850, issue of *Scientific American* (Scientific American 1850a:154):

> A pair of Georgia Burr Mill Stones have recently been received at a mill in Norfolk, Va., from Savannah. They have excited considerable attention; formerly such kind of stones were all imported from France but Mr. Hoyt, of Savannah, made the discovery of the Georgia Burr bed, which promises to supersede the foreign. These stones are fitted for a 4½ feet circle, weigh about 1600 pounds each, and are of a superior quality....
>
> We are happy to hear from our correspondent about the introduction of the Georgia Burrs into Virginia. We have had some specimens of the stone for a long time, in our possession, and we noticed it in our last volume. ed.

The Georgia Burr millstone was also mentioned in *Scientific American* in the April 27, 1850, edition (Scientific American 1850b:253):

> GEORGIA BURR VERSUS FRENCH BURR STONES
>
> The Schooner Hartford arrived at this port a few days since from Savannah with a lot of 5½ feet Georgia burr mill stones, to be used in the mills of Hackshall, Bro. & Co., at Richmond City, Va. It may somewhat astonish importers of the French burr to learn the fact, that the South will not only in the future quarry their own mill stones, but it will not be a year hence before they will be furnished for all new flouring mills that may be erected in this or the Western states. The Georgia stone, we have been informed by a manufacturer of this city, fully equals the best French; and he says, although he is engaged in the importation of the French burr, that the Georgia stone will inevitably take the place of the French in this country — "Sic transit Gloria Franice."

The final mention of the Georgia Burr millstone in *Scientific American* occurred in the May 25, 1850, issue in a story about the Wilson Patent Stone Dressing Machine (Scientific American 1850c:248): "The principle of action of the cutters is most effectual in dressing stone of the softest and hardest qualities. We have seen two of these machines in operation at the Empire Stone Works, Messr. Sherman & Howdayer, at the foot of 28th street East River. Marble, the hardest Staten Island granite, and even the now celebrated Georgia Burr Stone, have been dressed by one of the machines mentioned, in a period of time surprising to every body who witnessed the operation, and with but little wear of the cutters."

Dana (1857:360) reported that "a buhrstone occurs in Georgia, about 40 miles from the sea, near the Carolina line." The April 1876 edition of *Manufacturer and Builder* included the following information on the Georgia burr millstone:

> A quarry was discovered in Georgia a few years since, but so inferior in quality to the French stone, that it is now entirely abandoned.
>
> The stone is generally shipped in pieces varying in size from 1 to 2 feet in length and about 1 foot in width, and after being dressed to the required shape,— are placed in circular form, the cavities and joints filled with a patent cement made of refuse ground burr stone, calcined magnesite and chlorid of magnesium, and then backed by plaster of Paris, making it present the appearance of a whole stone, and if properly finished, will last, with constant use, for twenty years. This cement is also used in repairing old stones, and if applied with care, it nearly equals the stone itself, and bears the pick almost as well as the burr. For grinding coarse grain, a millstone dressed from one burr stone is preferred.
>
> When taking into consideration the scarcity and rarity of the burr quarries, and the labor

expended on the stone, the price seems nominal compared with its indispensable value, a perfect stone, 4 feet in diameter, selling for only $250.

The Straub Mill Co., of Cincinnati, O., has a very experienced foreman in Mr. R. Fuliman, who has spent several years among the quarries. The company have every facility for manufacturing the mill-stone; they are very courteous to strangers visiting their novel works, and an hour can not be spent more profitably in Cincinnati than in looking through their place of business, whether as a sight-seer or a purchaser.

Davis (1990a:6) provided the following information about the history of the Georgia Buhr:

Christopher Fitzsimons, Irish-born resident of Charleston, SC, was a successful merchant and pioneer American industrialist, owning manufactories, ships, warehouses and plantations. He acquired the Old Town Plantation, near Louisville, GA, in 1809, and beneath the soil there had been found buhr stone. On June 8, 1810, he sent a "Georgia buhr" to Philadelphia to Oliver Evans for comparison with the French buhrs. Evans must have been impressed, for northern workmen from Evans and Morton, under a Mr. Maskel, were in Georgia by January 1811 to acquire more stones. Some of the Georgia buhr stones were sent to flour mills at Brandywine, Pennsylvania. Evans' workmen were quoted as saying that the Georgia buhrs were as good or better than the French, a boast repeated many times in the years that followed....

More information is not available on Fitzsimons' and Evans' work with the Georgia buhrs. However, more than 1,000 buhr stones were reported taken from the Georgia quarries for sale in Georgia, Alabama, the Carolinas, and Virginia. Not all of them came from the Old Town Plantation. Jeremiah Miller and George Poythress in neighboring Burke County were quarrying buhrs by 1815. Miller brought workmen from New York to cut his stone.

Davis (1990a:6, 1990b:9) noted that the Georgia buhr declined after the War of 1812 when the French buhr was again available. By the mid 19th century the Georgia buhr was back (Davis 1990a:6): "By 1849 Georgia buhr had a major comeback, this time on Patrick B. Connally's 17,000-acre plantation near the Central of Georgia Railroad in Burke County. Rock from Connally's plantation was made into stones by S. & H. Hoyt's La Fayette Burr Stone Company of Savannah for sale in Georgia and through A. B. Allen & Co. of New York City. Georgia buhrs were made 15 inches to seven feet in diameter."

A very early ad concerning a millstone quarry in Burke County, Georgia, was published in the August 15, 1792, issue of *The Augusta Chronicle and State Messenger* (Hemphill 1999):

The subscriber has for sale, at his
Quarry in Burke County,
Mill Stones

Of the best quality, and on the most
Reasonable terms. The stone is
composed of seashells turned into flint,
very open and rough, and allowed by
good judges to be equal to the best
French burr. By adding the expense
of carriage, they will be delivered at
Mr. McFarland's store in Augusta,
where application may be made by those
that want to purchase, or to myself at
the Quarry, 5 miles above Louisville,
between Rockey Comfort and Ogeehee.

JOHN MURRAY

August 1, 1792

Hemphill (1999) noted that "John Murray's quarry was forty miles from Augusta as it is five miles north of Louisville which is 45 miles from Augusta. Louisville is now in Jefferson Co. which was cut off the old Burke Co."

Illinois

Beck (1823:194) provided the following information for Illinois: "Buhrstone millstones. In Illinois, near the junction of the Illinois river with the Mississippi." Another source mentioned flint millstones being made in Union County, Illinois (Anonymous 1905:198):

> To these the late Professor Worthen, for many years State Geologist, added, although with some hesitancy, the black shale formation of Illinois. Although these comprise an aggregate thickness of over 500 feet, their exposure is limited to a few isolated outcroppings along the bluffs of the Illinois, Mississippi and Rock Rivers. The lower division called "Clear Creek Limestone," is about 250 feet thick, and is only found in the extreme southern end of the State. It consists of chert, or impure flint, and thin-bedded silico-magnesian limestones, rather compact in texture, and of buff or light gray to nearly white colors. When decomposed by atmospheric influences, it forms a fine white clay, resembling common chalk in appearance. Some of the cherty beds resemble burr stones in porosity, and good mill-stones are made there from in Union County.

Indiana

A geological reconnaissance made for the state of Indiana in 1837 mentioned burr-stones in Harrison and Jennings counties (Owen 1859). For the burr-stone chert layers in Harrison County, Owen (1859:23) commented that "the cement of this burr-stone chert has usually too little tenacity to render it useful for millstones." In Jennings County, Owen (1859:27) stated that "[a]ssociated with these, at the forks of the Muscatatuk river, near Vernon and on the Big Sandy creek, about eight miles from that place, is a good variety of burr-stone. It is almost entirely made up of the remains of fossil coral-lines, cemented by a siliceous cement, with perhaps, locally, some calacreous matter also. When free from carbonate of lime, this porous, siliceous material affords a good mill-stone; some of our western mills have been supplied with this burr-stone, which makes a pretty good substitute for the French burr; it requires, however, to be more frequently dresses, because the cement is not so tough."

Finally, Owen (1859:28) observed, "The burr-stone mentioned in the first part of this report, which is wrought for millstones in Jennings county, is a member of the upper division of the preceding section."

Several years later, Holloway (1870:326) noted that "Buhr mill stone is also found in Jennings county, but it is hard to say what the value of the trade in it may be, or may be made."

Kentucky

Flint millstones were quarried in Franklin and Woodford counties, Kentucky (Hockensmith 2003c, 2006e). The following paragraphs provide the archival information encountered on these quarries.

Franklin County

The firm of Miller, Railsback & Miller advertised their flint millstones from a quarry in Franklin County in the August 9, 1821, edition of *The Argus of Western America*:

> LOOK HERE !
>
> To Mill-Wrights and Mill-Builders.
>
> THOSE who wish to purchase Mill-stones of the flint kind which have been cut for some years by Dudley and lately by Jeremiah Buckley, may now be had in Franklin county, Ky

eight miles above Frankfort, on the river, where mill-stones are cut by the undersigned, on the land of Henry Miller, or near the adjoining land of M. Johnston. We have as good rock and as thick as necessary, and will cut as cheap & do our work as good as any. As the quality of these flint rock has been proven to be good for many years, it is not necessary to recommend them. We will deliver millstones any where in this state or in the states adjoining, by the purchaser paying the common cost of carriage and as our rock are not inferior to any of the kind, we are determined the workmanship shall be the same. We wish every one that wishes to purchase rock, to review them for themselves and in particular those who are real judges of quality and workmanship. All millstones proving not good cut by us either shall be cut gratis for the purchaser that are good, and of the same size. All letters to the undersigned will be attended to. They must be directed to Lawrenceburg, Franklin county, Kentucky. The prices of our Mill-stones are annexed, but the purchaser having cash can almost make his own bargins as we intend to work cheap, and are possessed with experience and a mechanical eye.

5 feet	$150.
4 feet	$100.
3 feet	$50.

And all sizes accordingly.

MILLER, RAILSBACK & MILLER

May 23rd, 1821.

Jeremiah Buckley also advertised his flint millstones in the November 8, 1821, edition of *The Argus of Western America*:

MILL STONE QUARRY,

AT BUCKLEY'S FERRY, Franklin County. THE subscriber takes this opportunity of informing his friends and the public in general that he has on hand a good assortment of

MILL STONES,

and intends at all times to meet the calls of gentlemen that wish to get mill stones from him. As flint or grit is thought by the best of judges, to be superior to any other of the kind that has ever been discovered in the United States, for either corn or wheat when put in order, the French Burr not excepted it is hoped and sincerely requested by the subscriber that gentlemen in the west country wanting that article will inform themselves whether superiority of quality belongs to his Quarry or not; and for their information, he will refer them to a number of gentlemen who have his mill stones now in use: Col Robert McAfee, McCoum and Kennedy, Vandike and Keller, Robert Neal and Joseph Adams. These gentlemen have eight pair now in use on Salt River John Buford, Esq of Versailles and Mr. David Rice, on Clear Creek, Woodford county, Mr. James Rucker, of Caldwell county, one pair in use at this time, Mr. Gabriel Stansefer, on Main Elkhorn, has one pair now in use, also General George Baltzell and John Baltzell of this county, each one pair, Matthew Flourney, of Fayette, has also a pair in use. Mr. Flourney, Col McAfee and John Bufford, Esq. are at this time in the Legislature of our state, from whom, gentlemen living at a distance, through the medium of their representatives and others, may easily inform themselves.

The prices of my mill stones are as follows:

For Five Feet	$180
Four feet six inches	150
Four feet	125
Three feet nine inches	100
Three feet six inches	85
Three feet three inches	75
Three feet	60

Two feet nine inches 50
Two feet six inches 40

JEREMIAH BUCKLEY

October 18, 1821

Woodford County

The Frankfort newspaper, *The Palladium*, carried an ad in the February 27, 1800, issue for millstones quarried in Woodford County, Kentucky:

MILL STONES.

A Fair and impartial trial of a pair of millstones cut in my quarry, of flint quality, of four feet in diameter, now running in Colonel Robert Johnson's mill on North Elkhorn Creek, was made with his Burr stones, as stated in his certificate, 330 lb. of wheat being weighed into each mill, and the wheat weighing 59½ lb, per bushel, and both mills managed to the best advantage, and the flour bolted his superfine cloth immediately, that ground on the flint stones was first bolted, and made 243 lb. of flour, the Burrs 213 lb. leaving 30 pounds in favor of the flint stones, & of equal quality. I have here the Colonel's certificate for further information:

"I HAVE in my mill on North Elkhorn, a pair of French Burr Stones, four feet diameter; also, a pair the same size, which I purchased of Mr. John Tanner, cut in his quarry, in Woodford county. I have made trial of the flint mill stones I purchased of Mr. Tanner, and find them exceedingly good for manufacturing flour; and I further certify, that I am of opinion they are equal to any French Burrs of the same size in this state. Given under my hand, this 15th day of December, 1799."

A true Copy.

(Signed) Robt. Johnson.

I Have for Sale, the Quarry out of which the above stones were cut, and five acres of ground including the same; and will make the terms cash (?) to the purchaser, by taking the price in mill stones and the produce of the country. Seven hands unto of whom may be warrent will make the 100 £ per month, clear of all expenses, at my usual selling prices, and the probability is, that the stones will bear a higher price than usual, if their value could be generally known, and if a speedy trial could be made of those who incline to purchase of me. I will warrant them to be equal to the Burrs. I have on hand, a few pair, from four feet two inches in diameter to four feet eight inches, which I will warrant as above.

John Tanner

February 5th, 1800.

Geologist William Mather (1839:282) made the following observations concerning Kentucky millstones in his discussion of limestone: "A stratum of silicious matter which varies in its texture from hornstone to a porous material like buhrstone, is found in many places in the cavernous limestone. The compact variety of this material was long used by the early settlers for flints for their rifles. Another form of it is used for hones. The coarser varieties are frequently seen of such a quality and texture, as to fit them for millstones. The localities of these materials are frequently called flint knobs, and their surfaces are thickly strewed with silicious masses."

Massachusetts

Dana (1857:360) reported that "the quartz rock of Washington, near Pittsfield, Mass., is in some parts cellular, and makes good millstones." No other references to this rock were encountered in the sources consulted. Undoubtedly, other geological reports for Massachusetts mention this type of rock.

Missouri

The *Gazetteer of the State of Missouri* (Wetmore 1837) mentioned millstones in Carrol, Madison, Montgomery, and Washington counties. Of these stones, the rock in Madison County is compared to the French buhr (Wetmore 1837:109): "There is an abundance of limestone and sandstone in this county; and there is likewise a rock of suitable quality for millstones, and which is now in use for that purpose. It is believed that this stone, if properly wrought in the fashion of the French buhrstone, will be a valuable substitute for that costly material, and take place of it to a great extent. The experiments that have been made in grinding with this stone have proved the great value of the material, and placed it on the long list of the resources of Missouri."

The North American Review (1839:522) noted that "Buhr stone, said to be equal to the French, is in great abundance on the Osage and Gasconade Rivers."

Another statement about Missouri millstones was found in *The North American Review* (1839:521): "in a single county (Washington) are found ... millstones, resembling the French buhr."

The First and Second Annual Reports of the Geological Survey of Missouri (Swallow 1855:11) provided the following information on millstones in Moniteau County: "Some of the beds of silicious breccia, previously referred to in the south-eastern corner of the county, I was informed have furnished very good mill-stones. Judging from their extreme hardness, and cellular structure, it seems quite probable they would answer very well for such purpose."

In the reports on the *Geological Survey of the State of Missouri, 1855–1871*, Broadhead, Meek and Shumard (1873:11) stated, "In Maries and Osage counties, it occurs as a Silicious Buhr-stone in small fragments, and I almost invariably found it on slopes just above the Second Sandstone. Beautiful specimens of this rock are found in Webster county, where some varieties have been used with very good success as a mill-stone."

Broadhead, Meek and Shumard (1873:31) indicated that in Osage County, Missouri, "Lying at the base of the Second Magnesian Limestone, and over the Second Sandstone, are found silicious beds, generally cellular, which very much resemble the French Buhr-stone. The cells are somewhat oval or spheroidal in shape, their general size about that of an ordinary bean — some are larger and some much smaller — and, under a magnifier, their interior appears as if studded with minute quartz crystals."

For Ozark County, Missouri, Broadhead, Meek and Shumard (1873:201) noted that "[l]arge amorphous masses of silex, sometimes brecciated and containing cavities, are quite common on the uplands in different parts of the county, and these will often furnish a good variety of Buhr-stone."

In Wright County, Missouri, Broadhead, Meek and Shumard (1873:212) observed that the "large silicious masses which occur frequently on the tops of the high ridges in various parts of the county, often have the cellular and other appearances of a good quality Buhr-stone."

Finally, Broadhead, Meek and Shumard (1873:231) recorded a chert used for millstones in Pulaski County: "A better material is found near the Gasconade, in the vicinity of Bates' mill, examined by Mr. Engelmann. This is a bed of decomposing chert, traversed with seams of quartz, in the Third Magnesian Limestone. Mill-stones were obtained for several mills in the county at this locality, and they are said to answer a very good purpose."

In a discussion of different types of Missouri millstones, Broadhead (1874:55–56) stated that "[s]ome of the Chert beds have also been used, from rock obtained from interstratified beds of the Second and Third Magnesian Limestone. The Chert occurring among the Lower carboniferous rocks of South-west Missouri, in outward appearance closely resembles the French Buhr. Beds of this Chert are well exposed in Newton, Jasper, and Cedar counties. Some specimens from the last named county, under a strong magnifier, disclose minute crinoid stems. There would be danger of their breaking and becoming disseminated into flour, rendering it not very palateable."

North Carolina

Emmons (1856:218–219), in his *Geological Report of the Midland Counties of North Carolina*, provided the following discussion:

> Burrhstone or Pseudo-burrhstone.—This rock, upon its exterior, is exceeding rough and ragged, and as it is an extremely tough siliceous rock, it may probably possess the same valuable properties as the Paris burrhstone. But I do not speak confidently; for the fact can be known only by applying a stone suitably prepared to this use, that of a millstone. The material of the rock is in the required condition at the surface. It consist of a porous chert; or, originally, it was a porphyrized chert, the felspar having disintegrated, leaves rough cavities bounded by a tough sharp edged material, similar to that of the Paris burrhstone. One difficulty may materially interfere with the introduction of this stone for the purpose proposed, viz., the expense of cutting it, in consequence of the toughness of the material; and besides, the stone has not weathered deeply, except when detached from the parent bed. But the formation is extensive, and the varieties or kinds are quite numerous; and hence, it is expected that localities will occur suitable for the purpose I have proposed.
>
> The most extensive beds of the Pseudo-burrhstone occur in Montgomery county, near Troy. It is the fossilliferous mass which is intimately connected with the granular quartz. A rock possessing characters quite similar to the foregoing exists in large massess at the Flat Swamp mountain, in a part of Stanly and Davidson counties. So far as they have fallen under my observation, they are not so good as those of Montgomery county. The whole range too, from a point near Gold Hill to the Flat Swamp mountain is traversed by a siliceous porous rock, which possesses many of the characters required for good millstone.

In his "Agriculture of the Eastern Counties: Together With Descriptions of the Fossil of the Marl Beds," Emmons (1858:90) furnished the following information on stone marl used for millstones: "There are two varieties of stone marl, both of which deserve a special notice. The first consist of shells cemented strongly together, and which are usually from one to one and one and a half inches across, and very uniform as to size. They are very firmly cemented by silica, which seems to have penetrated the shells more or less. This rock has been employed for a long period for small millstones. Its valuable qualities consist in being easily wrought when first removed from the quarry, but subsequently becomes very hard and strong. Being made up of shells, it has a rough appearance, even when cut evenly; but this feature constitutes its recommendation."

He further noted (Emmons 1858:102), "The Eocene is known to exist at Wilmington, at Pollocksville, in Jones county, and underlies the whole country in the vicinity of Newbern, upon Neuse. In this information I include the consolidated beds which have been employed for mill stones, and which consist of a mass of the cast of shells, the most common of which is a small species of clam."

The 1875 *Report of the Geological Survey of North Carolina* stated that "in the eastern section, the shell rock is often partly or wholly silicified, forming a sort of buhr-stone, as in Georgia, and is well adapted to the same uses" (North Carolina State Geologist 1875:305).

Ohio

Flint was exploited for making millstones at Flint Ridge and Raccoon Creek in Ohio (Hockensmith 2003d, 2007a). At Flint Ridge in Licking and Muskingum counties, Ohio, millstones were quarried from a flint that was "very porous and fossiliferous, and very frequently mixed with calcareous or argillaceous material" (Mills 1921:96). Samuel Drumm made the first millstones (monolithic) at Flint Ridge and in 1827, Joshua Evans & Co. began to manufacture composite millstones similar to French buhrs while Richard Watkins was making millstones there (Garber 1970:81). Garber (1970:81) reproduced an ad from the May 15, 1827, issue of the *Mansfield Gazette* which is quoted below in a different format:

FLINT RIDGE BURR

MILL STONES.

> THE Subscribers inform the public that they have commenced the business of manufacturing Mill Stones of the above description, in Washington township seven mile South of Mansfield, where they have and intend keeping constantly on hand, Stones of various sizes. They have also an establishment at Flint Ridge, in Licking county where Stones of any size will be made on the shortest notice.
>
> Persons engaged in the milling business are invited to call and examine for themselves, the subscribers believe that they can furnish stones of as good a quality, and at as reasonable prices, as any establishment in the western country.
>
> JOSHUA EVANS & Co.
> Washington, May 15, 1827

Hildreth (1828:40) stated that "at the east end of the ridge, twelve miles from Zanesville, fine mill stones are made of cellular quartz, equal to, or better than French Burr."

Mather (1838a:33–34) described the Raccoon Creek buhr in his *First Annual Report on the Geological Survey of the State of Ohio*. W. W. Mather (1838a:33–34) made the following comments concerning the value and importance of the Buhr-stone:

> The importance of this article in a commercial and domestic point of view, may in some measure be estimated, when it is stated, by intelligent persons who have been long engaged in the manufacture of mill-stones, that the annual amount of the manufactured article is not less than 20,000 dollars; and that it may be safely calculated at this sum, for twenty years past. When to this, is added, the money saved to mill owners, from the use of the native, instead of the foreign buhr-stone, that amount will be nearly doubled. It came into use about the year 1807; and the first pair of stones constructed of this article, on the waters of Raccoon, was by Abraham Neisby, a native of Germany. He being familiar with the foreign, or French buhr, and seeing this rock so nearly resemble that in composition and aspect, was led to make trial of it. Henry Castle, also began to make them about the same time. Soon after this, our embarrassments with Great Britain, and other commercial difficulties, led the American people to establish manufactories of various articles, heretofore altogether brought from Europe. Amongst other things the enhanced value of the French buhr, led to the search of a material of a similar quality at home, and no doubt brought the domestic article much sooner into general use in the Western States, and especially in Ohio. The early manufactured millstones were made of a single piece; but these often proving to be of unequal density, and not making good flour, were abandoned, and staves constructed of separate blocks, cemented with plaster, and confined together with iron bands. Where these blocks are selected with care, by an experienced workman, the flour is said to be equal in quality to that made by the French stones.
>
> From the 1814 to 1820, the price of a pair of 4½ feet stones was $350, and a pair of 7 feet, sold for $500; while the foreign article sold for a still higher sum. The 4 feet stones now sell

> for $150. In the townships of Richland, Elk and Clinton, a large number of the inhabitants are engaged in the dressing of blocks, and in the construction of mill-stones. The buhr-rock is a mine of wealth to the inhabitants, and has contributed largely to the prosperity and the independence of this whole region of country. The manufacture of mill-stones is not confined to the waters of Raccoon, but is also carried on to considerable extent in Hopewell township, Muskingum county. The quality is apparently inexhaustible, and new quarries will be opened, at points where it is not at present looked for, and probably of a more even and compact texture than that now obtained. Few or no quarries have been yet worked by drifting under the sides of the hills, but the rock is generally procured by what is technically called "stripping," or excavating the superincumbent earth, near the top of some ridge or hill, where it is easy of access.
>
> The characteristic excellence of the best mill-stone rock consist in its uniform texture; composed of quartz; free from lime and oxyde of iron, which more or less pervade the larger portion of the deposit; color light grey; structure open and full of cells; the fragments giving a clear metallic sound when struck with the hammer.... Although an intensely hard rock, yet its structure is similar to that of all other stratified rocks; and it has a regular horizontal division, as well as cross fracture. The bed, or horizontal surface, is the one which contains the most cells, and is selected for the face of the mill-stone. In nearly all the quarries, the rock is found naturally broken into rectangular masses of various dimensions. As the larger portion of the quarries contain more or less of petrified shells, those the most free of foreign matters are considered the best. While the fossils in the French rocks are all of fresh water species, those in the Raccoon buhrs are altogether of marine origin; and as this deposit is a member of the coal series, it is a much older rock than the French, which belongs to the tertiary group.
>
> In the composition and the chemical affinities of the Raccoon deposits, there is a close resemblance to those of the Paris basin, as they both abound in calcareous materials, as well as silicious. In several of the Raccoon beds, there is a stratum of lime-stone, two or more feet in thickness, both above and below the buhr-stone. In such quarries, the portion suitable for mill-stones is only from two to four feet in thickness. In others, the quartzy portion is below, and the silicious lime-stone above. The greatest thickness of the deposit in any one bed, is nine feet."

In his *Second Annual Report on the Geological Survey of the State of Ohio*, W. W. Mather (1838b:90–91) offered the following comments concerning character of the buhr stone:

> This is the result of a chemical precipitate — rather than a mechanical deposit — of silex; although we are unacquainted with the process by which it was effected over so large an area. That silex is soluble in boiling water, is evident from the deposits of the geysers of Iceland and Bohemia; nor is a high temperature absolutely necessary, since it enters largely into the composition of most of our canes and rushes.
>
> *External Characters.* — Greyish or yellowish white — also with a greenish tinge; opaque — sometimes passing into hornstone, when it becomes translucent. It contains numerous cavities — bearing some resemblance to amygdaloidal trap. These cavities, in some cases, are formed by the mold of small infusoria, myriads of which are contained in a cubic foot.
>
> *Mineral Contents.* — Quartz in beautiful druses, lining the oblique fractures of the buhr — also in six-sided pyramids-occasionally smoky. Chalcedony in a confused aggregation of crystals, in the cavities of the buhr, as though deposited by infiltration before the consolidation of the surrounding mass was complete. Calc spar, in rhombic prisms of pearly luster — translucent, or nearly transparent. Heavy spar, or sulphate of barytes, is also found.

Concerning the range and extent of the Ohio buhr stone, Mather (1838b:91) noted: "Beginning about a mile west of Somerset, Perry county, it ranges near the dividing line between Muskingum and Licking, passing through the townships of Hopewell, Hanover, Licking and Jackson, crossing the national road near Gratiot. Its average breath is from 8 to 10 miles. This rock is so liable to disintegration, as to render it difficult to ascertain the line of junction. Its examination, therefore, is attended with some degree of difficulty. To specify all the localities at which it has been observed, would be unnecessary as they are indicated by symbols on the map."

Mather (1838b:91) made the following statements about the use of a buhr stone (2 to 6 feet thick): "It is quarried and wrought into mill stones. The principal quarries are those of Samuel Drumm and S. Henslee, on sec. 15, Hopewell township, Muskingum county; and Adam Drumm and Joseph Baird, on sec. 11, Hopewell township, Licking county. It is inferior in toughness to the Raccoon buhr, and therefore less valuable for millstones."

Another early discussion concerning the use of Ohio buhr stone for millstones was published by Caleb Atwater (1838:16):

> The burghstone, of which millstones are made, in considerable numbers, in the counties of Muskingum, Hocking, Jackson and Gallia, occurs in amorphous masses, partly compact, but this rock always contains in it, more of less irregular cavities. These holes are occasioned sometimes, by the seashells which originally filled them having fallen out of their places in the rock. The aspect of this millstone is somewhat peculiar, resembling paste, which had been in a state of fermentation, when moist, and warm; but when the heat had ceased to act, the mass became dry, hard and compact, with all the marks of fermentation remaining in it. The cavities are sometimes, filled with crystals of quartz. The fracture of this burghstone is commonly dull, and its colour is whitish or redish brown. Its hardness and cavities, when not too numerous, render it very useful for making millstones, many of which are manufactured, and sent all over this state, and to the western ones generally.

Millstones were made to order by Richard Watkins at the quarry in 1841 (Garber 1970:81). It was a common practice for mill owners to specify the diameter and thickness of the millstones they ordered.

In an 1851 discussion on Muskingum County, Ohio, Howe (1851:381) noted the presence of a "burr-stone or cellular quartz, suitable for mill stones." Howe (1854:586, 1875:586) also provided the following description of a stone in southern Ohio: "In the 1st Geological Report of Ohio, p. 28, Dr. Hildreth notices the 'calcareo siliceous,' or 'buhr stone rock,' of the coal series of Ohio, which resembles very closely the French buhr, used in this country for mill stones, and imported from France. On Raccoon's creek, and at other places in the south, near the Ohio river, this rock is wrought into mill stones to a considerable extent; but millers, as yet, prefer the foreign buhr, at a considerably higher price."

Brainard (1854:12) in his "Origin of the Quartz Pebbles of the Sandstone Conglomerate and the Formation of the Stratified Sand Rock" observed: "On what is called Flint Ridge, in Ohio, there is a bed of compact quartz, with cavities beautifully studded with crystals. This stratum belongs to the carboniferous formation. It is generally known as the buhrstone of Ohio, and is used to some extent in the manufacture of millstones. Like the buhr stone of France, this has been formed from solution, which fact is most satisfactorily shown by an examination of the specimens."

Bishop (1866:185) made the following comments about the Raccoon Creek burr millstones: "A domestic supply of 'Burr' millstones, for the western country, was opened near the head of Raccoon creek, Athens co., Ohio. It was considered identical in composition with the French curb stone. The first pair were put in the steam flour mill of the Mariette Mill Company, started in January by Messrs. Gilman, Barber, Skinner, Fearing & Putnam, who afterwards added woolen machinery."

In his *The American Miller, and Millwright's Assistant*, Hughes (1869:73–74) shared the following information about the Raccoon Burr Stone:

ON THE RACCOON BURR STONE

> This description of stone is of American production, and its geological nativity is confined to the State of Ohio, not being known elsewhere. Its locality is in Muskingum and adjoining counties, known by the name of the "Flint Ridge." This stone is a description of burr, and makes a very good substitute for the imported or French burr. During my residence in the State of Ohio, I was employed by the Messrs. Adams, of Muskingum county, who do a large business in flouring, being the most extensive millers in that part of the State. One of their mills, in which the author was employed, was of six run of stones, all of them of Raccoon burr, and having dressed them, the only conclusions I drew, from the work the stones made, was that they required to be dressed oftener than the generality of the French burr. The reputation of this mill then stood high in New York for making a good article of superfine flour. The difference in the price between the Raccoon and imported being from 35 to 45 per cent cheaper. They are put together in blocks and fitted up as the French burr, and will answer a good purpose for grist mills, or for grinding coarse grains, such as grist grinding generally consist of, for the use of the farmer.

An 1874 report entitled *The Agricultural and Mineral Resources of Muskingum County* (Zanesville, Ohio Board of Trade 1874:17) contained the following comments concerning the Ohio buhr stone:

> Buhr.— What is called "Buhr Stone" is found in the Western section of the County. It exists on both sides of the line dividing Muskingum and Licking Counties, and extends into the North-east corner of Perry County. The stone is a grayish or yellowish white, sometimes passing into hornstone, exists in beds from two to six feet in thickness, is fine grained and compact, and well calculated to give a fine edge to cutting tools or implements. The Indians used the compact hornstone for arrowheads. This Buhr was, years ago, quarried to quite an extent, and made into millstones, but as the material lacked tenacity they were not regarded with equal favor with the "French Buhr." Their manufacture has of late been abandoned, and at present the Buhr is not sought.

A quartz millstone quarry was established in about 1805 by a man named Musselman in Vinton County, Ohio (Garber 1970:78). These millstones became known as "Raccoon buhrs" since the quarry was near Raccoon Creek (Garber 1970:78). Blocks of stone were quarried and fitted together to form composite millstones (Cinadr and Brose 1978:59). By 1860 the Raccoon Creek quarry had apparently ceased production (Garber 1970:78).

George Ogden (1823:77) in his *Letters from the West* provided the following information about the Raccoon Creek millstone quarries: "On Racoon [*sic*] Creek, in Athens county, are extensive quarries of stone, from which burr mill stones are made, equal to the best English burrs; they are so good that all the western states are supplied with them."

Andrews (1870:102) noted that "the flint or buhr stratum, on Flint Ridge [Licking County], is not found to correspond in stratigraphical position with the other layers of flint found in the district especially examined." Further, he made the following comment about Flint Ridge (Andrews 1870:102): "The buhr is of a variable thickness, its maximum being perhaps 8 feet. Formerly, mill-stones were made from the rock, but the quarries have been of late years abandoned."

William C. Mills (1921:96, 98) provided the following information on historic millstone quarrying at Flint Ridge:

> The flint at the west end of the "Ridge," in Licking County, was especially useless to primitive man, but the early white settler found it well adapted to the making of buhr-stones, used in grinding grain into flour. Near the western edge of the outcrop of the flint, several partly formed buhr-stones, each weighing a ton or more, may be seen where they were quar-

> ried upon the farm of Mr. William Hazlett, near the only large mound located upon the "Ridge."
>
> The flint at the eastern end of the "Ridge" is likewise unfit for implement making but well adapted for buhr-stones. In the early pioneer days of Ohio, Mr. Samuel Drumm quarried the flint in suitable blocks and fashioned them into small hand buhr-stones....
>
> The farm upon which the quarry is located is owned by Mr. George Fisher, who kindly presented to the museum a fine sample of [a] ... partly shaped buhr-stone ... as well as a buhr-stone sent from France and used as a sample stone.
>
> The manufacture of these small buhr-stones during the early settlement of the country was a very great convenience to the people, as water mills for grinding grain could only be constructed where proper conditions prevailed, and often long distances would be traveled to find such a mill; consequently the small hand mill made from Flint Ridge flint was very desirable, and the manufacture of the buhr-stones proved to be a very lucrative industry. The buhr-stones manufactured at the Drumm site were sent to a point on the Old National Road, three miles to the south, where they were transported by ox teams as far west as the Mississippi River and as far east as Pittsburgh.

In his discussion on the growth of industries, Utter (1942:241) noted that the: "millstones were imported from the East, even from England, until suitable stone for their construction was found in Athens and Jackson counties."

Howe (1888:427) provided the following information on the millstone industry in Vinton County, Ohio:

> When we first came here there were perhaps fifty families in and around this settlement, most of them quarrying and making millstones. There was no person making a business of farming. All had their little patches of garden, but making millstones was the principal business. Isaac Pierson, father of Sara Pierson, of Chillicothe, had the principal quarry. Afterward Aaron Lantz and Richard McDougal had large quarries. A man named Musselman first discovered the stone in 1805 and in 1806 employed Isaac Pierson to work for him. This was on section seven. There were no white people here at that time and the two camped out. Musselman quit, but the next year Pierson, finding the business to be very profitable, moved out, built the first cabin and made the first permanent settlement.
>
> He employed hands to help him, and soon the settlement grew. The business was very profitable, and all engaged in it would have become independently rich but for one thing—whiskey! Most of them drank; and nearly every pair of millstones that was sold must bring back a barrel of whiskey, whether it brought flour or not. If the flour was out they could grind corn on their hand-mills but they made it a point never to get out of whiskey.

Fess (1937:290) said of Vinton County, Ohio, "The tradition is that the first settler of the county did not locate here for farming, but because he found here that particular type of limestone most suitable for making the burrstones or millstones used in the old grist mills. Most of the early settlements were made along the Elk fork of the Raccoon in the vicinity of McArthur and Zaleski."

The Maysville Eagle, published in Maysville, Kentucky, carried the following ad for Raccoon Burr millstones in the November 20, 1834, edition:

> MILL STONES.
>
> RACOON BUR MILL STONES, manufactured by David Richmond, of various sizes, warranted. Can be had at short notice, by application to
>
> JANUARY & HUSTON, Agents
>
> Maysville, May 3, 1834

James L. Murphy (2004) indicated that he had seen advertisements for the "Raccoon buhrstones" in early Ohio newspapers. He stated that these millstones were made from the Zaleski flint of Vinton County, Ohio. He thinks that these quarries were located west of McArthur, Ohio, but were destroyed by strip mining. Steve Parker (2005) located one of the Raccoon Creek Buhr millstone quarries near McArthur, in Vinton County, Ohio, during March 2005. The field trip revealed 12 quarry pits, piles of flint debris from shaping millstones, and one squared piece of flint (12 to 14 inches across) with tool marks. He also noted that a larger quarry had been destroyed by strip mining for coal and that several unfinished millstones were reportedly buried under the slag piles (Parker 2005).

South Carolina

Millstones were quarried from a flint-like stone at two locations in South Carolina. The first location is near Barnwell in the southwestern portion of the state where "beds of silicified shells" were quarried (Tuomey 1848:290). Tuomey noted that "pieces may be found that agree exactly with the French buhrstone, but those who have attempted to procure millstones at this locality, have committed a great mistake in trying to get them in one piece. Everyone knows that French burr mill stones are made up of from 16 to 20 pieces, cemented and bound together with iron hoops." The second quarry was located in central South Carolina near Orangeburg and where millstones were made from "a bed of close grained silicious rock" (Tuomey 1848:290). Finally, Foster (1869:320) noted that "in South Carolina occurs a cellular Buhr-stone, adapted to mill-stones."

Tennessee

Flint millstones were manufactured in Clairborne, Jefferson, Knox, Sumner, and Williams counties, Tennessee (Ball and Hockensmith 2005, 2007d; Hockensmith 2004d). The earliest mention of buhrstone millstones in Tennessee is found in Haywood's (1823:14) history: "Some of the inner mountains above described; and particularly one lying fifteen miles to the north of Knoxville, are mountains of Buhr stone, which is acknowledged by the best Judges, to exceed all of the like kind in the world."

Safford (1869:221) in his *Geology of Tennessee* provided the following information on flint millstones in Tennessee: "The chert so characteristic of the upper part of the Knox Dolomite is manufactured, at several points, into excellent millstones. Layers of it, having a suitable cellular structure, occur in Claiborne, Jefferson, Knox, and other counties. It is generally the weathered outcropping portion of these layers that is used. After getting to a certain depth, the cavities are found to be more or less filled with crystals of dolomite and other matter."

In further discussion of the Knox Dolomite and the flint millstones made from it, Safford (1869:511) noted, "This has been made into millstones at a number of points in East Tennessee. At Big Spring, in Claiborne County, Col. Hugh Jones manufactured, during his life time, not many less than 100 pairs of stones from this chert. They were quite in demand, and were considered to be equal to the French buhrstone."

A bed of silicified shells was used for millstones in Sumner County, Tennessee. Safford (1869:283) described this stone and the millstones made from them as follows:

> In Sumner County, a few miles north of Hartsville, immediately below the Black Shale, is a bed, from which millstones were formerly extensively manufactured. This bed is a mass of shells, closely packed, and silicified. The bed is several feet thick, and contains Nashville species.
>
> The shells are so packed as to make the rock, in due degree, cellular. The weathered portions, near the outcrop, are preferred, for the reason that, within, the spaces between the

shells are filled with calcareous matter, which, by exposure, is leached out. The millstones manufactured here, were highly esteemed. I do not know that any have been made of late years.

Safford (1869:511) also reported on some additional chert millstones that he had not personally seen: "Dr. Troost, in his Third Report, speaks of a superior kind of 'siliceous millstone' near Harpeth River. I have not seen this, but suppose it to be some layer of chert, in the Siliceous Group."

Flint millstones were manufactured in Claiborne County, Tennessee, at Big Spring (Killebrew 1874:260). The raw material at this quarry was described as a "a flint mass, filled with cellular cavities" (Killebrew 1874:260). The quarry was in business prior to the Civil War and the millstones were regarded "as equal in every particular to the French buhr" (Killebrew 1874:260).

During 1989, the author was made aware of a millstone quarry in Williams County, Tennessee (Hockensmith 1993c:5; Rogers 1989). This small quarry exploited a Mississippian age Fort Payne Formation which consisted of siltstone with shale and dolomite (brownish gray to olive gray) containing lenses and blebs of brownish black chert and small quartz geodes (Rogers 1989).

Virginia

Dealing with the period of 1835 to 1841, Rogers (1884:173) described the Montgomery Buhr deposits of Floyd County in southwest Virginia: "In the same region, and associated with the valley limestone, occurs that interesting and valuable siliceous deposit, the Montgomery Buhr. Varying from a greyish and yellowish white to a deep orange brown, and presenting a cellular texture and great hardness and sharpness of grit, this unique material possesses qualities which admirably adapt it to the formation of mill-stones; and is accordingly, though to a much less extent than could be wished, used for this purpose."

An early gazetteer of Virginia (Martin and Brockenbrough 1835:218) mentioned millstones equal to the French burr in the section on Smyth County in the western part of the state: "Preston's Saltville land contains a description of millstones, easily quarried, which are equal to the best French burr stones for flour mills; and in Russell County are quarries of various marbles."

Another reference to millstones in southwest Virginia appeared in the February 1869 edition of the *Manufacturer and Builder* (Fontaine 1869:47): "Here may also be found deposits of 'buhr-stone,' equal to the French, and 'pebble-stone,' suited also for millstones."

Quartzite Quarries

Quarries producing millstones from quartzite or quartz have been reported in Alabama, Connecticut, and North Carolina. Gary, McAfee, and Wolf (1974:582) defined quartzite as both a metamorphic and a sedimentary rock: "A granoblastic metamorphic rock consisting of mainly of quartz and formed by recrystallization of sandstone or chert by either regional or thermal metamorphism. ...A very hard but unmetamorphosed sandstone consisting chiefly of quartz grains that have been so completely and solidly cemented (diagenetically) with secondary silica that the rock breaks across or through the individual grains rather than around them."

Alabama

Jones (1926:59) stated that the Weisener Quartzite in Alabama was capable of producing good millstones. No other comments were encountered for millstones being made from these deposits. However, it should be noted that time was not available to search the early reports on Alabama geology.

Connecticut

Shepard (1837:68) described a Connecticut quarry as follows: "a cellular drusy quartz found on Whortleberry-hill in the south part of Canton, was used during the last war for millstones. Several acres near the top of this hill are strewed over with large masses of this rock."

North Carolina

A North Carolina millstone quarry exploited a quartzite along the Laurel River in Madison County (Schrader, Stone, and Sanford 1917:234). An early account (North Carolina Land Company 1869:106) noted that: "On Laurel River in Madison is a peculiar cherty splintered whitish quartz rock which Mr. George Gehagan has manufactured into millstones, which are described as nearly equal in performance to the French buhrstone."

The 1875 *Report of the Geological Survey of North Carolina* (North Carolina State Geologist 1875:305) stated that: "In Madison county, in the Huronian slates on Laurel River, there is an irregular laminated whitish quartz, occurring in large veins, which is used for millstones, which are reported to be a good substitute for buhr-stone."

Gneiss Quarries

Millstones were manufactured from gneiss in Alabama, Massachusetts, New York, North Carolina, and Tennessee (Hockensmith 2007b). Gary, McAfee, and Wolf (1974:303) define gneiss as a "foliated rock formed by regional metamorphism in which bands or lenticles of granular minerals alternate with bands and lenticles in which minerals having flaky or elongate prismatic habits predominate.... Although a gneiss is commonly feldspar- and quartz-rich, the mineral composition is not an essential factor in its definition (American usage). Varieties are distinguished by texture (e.g. augen gneiss), characteristic minerals (e.g. hornblende gneiss), or general composition and/or origins (e.g. granite gneiss)."

Alabama

Dean (1996) stated that "near the historic Chewacla lime works in Lee County a porphyritic gneiss was used for millstones." No other information is currently available concerning these millstones.

Massachusetts

Hitchcock (1841:218) indicated that a singular porous quartz derived from gneiss was quarried on a hill in Washington, a few miles from Pittsfield, Massachusetts. Other early geological reports may shed light on this quarry but they were not readily available.

New York

A gneiss millstone quarry was located at Antwerp in Jefferson County, New York. Emerson (1898:452–453) noted that "Antwerp is the extreme eastern town of the county, and is bounded north by the St. Lawrence and east by Lewis County." Haddock (1895:434) stated that "Antwerp possesses stone of many kinds, whose commercial value is yet practically to become known, although David Coffeen and James Parker, in 1805, quarried over 100 pairs of mill-stones at $100 per pair, from a ledge of gneiss on the old State road, between Mr. Cook's place and Ox-bow."

Emerson (1898:453) provided additional information on these quarries: "The ridges are composed of masses of gneiss, a product which was an early source of business and profit to the inhabitants, for between the years 1805 and 1828 about one hundred pairs of millstones were manufactured from that rock in the town."

Oakes (1905:1125–1126) shared useful biographical information about James Parker and his involvement in the millstone industry: "James Parker was born September 29, 1764, in Richmond, New Hampshire.... He died January 26, 1828, in the town of Watertown, where he settled in 1801.... His farm embraced three hundred and ten acres, on the Brownsville road, in Watertown, where a grandson now resides.... He was also one of the pioneers in stone quarrying and as early as 1806 open the Parker ledge on the state road between Lee's tavern and Ox Bow. It was from this quarry that one hundred pair of millstones were manufactured, and the old Church mill at Antwerp village was also built with the product of the Parker ledge."

In *A History of Lewis County, in the State of New York*, Hough (1860:145) commented that a mill at Lowville, in operation on September 22, 1799, had stones "dressed from a boulder of gneiss rock by James Parker, the well known mill-stone maker of Watertown."

James Parker's son Alexander was also involved in the millstone industry (Oakes 1905:1129–1130): "Alexander Parker, eldest son of James and Martha (Houston) Parker, was born September 3, 1787, in Ackworth, New Hampshire. He was educated in the common schools, and came to Watertown, New York, in March, 1801, with his father.... In company with his father he became interested in the manufacture of the solid rock millstones, then the only stones used in grinding corn and in use throughout the whole northern country, which eventually superseded the French burrstone.... Mr. Parker died in June, 1871, aged eighty-four years."

North Carolina

The 1875 *Report of the Geological Survey of North Carolina* (North Carolina State Geologist 1875:305) stated, "The coarse porphyroidal granites and gneisses which are scattered over so large a part of the State, are ... the most common material for mill-stones."

Tennessee

Safford (1869:178–179) provided the following information on gneiss millstones in Tennessee:

> At several points the hard quartose gneiss has afforded tolerably good millstones. In Johnson County, a few miles east of Taylorsville, stones have been cut out of such material, and have answered a good purpose, especially for grinding corn. In Carter, on Little Doe river, about twelve miles from Elizabethtown, a syenitic gneiss, or granite, was formerly worked into millstones, some of which were of large size, and were, at one time highly esteemed, both for grinding corn and wheat. Of late years, however, they have not been manufactured, owing to the introduction of foreign buhr-stone. White quartz rock, too, from veins in this Group, has been cut into millstones, which have been used with very satisfactory results.

Killebrew (1874:260) noted that a "hard gnessoid rock near Taylorsville, in Johnson County, has been used for millstones. Those made of this rock are inferior to the last mentioned, [flint with cellular cavities] and are unsuited for the grinding of wheat though they answer tolerably well for corn."

Sandstone Quarries

Sandstone millstone quarries were reported in Alabama, Arkansas, Massachusetts, Mississippi, Missouri, Ohio, and Texas. Gary, McAfee, and Wolf (1974:628) define sandstone as a "medium-grained, clastic sedimentary rock composed of abundant and rounded or angular fragments of sand size set in a fine-grained matrix (silt or clay) and more or less firmly united by a cementing material (commonly silica, iron oxide, or calcium carbonate): the consolidated equivalent of sand, intermediate in texture between conglomerate and shale. The sand particles usually consist of quartz and the term "sandstone," when used without qualification, indicates a consolidated clastic rock containing about 85–90 percent quartz sand (Krynine, 1940)."

Few sandstone millstone quarries were reported in the literature. This is not surprising in light of Garber's (1970:82) comments about problems with sandstone: "Buhrs made of sandstone were seldom used in waterpowered gristmills to grind meal for human consumption. These millstones were solid, inexpensive, and easy to make, but grains of sand rubbed off in grinding. They were satisfactory for crushing corn for mash to make whiskey and were occasionally used in a corncracker operated in conjunction with a distillery because the solids were discarded in making whiskey."

Alabama

A millstone quarry utilizing a Pennsylvanian age sandstone was located near Dutton in Jackson County, Alabama (Katz 1916:554, 1920:217). It was still producing a few millstones during 1917 (Katz 1920:217).

Arkansas

In his publication "Journal of Travels into the Arkansa Territory, During the Year 1819," Nuttall (1821:243) reported that "[in] the adjoining heights, a coarse-grained sandstone occurs, answering the purpose of mill-stones."

Massachusetts

A millstone quarry is located on the Stonehill College campus at Easton, Massachusetts. This quarry was briefly described in an article entitled "Millstones" on ScienceMusings.com (*http://www.sciencemusings.com/blog/2005/12/millstones.html*). According to Chet Raymo (2005) the 200-year-old millstone quarry is located adjacent to a campus parking lot. Three abandoned millstones were mentioned in the article. The largest stone was five feet in diameter and one foot thick, and the other two millstones are unfinished (Raymo 2005).

Greg Galer (personal communication 2006), Curator of Stonehill Industrial History Center at Stonehill College, shared two quotes that mentioned the millstone quarry. The first quote is from a booklet entitled *King Philip History Trail* (Stonehill College n.d.:2–3):

> Although the total number of millstones (grindstones) taken from this site is unknown, three millstones from colonial times remain. During the 1960s, the Stonehill College Archeology Club determined that these particular stones were most likely used to grind acorns to make pig feed. They based this belief on their knowledge that sandstone, the type of rock from which the stones are carved, is incapable of producing a fine grain that is necessary when grinding wheat or corn.
>
> Some stones remain partially chiseled from slabs of rock in the ground, but only one appears to be near completion as its center hole as already been carved. We have been unable to confirm who owned the stones or exactly which mill commissioned their constructions,

but deed research indicates J. O. Dean as the leading possibility. Dean was the owner of a central gristmill in Easton, and owned this piece of Stonehill in the 1800s.

The second quote found by Galer (personal communication 2006) was from *Easton's Neighborhoods* (Hands 1995:21):

> Although the Wamsutta Formation is estimated to be 500 feet thick, very little of it appears on the surface. One place where there is a prominent outcropping is near Stone House Hill where someone quarried the sandstone to make millstones. This quarry has stones five feet across in various stages of completion. No one knows for certain who started the quarry or when it began, but the fine grained stone might have made an effective grindstone in an area where good mill stones had to be imported from Europe. Recent research by Karl West of Foxboro shows that John Dailey, who came to Easton before 1708, sold land in this area to John Randall, miller, in 1736. Chaffin noted that this branch of the Randall family erected a gristmill in North Easton village before 1760, so perhaps Randall is the creator of this mysterious quarry.

Mississippi

A quarry was located in Claiborne County in southwest Mississippi. Wailes (1854: 216–217) noted that "in early times [Grand Gulf sandstone, part of the Catahoula formation] was frequently used in neighborhood for millstones."

The following information was provided on millstones in the *Preliminary Report on the Geology and Agriculture of the State of Mississippi* (Harper 1857:169):

> A very fine sandstone, a real buhr stone, as good as it is found anywhere in the United States, is to be found near Marion, Lauderdale county. It is a very hard silicious stone, containing a great many silicified shells, which give the stone a rough surface, and render it very fine for mill stones.
>
> A similar kind of stone, perhaps not quite as hard, is found near Chunkey creek, in Newton county, and in Clarke county, on Sec. 33, T. 5, R. 15.
>
> A tolerably good stone for mill stones is found in Tippah county, in the northern eocene, on Sec. 9, T. 5, R. 2.

Missouri

In Laclede County, Missouri, Broadhead, Meck and Shumard (1873:231) provided the following information on sandstone millstones: "Very often we find the Second Sandstone composed of grains firmly cemented with a silicious paste, and this variety may be employed for mill-stones. At Cherry's grist-mill, on the Osage Fork, I saw a set of stones, made from this rock, that have been in constant use for upward of twenty years, and I am assured by Mr. Cherry that the farmers prefer the meal made by these stones to that made by the French Buhr. They were quarried in T.34, R.14, Sec. 16. The same rock exist also at a number of points on the Gasconade and Osage Fork Rivers."

Broadhead, Meek and Shumard (1873:231) discussed a particular sandstone used for millstones in Pulaski County: "The compact layers of the Second Sandstone, sometimes form a good substitute for mill-stones, and they may nearly always be obtained in the districts where this formation prevails. They are, however, objectionable on account of the frequent dressings they require."

Ohio

The best described quarry exploiting sandstone was located south of Peninsula, Ohio in the Borton township within Summit County (Katz 1920:217; Ladoo 1925:9–10; Read

1883:478–479; Williams 1885:355). One variety was used to manufacture grindstones (Read 1883:478). The white variety of the Berea grit was primarily used for grinding oatmeal and pearling barley, especially in the Akron area (Katz 1920:217; Read 1883:478–479).

Read (1883:478) made the following comments about the appearance and distribution of Berea grit: "Its exposure in Summit county is confined to the valley of the Cayahoga river and the bluffs bordering it. It appears at the bottom of the valley near the south line of Borton township, and from thence rises in the bluffs to the north line of the county. It is in this county a hard, evenly-bedded, compact rock varying in color from nearly white to a dark brown, but in places so irregularly colored as to detract somewhat from its value as a building stone."

Near Akron, Ohio, millstones were manufactured from Berea grit as reported by Read (1883:478–479):

> Mr. Fred Shumacher, of Akron, is the founder of the oatmeal industry in this country, and has also been extensively engaged in the work of pearling barley. He found considerable difficulty in obtaining millstones suitable for this work, and after protracted tests of all available material, selected the white variety of this Peninsula stone as the best that he could find. Its great hardness and very sharp grit especially fitted it for this use. A layer, the bottom of which was 3 or 4 feet above the underlying shale, proved to be the best, that which is of strictly first-class quality having been yet found only in this quarry. He has used for this purpose about 250 blocks ranging in diameter from 4 to 5 feet and in thickness from 14 to 18 inches, and has sold to others about 30 such blocks. These blocks he values at from one to two dollars per cubic foot. In his mills at Akron he is now using 28 runs of these stones, and producing 80,000 barrels of oatmeal and 50,000 barrels of pearl barley per annum. These stones used for pearling barley are placed upright in iron shells revolving slowly in a direction opposite to that of the stone. Mr. Shumacher is now putting the machinery into a new mill, which will increase the annual production of oatmeal to 240,000 barrels, of the value of $1,500,000.
>
> The success of this industry is largely the result of the peculiar character of a small part of the stone from this quarry. The work was commenced in 1856, and has steadily increased until the present time. Its dependence upon the character of this rock, in the estimation of Mr. Shumacher, is indicated by the fact that he has purchased the entire quarry and is now putting it in shape for large production of building stone, expecting, as this is removed, to uncover and make available a full supply of that which is especially fitted for milling purposes.

Texas

According to a "History of Texas," sandstone was used for millstones in Texas (Anonymous 1894b:171): "Buhrstones.— In the Fayette sands are found stones of excellent quality for use as millstones. In Jasper and other counties millstones which have given perfect satisfaction in use have been cut from certain horizons of these sands."

Dolomite Quarries

A single dolomite millstone quarry was encountered in the literature. Gary, McAfee, and Wolf (1974:206) define dolomite as follows a "carbonate sedimentary rock consisting chiefly (more than 50 percent by weight or by areal percentages under the microscope) of the mineral dolomite or approximating the mineral dolomite in composition, or a variety of limestone or marble rich in magnesium carbonate; specif. a carbonate sedimentary rock containing more than 90 percent dolomite and less than 10 percent calcite, or having a Ca/Mg ratio in the range of 1.5–1.7 (Chilingar, 1957), or having an approximate MgO equivalent of 19.5–21.6 or magnesium-carbonate equivalent o 41.0–45.4 percent (Pettijohn 1957, p. 418)."

At Knoxville, Tennessee, dolomite was utilized for millstones (Rule, Mellen, and Wooldridge 1900:16): "But the Knox dolomite is the most important and massive of the three divisions of the Knox group, the thicker layers being often worked into millstones, and in the upper strata of this division there are cuts of dull, variegated dolomite, which are worked as marble and used as building material. In color it is light gray, variegated with brownish red clouds, and is rather fine grained."

Sienite Quarries

One millstone quarry was listed as exploiting a sienite. Gary, McAfee, and Wolf (1974:657) do not list this rock type which suggests that sienite is an obsolete geological term. In Danvers, Essex County, Massachusetts, millstones were quarried from sienite between Boston and Lowell (Hanson 1848:9): "There is an exhaustless supply of sienite from which the choicest millstones are manufactured, equal to any in the world, and some fine specimens of quartz have been found, of which No. 1312 in the State College is a sample."

Unspecified Materials

Millstone quarries of unspecified raw materials were mentioned in the literature examined for Connecticut, Georgia, Illinois, Kentucky, Missouri, New York, Pennsylvania, Tennessee, Virginia, and West Virginia. These sources are discussed below alphabetically by state.

Shepard (1837:68) made the following comments about stone suitable for millstones at New Milford, Litchfield County, in southwest Connecticut: "No material however, appeared to possess such advantages as a carious jasper found disseminated through the bed of porcelain-clay in New Milford. Provided they occur in sufficient quantity and in masses of a suitable size, they will probably be found capable of being substituted for the French buhr-stone."

Another reference to a millstone quarry in Connecticut was reported in the August 13, 1883, edition of *The New York Times* (on ProQuest). The brief article reported on the murder of Peter Nelson who was employed at the New London, Connecticut, millstone quarry. Nelson was visiting New York with friends and was murdered during the trip. New London is located near the coast in southeast Connecticut.

Brief references to millstone quarries of unknown material types were found for Delaware and Georgia. A deserted millstone quarry was reported as being located in White Clay Creek State Park in Delaware (Brown 2000). In Georgia, millstones of unknown types were quarried in Bulloch, Burke, Early, Jefferson, and Screven counties (Schrader, Stone, and Sanford 1917:107).

Allen (1949:65) provided the following information on Millstone Knob in Pope County, Illinois: "Millstone Knob is shown on the 1876 map of Pope County. It is said to have taken its name from the fact that millers once got mill stones from a quarry there. A number of the early mills in Pope County are said to have been equipped with millstones from this source. According to another explanation, it was so named because of its resemblance it bears to a millstone, when viewed from a distance."

Millstones were also made in Saline County, Illinois (Anonymous 1967:156): "Zadock Aydolett was a Frenchman, who put up a horse mill for the grinding of corn. The millstones were made from the millstone grit in the mountain in Somerset Township, near which he lived, and they were propelled by means of wooden gear machinery, and a long sweep to which the horses were attached."

In 1816, Benjamin D. Price (South Union, Logan County, Kentucky) wrote to John L. Baker (New Harmony, Indiana) concerning millstones (Arndt 1975:234): "I promised to write you when we could be satisfyed in regard to the mill stones that are obtained in this country. We have got two pair of them at home and our people esteem them to be the best they have seen. There will be about six miles of land carriages to cumberland river which will make it convenient for you as they can conveyed by water to the place you want them."

Two years later (September 4, 1818), Joseph Allen (South Union) wrote Frederick Rapp (New Harmony) about providing millstones (Arndt 1975:565–566): "We have imployed a man to cut the four feet mill stones for corn, and as soon as it is practicable will send them to you, also the stone for your Hemp Mill with the other stones.... The man we Have imployed will undertake to git out as many pare as you want, if you find they will answer your purpose."

On February 25, 1819, Samuel G. Whyte (South Union) wrote Frederick Rapp to advise him that the millstones were finished and ready to be shipped (Arndt 1975:656). Benjamin S. Youngs (South Union) wrote to Frederick Rapp on May 20, 1819 to apologize for delays (Arndt 1975:712): "We have felt considerable concern for you on account of the Mill Stones, tanners tables & Hemp stone which you have expected from here. They have been all finished & taken to the bank of the Big Barren River about three months ago — but our Brethren have not been able as yet to get them a safe conveyance to Shawnee town."

Samuel G. Whyte sent Frederick Rapp a bill for the millstones on September 1, 1819, and referred to them as "Goose Creek Mill Stones" (Arndt 1975:767).

The Goose Creek millstones were probably quarried in the general vicinity of South Union, Kentucky. The proposed shipping routes mentioned in the correspondence provide some clues. The first route mentioned 6 miles of land transport to the Cumberland River which would have been to the south in Tennessee. The second reference mentions the Big Barren River (Barren River on later maps) which would have been to the north in Warren County. Using these references, the quarry could have been in Logan, Simpson or Warren counties, Kentucky. The final clue provided was the name "Goose Creek" that does not occur as a named stream in that area today. Arnow (1984:283) mentioned that a pair of "Goose Creek burrs" were used in the Croft Mills of General James Winchester on Bledsoe Creek when it was sold in 1802. Bledsoe Creek is a tributary of the Cumberland River in southwest Russell County, Kentucky. This suggests that these millstones were being manufactured prior to 1802.

A possible millstone quarry was reported to the author by Tom Sussenbach (personal communication 1990) as being located along Indian Creek in McCreary County, Kentucky. Sussenbach was conducting an archaeological survey in the vicinity when the landowner mentioned the millstone quarry. Sussenbach observed an unfinished millstone in Indian Creek. A plan to visit the quarry in January of 1991 did not occur. Fortunately, the author has a location for the quarry so that it can be investigated in the future.

Two early references mention millstone quarries in Maryland. The October 25, 1753, edition of the *Maryland Gazette* carried an ad by Herman Husbands "that he was making millstones near Jacob Giles' at the mouth of the Susquehanna" (McGrain 1991). John McGrain (1991) discovered a deed in Baltimore County, Maryland (WG 126:63) dating to October 29, 1813, that mentioned "John M. Gorsuch and Dickinson Gorsuch agree to share 'parcels of Burr stones or stone suitable for making millstones' ... also to share Big, Little Cold Bottom

Springs ... near 19th mile stone on York Turnpike." He also indicated that "Scharf states that Dickinson Gorsuch built a gristmill on Forest Farm on York Road and built the Gorsuch Tavern or Homestead and also made his own millstones of local rock, HBCBC, p. 881."

Several sources mentioned millstone manufacture in Missouri. Wetmore (1837:53) noted that in Carrol County, Missouri "a mill-stone quarry has lately been discovered." In Montgomery, County, Missouri, "the rock called millstone grit, or the lost rock, is found in detached masses, apparently rounded by attrition" (Wetmore 1837:124). In Washington County, Missouri, "millstones ... are now in use, and what is supposed the French buhr, are all very abundant, though no actual experiment has been made with the later" (Wetmore 1837:233). Also, in Washington County, a history of the area (Anonymous 1888a:461) mentioned superior quality stone for "millstones or buhrs."

A historic millstone quarry is mentioned along a hiking trail in northeast New Jersey in Bergen County according to two articles in *The Record* by Daniel Chazin (2003, 2005) (on ProQuest). The quarry is near Todd Lake in Ramapo Mountain State Forest (Chazin 2005). Chazin (2005:G31) provided the following description for the millstone trail: "Here, several abandoned millstones in various stages of completion may be seen to the left of the trail. This area was once the site of a millstone quarry, and the stones that you see were either damaged during quarrying or abandoned when the quarry operation shut down. Another millstone, in nearly perfect condition, is a short distance down the trail, about 25 feet off the trail to the right."

In Allegany County, New York, the following information was presented on the source of millstones (Shear 1960:88): "The kind of stone used for millstones was not native to this section. Some were obtained from Auburn and there was some place from which they could be obtained near Nunda. Information on where and how millstones were brought in is much desired. Experts in chiseling the stones were available. It has been suggested that the same men who did engraving on tombstones were the men who 'cut' millstones."

A New York millstone quarry was mentioned in three stories appearing in the April 18, April 19, and April 25, 1900, editions of *The New York Times* (The New York Times 1900a, 1900b, 1900c on ProQuest). The stories refer to work on the new Cornell Dam with references to the old Croton Dam and the "Millstone Quarries back of Peekskill." The community of Peekskill is north of New York City, in southern New York State, on the east side of the Hudson River. Because of a strike of workers, the New York National Guard was sent in as "a precautionary measure to guard the old Croton Dam and the Millstone quarries" (The New York Times 1900a). These three stories follow the strike and the effort to protect the dam and quarries.

Millstones were manufactured in modern day Lawrence County, Ohio, northwest of Ironton along the Ohio River. On September 30, 1803, Thomas Rodney made the following observations during a trip down the Ohio River (Smith and Swick 1997:88, reprinted with permission of Ohio University Press, Athens, Ohio):

> We have had a pretty strong current all the way today as far as the two creeks and some distance below them. A mile below this we came to another castle of rocks on the face of a mountain on the Ohio State shore. We came too; and the Major and I went on shore in the skiff. Between the river and foot of the mountain we met with several old woods camps. At the foot of the mountain the people bled all the sugar trees; and others had been making millstones out of the monstrous blocks of rock that had falln off and rolled d[o]wn from the mountain. I passed them and assended about 200 feet higher to the foot of the castle. The rocks assended perpendicular from this base to apparently about 400 feet, like the wall of a

> castle; all the rocks on that side facing the river having decayed or falln off and falln down to the foot of the mountain.

In Forest County, Pennsylvania, "burr stone, well calculated for mill-stones, is found in various parts of the county" (Egle 1876:734). In his summary of millstone production during 1883 and 1884, Williams (1885:712) stated that "stones more or less suitable for coarse work are, or have been, quarried in parts of Virginia, North Carolina, Georgia, Alabama, Missouri and Arkansas."

Millstone quarrying in west Tennessee, in Tipton County, was described by Williams (1873:142):

> The topographic features of the county differ but little from the other counties in West Tennessee, noted only for its beautiful western front, overlooking the great river. The "Mill Stone Mountain," an interesting feature, found among the range of hills bordering on the Big Hatchie, near its mouth — a novelty of itself— is more interesting for its being a solid mass of concrete rock, from which is wrought the best mill-stones in use; said to be equal, if not better, than the celebrated French burr. Less than a half mile in diameter at its base, it raises in cone shape from the banks of the Hatchie, towering above the tallest forest trees, its apex perfectly level, overlooking the surrounding country.

McGrain (1991) cited Beitzell (1968:194) concerning millstones from Aquia Creek, Virginia, that were distributed in the Tidewater of Chesapeake Bay area. Grimshaw (1882:288) mentioned a millstone quarry in West Virginia: "There is a quarry of valuable material in Randolph County, W. Va., near the great Cheat Mountain limestone belt, sixty miles south of Grafton. This stone is pronounced good for flour and corn, and unusually hard."

Limestone millstones are poorly represented in the literature but the author has seen a number of these in Kentucky. The limestone specimens are primarily large edge runners used for crushing materials. A millstone quarry was described by Riedl, Ball, and Cavender (1976:125) in Coffee County, Tennessee, as "a bed of Leipers or Catheys limestone, which as late as 1975 still held a partially shaped and detached millstone."

Querns were also manufactured in the United States but little has been published on this topic. Labs (1917) and Mercer (1917) discussed the survival of querns in the southeast United States for grinding corn. Swain (1917) described a hand quern from Georgetown, South Carolina. The American millstone literature mentions small stones that obviously would have functioned as querns. Typically, these small grinding stones would have been used by individuals or families to grind grain for their own consumption. Most of the research to date has focused on the large millstones used in commercial grist mills. Undoubtedly, the literature on early American life and history contains many references to querns that could be compiled. Hopefully, future researchers will take a more in-depth look at querns.

Boulders Used for Millstones

Boulders were used for millstones in several states where suitable stone was not available or boulders were dropped by ancient glaciers. In Parke County, Indiana, millstones were made from "granite boulders or nigger-heads such as we find scattered over many farms" (Branson 1926:24). Another example was found for Montgomery County, Indiana (Beckwith and Kennedy 1881:328): "It is becoming that an acknowledgment be made of the influence of the Rev. Samuel Van Cleave upon the early society of Brown township. He was among the settlers of 1827, in which year he built the Van Cleave mills, on Indian creek, by his own genius and efforts, manufacturing the mill-stones from a large gray boulder, which he split in halves and dressed them to a true face, which did effective work while the mill stood."

Millstones were also made from boulders in Illinois. In McLean County, Illinois, it was noted that "sometimes they constructed sort of mills with millstones, cut from the lost rocks found on the prairies" (Anonymous 1879:212). Prairie boulders were used for millstones in De Witt County, Illinois (Anonymous 1882:39, 1910:88): "The first utilized rocks were the prairie boulders in the construction of millstones, a use long since abandoned before the introduction of superior burr stone rock from other sections of the country."

In Will County, Illinois, Charles Reed used a boulder for millstones in his ca. 1833 mill near Joliet City (Anonymous 1898:272). According to the account (Anonymous 1898:273): "Reed's millstone, we remember was made from a large hard-head, or niger-head, as they are sometimes called, and for a long time lay upon the old mill yard."

Peck (1840:120) made the following comments concerning granite boulders being used for millstones in Illinois: "Those singular boulders of granite, which are found throughout the West, are frequently seen on the surface in this State. They are of all sizes, from a few pounds to as many tons, and their structure is so hard, that they are often wrought into millstones."

Boulders were shaped into millstones in Iowa and Michigan. Swisher (1940:59) indicated that "Iowa millstones were often made from prairie boulders, sometimes called 'niggerheads.'" In Oakland County, Michigan, millstones were made from boulders as documented by the following quote from Seeley (1912:287): "In 1819–20, the flour mill was finished — the first in the county. It contained one or two burr stones, one run of common stone made from native boulders."

Boulders were also used for millstones in Missouri, New York, and Oregon. Drift boulders from north Missouri were used for millstones (Broadhead 1874:55). In Chautauqua County, New York, boulders were used to make millstones for a mill built by Abel Cleveland and David Dickinson on Silver Creek during the early 19th century (Young 1875:409). Elisha I. Applegate made several sets of millstones from glacial erratic during the mid 19th century in west central Oregon (Harrison 1996a:2–3, 1996b). These glacial boulders were carried to Oregon from southern Canada by icebergs during great floods associated with the melting of glaciers (Harrison 1996a:2–3, 1996b). Finally, Craik (1870:293) noted that "[f]ormerly mill stones were made of granite, or some other flinty conglomerate rock. Not unfrequently from boulders found in the vicinity where the mill was built, and in those days the millwright and miller's trade included the making of these."

3

Millstone Makers and Urban Factories

Several millstone makers operated businesses in urban areas. Many of these companies shaped and assembled imported blocks of stone from France into millstones. These millstones were called burrs or buhrs, with both spellings being in common usage. This chapter compiles some of these sources into one section for convenient use. A comprehensive listing of urban millstone manufactories would require years of research. See Ball and Hockensmith's (2007a) "Preliminary Directory of Millstone Makers in the Eastern United States" for a much more comprehensive listing of urban millstone makers. Information is presented for the following states: Connecticut, Illinois, Kansas, Kentucky, Maryland, Massachusetts, Missouri, New York, North Carolina, Ohio, Pennsylvania, Tennessee, and Wisconsin.

Stoner (1947:424) made the following comments about millstone factories: "So profitable had the millstone business become that factories for making them were established in Philadelphia and Baltimore. Stones in the rough were received in these factories and by skilled workmen dressed into millstone shape. Even stones from France were shipped over here in small pieces and dressed, fitted and assembled ready to run. Later, quarries suitable for millstones were opened up in many places in the United States with the result that neither the English nor the French had a monoply of the millstone business."

Connecticut

One millstone manufacturer in Fairfield County, Connecticut, was listed in the U.S. Census for 1860 with $500 of capital invested and $1,800 worth of raw materials. The manufacturer employed three men, paid $1,200 in annual labor, and produced $3,500 worth of products annually (Kennedy 1864:37).

Illinois

The U.S. Census for 1860 mentioned one millstone maker in Rock Island County, Illinois, with $10,000 of capital invested and $3,875 worth of raw materials, who employed five men, paid $600 in annual labor, and produced $5,230 worth of products annually (Kennedy 1864:103).

Kansas

A millstone manufacturing firm was located at Atchison in Donipan County, Kansas, in 1868 (Smith 1868:9). The first part of a detailed ad called attention to "Plamondon & Maher, Practical Millwrights, Manufacturers of and Dealers in French Burr Millstones, Portable Mills, and all kinds of Mill Machinery..." (Smith 1868:9). The company owners were P. Plamondon and N. A. Maher (Smith 1868:9).

Kentucky

Louisville, Kentucky, was a major source of millstones during much of the nineteenth century. Mill supply houses sold millstones along with a wide variety of mill related equipment and supplies (Hockensmith 2008a). Perhaps the best known of these companies was Herbert & Wright who made millstones. Other supply houses in Louisville that only sold millstones were W. T. Pyne, Jabez G. Kirker, Wilkes & Dillingham, William Dillingham, and John Braun & Company. With the exception of Herbert & Wright, their ads did not mention the types of millstones sold but it is reasonable to assume that at least some of the millstones for sale were French Burr. Such supply houses were convenient places where millers could obtain equipment and supplies for furnishing new mills or for maintaining older mills.

The firm of Herbert & Wright was the primary manufacturer of French Burr millstones in Louisville. An 1844–1845 directory for the city of Louisville (Poor 1844:98) included the following information:

> FRENCH BURR MILL STONE MANUFACTORY
>
> This branch of business is carried on very extensively by Messrs. Herbert & Co., on Water, between Second and Third. They construct them of all sizes, from the largest size flouring stones, down to 18 inches diameter, for plantation use.

Jegli (1845:266) also listed G. W. Herbert & Co. as French Buhr millstone manufacturers for 1845–1846. In 1848, Herbert & Wright, Mill Furnishers in Louisville, were offering all sizes of French Buhr millstones for sale "made from best quality Buhr Blocks" (Collins 1848:362). Again, in 1850, Herbert & Wright were selling French Buhr millstones in Louisville (Jegli 1850:85). A reference to burr stone manufacture during 1850 was also made by Casseday (1852:244) who noted that there was one factory in Louisville (undoubtedly Herbert & Wright) that employed 8 hands and earned $12,000 annually. Deering (1859:72), in his book *Louisville: Her Commercial, Manufacturing and Social Advantages*, published the following information on the millstone factory: "There is one establishment where portable and plantation mills, mill-stones, bolting cloths, etc., are manufactured. This shop has been long established, and is favorably known. They employ $20,000 capital, 20 hands, and $25,000 worth of business is done annually."

Herbert & Wright continued manufacturing French millstones for many years (Jegli 1851:337; Hurd and Burrows 1858:296; Hawes 1859:385; Edwards 1864, Hodgman & Co. 1865:423; Edwards, Greenough, and Deved 1866; Southern Publishing Company 1868; Southern Publishing Company 1869:470; Edwards and Deved 1870:460; Caron 1871:556; Caron 1872:604; Caron 1873:631–632; Caron 1874:650; Caron 1875:671; Caron 1876:746; Polk and Danser 1876:662; Caron 1877:655). In 1878, J. C. Wright, successor to Herbert & Wright, was manufacturing French Burr Mill Stones in Louisville (Caron 1878:721, 797; Caron 1879:782; Polk and Danser 1879:668). By 1880, J. C. Wright had a very small ad (Caron 1880:791) and he did not reappear in the subsequent Louisville directories.

George W. Herbert, of Herbert & Wright, was born in Leicestershire, England, on January 1, 1808 (Armstrong & Co. 1878:636). The son of a miller, Herbert came to America in 1832 (Armstrong & Co. 1878:636). George Herbert died on September 8, 1872 (Armstrong & Co. 1878:637). His biographical sketch noted (Armstrong & Co. 1878:636): "He came to Louisville, Kentucky, in the year 1835, and entered into the mill furnishing business with a Mr. Frazee, as clerk and general stock-keeper. After three years' service in this capacity, Mr. Frazee sold his stock out to him."

The U.S. Census for 1860 listed one millstone manufacturer in Kenton County, Kentucky, with $3,500 of capital invested and $4,000 worth of raw materials, that employed seven men, paid $3,372 in annual labor, and produced $10,000 worth of products annually (Kennedy 1864:181). This company was probably operated by Nicholas Spanager, a 48-year-old Prussian, listed as a millstone maker in Kenton County in the 1860 Population Census (Wieck 1983).

Besides Spanger, the Population Census records for 1860, 1870, and 1880 for Kenton County, Kentucky, list other individuals as millstone makers and millstone dressers. The 1870 Census listed 25-year-old Herman Schulte and 58-year-old Nicholas Spanier (probably a different spelling of Spanager), both born in Prussia (Wieck 1986). Both men were listed as millstone dressers (Wieck 1986). Schulte, now 34, was listed again in the 1880 Census (Wieck 1996). It is suspected that a number of other men involved in millstone making and dressing are listed in the U.S. Population Census records for other Kentucky counties. However, no one has undertaken the monumental task of searching all these records. The original handwritten records are only available on microfilm. Fortunately, some individuals have transcribed and published census information for select counties and years.

Maryland

Two early millstone factories were mentioned in Baltimore, Maryland. A millstone manufactory in Baltimore during 1810 was valued at $6,000 (Coxe 1814:28, 84). Morris & Trimble of Baltimore advertised French Burr Millstones and Burr Blocks in 1855 (Edwards 1855:290). Their attractive ad showed men working on millstones in the factory (Edwards 1855:290). The ad noted that they were "Manufacturers of French BURR MILLSTONES, warranted of superior quality and workmanship, being made from Burr Blocks of their own importation and selected from the best quarries in France" (Edwards 1855:290). In addition to producing French Burr millstones, the company also sold Cologne, Cocalico, and Esopus millstones (Edwards 1855:290). The Cologne stones were made in Germany, the Cocalico stones were produced in Pennsylvania, and the Esopus millstones were manufactured in New York State. Similar information on the Morris & Trimble Company was published in 1869 (Hughes 1869:226–227):

> Morris & Trimble, West Falls Avenue, near Pratt Street Bridge, Baltimore, manufacturers of French Burr Mill-Stones, warranted of superior quality and workmanship, being made from burr-blocks of their own importation, and selected from the best quarries in France. Constantly on hand a general assortment of Cologne, Cocalico and Esopus Mill-Stones, Burr-Blocks, Bolting Cloths, Calcined Plaster, &c. Orders from any part of the country promptly executed.
>
> The mill-stones of this firm are taken entirely out of wind, and ready for dress, making a savings in the expense of putting them in working order. They manufacture their mill-stones from burr-blocks of their own importation; and from a long experience in the business of making mill-stones, being established since the year 1815, and possessing many of the important improvements for building mill-stones, they offer many inducements to millers requiring a selection of all varieties. The original founders of this house were Messrs. Morris & Egerton, and in continual succession to the present parties. Their mill-stones are sent to various States, namely: Virginia, North Carolina, Tennessee, and the adjoining counties of Pennsylvania, and to South America. Their mill-stones are finished with cast-iron eyes of the most durable construction, being set into the eye-blocks with ears well fastened and secure.

The U.S. Census for 1860 listed one millstone manufacturer in Maryland with $3,420 of capital invested. He had $1,910 worth of raw materials, employed five men, paid $1,920 in annual labor, and produced $5,486 worth of products annually (Kennedy 1864:230).

Statistics compiled from the 1870 U.S. Census listed two companies manufacturing millstones in Baltimore. These companies employed 22 men, had $23,400 of capital, paid $31,557 in wages, spent $34,700 on materials, and produced $35,530 of products (Patapsco Land Company of Baltimore City 1874:31).

B. F. Starr & Company of Baltimore, Maryland, advertised their "Baltimore French Burr Mill Stone Manufactory and Mill Furnishing Establishment" in 1873. In addition to selling French Burr millstones, B. F. Starr & Company also sold Esopus, Cocalico, and Cologone millstones (Howard 1873:242).

Massachusetts

A publication containing statistical information on certain industries in Massachusetts listed two French Burr Mill Stone Manufactories at Boston in 1855. These companies were valued at $12,500, had a capital of $8,800, and employed 11 men (Massachusetts Office of the Secretary of State 1856:461).

The U.S. Census for 1860 mentioned one millstone manufacturer in Sulfolk County, Massachusetts, with $20,000 of capital invested and $7,000 worth of raw materials, who employed six men, paid $3,000 in annual labor, and produced $12,000 worth of products annually (Kennedy 1864:246).

Missouri

One millstone company was operating in St. Louis, Missouri, during 1839 (Edwards and Hopewell 1860:365). Twenty years later, J. Todd was listed as a burr millstone manufacturer (Edwards 1860:365). The U.S. Census for 1860 included one millstone manufacturer living in St. Louis County, Missouri, with $25,000 of capital invested and $10,450 worth of raw materials, who employed 25 men, paid $9,600 in annual labor, and produced $30,000 worth of products annually (Kennedy 1864:311).

New York

Several millstone manufacturers were listed in New York City, Buffalo, and Utica, New York. The "Census of the State of New York, for 1855" listed six millstone manufactories (New York State Secretary's Office 1857:413). These manufactories employed 67 men, had $28,500 invested in real estate, invested $8,910 in tools and machinery, spent $39,725 on raw materials, and manufactured $71,200 worth of millstones (New York State Secretary's Office 1857:413).

The Columbian Foundry and Burr Mill-Stone Manufactory was located on Duane Street in New York City (Hughes 1869:224):

> For the manufacture of Steam Engines, Boilers, Sugar Mills, Iron and Brass Castings, Wrought Iron work, Iron Columns and Pipes, Screws of all kinds, and machinery of every description.
>
> Constantly on hand all kinds of Burr, Holland, and Esopus Mill-Stones. Burr Mill-Stones made to order, and warranted to be of the best quality. Burr Blocks for sale.

In Erie County, New York, millstones were manufactured by John T. Noye (Horton, Williams, and Douglass 1947:229): "Into far places also went the product of the company founded by John T. Noye to manufacture millstones. In '84 the company moved from its old site in lower Washington Street to the west side block bounded by Fourth, Lake View, Pennsylvania and Jersey, where in more commodious quarters it went on with the manufacture not of millstones, but of roller mills, the modern equivalent, for which it found lucrative vent in Australia and Argentine."

The following excerpt from John T. Noye's biographical sketch provides useful background information (Smith 1884:61–62):

> **JOHN T. NOYE.**—John T. Noye was born on the 21st of March, 1814, in the city of New York. He died at his residence in Buffalo, April 6, 1881. His parents originally came from England; his father Richard Noye, came to this country when eighteen years old, and is said to have been the first merchant miller in the State of New York. Mr. Noye passed the early years of his life in the various towns of Westchester county, and when seventeen years old took charge of the flouring mill at Rye, N. Y., where he remained until the spring of 1835, when he came to Buffalo and was employed in the Frontier Mills, at Black Rock. A few months later he was engaged to take charge of the mills at Springville, but returned to Buffalo in a short time and accepted a position with Elisha Hayward, who was at that time interested in the flour and grain trade and carried on in a small way the manufacture of millstones. From that small shop, established in 1828, has grown the immense business of the John T. Noye Manufacturing Company, where some three hundred and fifty men find employment, the name and reputation of which is known wherever wheat is made into flour.
>
> After a brief apprenticeship as clerk with Mr. Hayward, Mr. Noye became a member of the firm, the name of which, after the death of Mr. Hayward, in 1846, was Hayward & Noye, Nelson Hayward acting as trustee for the widow of Elisha Hayward and representing her in the business. The office was at that time on Hanover street. This partnership lasted until 1850, when Mr. Noye assumed the entire control and ownership of the business, and removed to the location on Washington street with which the business has so long been identified. In 1883 the Lehigh Valley Railroad Company purchased a part of the property and the works have recently been removed to their new location on Lake View avenue.
>
> Up to the year 1850 Mr. Noye had not given up the grain business, but carried it on extensively and profitably.... About the year 1856 Mr. Noye turned his attention almost exclusively to the manufacture of machinery for flour mills, which business grew rapidly under his energetic and intelligent management.

John T. Noye of Buffalo advertised his Millstone Manufactory and Mill Furnishing Establishment in 1860 (Hawes 1860:910). In the same gazetteer and business directory, Z. G. Allen of Buffalo placed the following ad (Hawes 1860:910): "ALLEN, Z. G., manufacturer French Burr Millstones, Portable Flouring and Grist Mills, and all kinds of Mill Machinery, also dealer in Dutch Anker Bolting Cloths, etc., Miami, near Central R. R. Freight Depot."

In Utica, Oneida County, New York, a buhr millstone manufactory once operated (Bragg 1892:594):

> The manufacture of buhr-millstones was commenced by Alfred Munson in 1823, on the corner of Hotel and Liberty streets, in the basement of the Kirkland block. He soon removed to the east side of Washington street, where it is crossed by the canal. Martin Hart became associated with Mr. Munson in 1830, under the firm name of Munson & Hart. This continued for a number of years, when the firm dissolved, and Alexander B. Hart (a son of Martin) and Edmund Munson (a nephew of Alfred) became associated under the name of Hart & Munson. Edmund Munson has already been noticed as an excellent mechanic and inventor, and under his direction the business rapidly increased. The firm of Hart & Munson did an extensive business, employing at times as many as 100 hands. In 1868 this firm dissolved and a new one was formed, under the title of Munson Brothers. Similar manufactures have since been established in Buffalo, Indianapolis, Richmond, and other localities, creating competition. The raw material (French buhr-stone) come from a locality near Paris, France, and is imported through New York houses. The trade covers all parts of the United States and Canada. In 1888 a foundry was added to the plant, an unused portion of the west building being utilized for this purpose. The firm employs at present fifty hands, and manufacture portable grain mills, roller mills, water wheels, and all kind of corn and flour mill machinery. Also dealers in German, English, and domestic cements, plaster. etc.

Hughes (1869:196) provided the following information on the Utica French Burr Mill-Stone Manufactory in New York:

> HART & MUNSON, successors to M. Hart and Son, in the above establishment, are now prepared to furnish French burr mill-stones of the best quality and greatly improved workmanship and finish; together with the best quality of bolting cloths, screen wire, hoisting screws, lighter screws, dansells, and mill picks.
>
> Mr. Munson, who is a practical miller and millwright, has recently invented and patented a machine, on which the mill-stone, after it is blocked up, is suspended upon its centre, where it is balanced in the course of filling up and finishing, instead of filling up the same without the means of testing the accuracy of its balance, leaving that to be done by the millwright, (as is usually the case,) in hanging the stone for actual use in the mill.
>
> In order that the great superiority of mill-stones finished in this way over all others, may be seen at once, a brief description of the machine and manner of finishing, is herewith given.

After describing the design and use of Munson's machine for balancing millstones, Hughes (1869:198) went on:

> The author is aware of the importance of millers being made fully acquainted with their true interests as regards the building of mill-stones and mill-furnishing generally. A mill-stone is the most essential part of the mill, for without a good quality of mill-stones, all other expenses, however costly, are of no use in getting up a mill. If the stones are either too hard or too soft, too close or too open, made out of blocks uneven in temper, they are not fit for making flour, and are sure to injure the reputation of any mill. Where they are to be obtained is therefore of great importance.
>
> At this establishment may be found mill-stones not to be equaled in the United States in any particular, as no improvement has been allowed to pass their notice. And it is gratifying, and highly recommends their work, to know, that if France alone produces the Burr in its pure native state, the honor is left to Messrs. Munson & Hart to form and fashion it into those mill-stones which so far supersede other manufactures in the beauty, excellence, and superiority of their workmanship.

In 1872, *Boyd's New York State Business Directory* (Boyd 1872:14) contained the following ad for the Munson Brothers:

> Central New York Burr Mill Stone Manufactory
> **MUNSON BROTHERS,**
> Proprietors and Manufacturers of
> **FRENCH BURR MILL STONES,**
> E. MUNSON'S PAT. PORTABLE MILLS.
> And all Kinds of
> Mill Furnishings at the Lowest Cash Prices,
> **UTICA, N.Y.**
> We take pleasure in announcing to our friends and to the public generally, that E. Munson, of the late firm of Hart & Munson, has assumed the entire Superintendence of our Works, both in the selection of Stock and manufacture of Mill Stones.

The Munson Brothers published catalogues promoting their products. Three editions of their catalogues were found during an internet search of WorldCat's library site. The earliest catalogue (ca. 1860) contained the following information on the cover: "Munson Brothers, proprietors and manufacturers of French burr mill stones ... machine finish for mill stones, pat. cast iron eyes ... portable and stationary engines and boilers ... and all kinds of mill furnishings..." (Munson Brothers 1860). In 1875, the Munson Brothers (1875) published their

"Illustrated Catalogue and Descriptive Pamphlet from Munson Brothers' Central New York Burr Mill Stone Manufactory, and Mill Furnishing Establishment...." The title of the 1886 catalogue was "Illustrated Catalogue of Munson Brothers, Mill Furnishers. Specialties: Munson's Patent Eyes and Spindles, Shafts, Gearing, Pulleys, and all Kinds of Mill Furnishings, at the Lowest Prices, Utica, N. Y." (Munson Brothers 1886). The 1886 catalogue contained a note on the cover that it was an "Annual catalogue of portable mills and mill machinery, made by Munson Brothers, Utica" (Munson Brothers 1886).

Three counties were listed in the U.S. Census for 1860 as millstone producers. In Erie County, New York, three millstone and mill furnishing companies were listed with $155,000 of capital invested, with $90,400 worth of raw materials, who employed 89 men, paid $15,480 in annual labor, and produced $217,000 worth of products annually (Kennedy 1864:366). New York County, New York, had two millstone burr companies with $15,000 of capital invested, with $7,760 worth of raw materials, who employed 19 men, paid $10,320 in annual labor, and produced $24,500 worth of products annually (Kennedy 1864:382). In Erie County, New York, three millstone and mill furnishing companies were listed with $155,000 of capital invested, with $90,400 worth of raw materials, who employed 89 men, paid $15,480 in annual labor, and produced $217,000 worth of products annually (Kennedy 1864:382). One burr millstone company was in Ulster County, New York, with $200 of capital invested, with $250 worth of raw materials, who employed three men, paid $720 in annual labor, and produced $1,120 worth of products annually (Kennedy 1864:405). In Troy, New York, two buhr millstone manufactories were operating in the city (Weise 1891:103).

Two New York millstone makers were mentioned in early documents consulted. First, a Mr. Webb who was "a burr-millstone maker" was mentioned in New York in 1776 (Burdett 1860:389, Burdett 1865:389). The second man, Alexander McClung, was mentioned in the will abstracts for New York on November 30, 1795 (New York Historical Society 1868:329). McClung was a native of Killafady, Ireland, and listed as a millstone maker (New York Historical Society 1868:329). It is not known whether these men were self employed or worked for companies.

North Carolina

The U.S. Census for 1860 listed one millstone company in Moore County, North Carolina, with $25 of capital invested, with $120 worth of raw materials, who employed one man, paid $300 in annual labor, and produced $1,200 worth of products annually (Kennedy 1864:429).

Ohio

Several millstone factories operated in Cincinnati, Ohio, at different periods. A burr millstone factory was operating in Cincinnati by 1819 (Flint 1822:240).

The January 23, 1828, edition of *The Maysville Eagle* contained the following ad for French Burr millstones produced in Cincinnati by a branch office of the Morris & Egenton company of Baltimore:

> FRENCH BURR MILL STONES.
>
> MORRIS & EGENTON, *French Burr Mill-stone Manufacturers*, of Baltimore, respectfully inform the Mill owners and millers of Cincinnati, and its vicinity, that they have established a branch of their business in Cincinnati, East Front Street, opposite the steam flour mill, and near the City Hotel, under the superintendance of ROBERT SMITH, in whose capability they have every confidence, and who has for several years past served them in Baltimore, in

> the capacity of a foreman. One of the firm (Wm. Egenton) being constantly residing in France, for the expressed purpose of procuring the first quality of Burr Blocks, from the best quarries, this arrangement cannot fail to afford them every opportunity of giving general satisfaction, and gives them a decided advantage over all others in the same line of business. All orders in the above line, by mail or otherwise will be thankfully received and carefully attended to, on the most moderate terms. Calcined PLASTER PARIS for sale as above. Refer to Barr, Lodsick & Co. in Cincinnati.
> January 8, 1828.

Limited information is available for the decade between 1841 and 1851. In 1841, two burr millstone factories in Cincinnati employed 15 hands and produced $10,500 of products annually (Cist 1841:56). During 1851, information was provided about French burr millstones being manufactured in Cincinnati (Cist 1851:182):

> *Burr Millstone makers.* Four factories.—Nineteen hands; value of product, twenty-four thousand dollars; raw material, 65 per cent.
>
> James Bradford & Co., 65 Walnut street, manufacture yearly, seventy-five pairs burr millstones.
>
> The burrs, of which the millstones are composed, are imported from France, in cubes of about twelve inches average. We have the same material in our own west, but it is not hard enough for service. The burrs are cemented with plaster of Paris, which is received from Nova Scotia and the Lake Erie region; and each stone is secured with four bands of iron, which being put on hot, as they shrink in cooling, serve to confine the whole under any amount of strain to which it may be exposed.

Additional information was provided by Cist (1859:243) for 1859:

> James Bradford & Co., northwest corner Elm and Second streets, office 65 Walnut Street, manufacturer French burr millstones, portable mills, corn and cob crushers, smut machines, hoisting screws, tempering screws, mill spindles, screen wire, mill castings, and damsel irons. Plaster paris, land plaster, hydraulic cement constantly on hand.
>
> Their portable mills are in successful operation in this city, competing with merchant mill-work, making as good flour, and as great yield, as can possibly be done upon any four, or four-and-a-half feet, that are running and capable of making fifty to sixty barrels superfine flour in twenty-four hours. Bradford & Co. warrants their portable mill to be equal, in quality and performance, to any portable mill in use. They have averaged three hundred and sixty pairs large millstones per year, beside one hundred portable mills. Sales, fifty thousand dollars annually; employ thirty hands; raw material, burr millstones, 45 per cent; portable mills, 25 per cent. An improvement has been made in the construction of millstones, by substituting for the four bands formerly used, one broad one, covering the entire edge, excepting the wearing surface.

James Bradford & Co. were listed as manufacturers of French Burr Mill Stones in *George Hawes Kentucky State Gazetteer and Business Directory for 1859 and 1860* (Hawes 1859: page opposite title page). The *Ohio State Gazetteer and Business Directory for 1860-'61* (Hawes 1860:viii) carried a full page ad for James Bradford and Company who were manufacturers of "French Burr Mill Stones" and sold many other mill related items. In 1861, T. Bradford and Company of Cincinnati were selling French Burr millstones and many other supplies for mills (Williams and Company 1861:480). James Bradford and Company manufactured "French Burr Mill Stones" for many years and advertised in the *Kentucky State Gazetteer, Shippers' Guide and Business Directory* for 1865 and 1866 (Hodgman & Co 1865:374). During 1881–1882, the Bradford Mill Company was advertising French Burr Mill Stones in Cincinnati at 8th and Evans streets (Polk and Danser 1881:659).

Other millstone manufactures produced millstones in Cincinnati, Cleveland, and

Cochran between 1837 and 1860. During 1837, MacCabe (1837:50) noted that "two French Burr millstone manufactories" were in operation in Cleveland, Ohio. In Dayton, Ohio, one "Burr Millstone Factory" was listed among the many manufacturing establishments (Cincinnati Gazette Company 1852:76). Robert Cochran owned another French Burr mill stone manufactory and also produced portable mills at 44 West Front Street in Cincinnati (Hawes 1860:111). C. S. Decker & Company manufactured French Burr millstones at Concord, a small community 30 miles northeast of Cleveland (Hawes 1860:284, 299).

Two counties were listed in the U.S. Census for 1860 as millstone producers in Ohio. In Hamilton County, Ohio, two burr millstone companies were listed with $8,000 of capital invested, with $4,500 worth of raw materials, that employed 12 men, paid $4,800 in annual labor, and produced $16,000 worth of products annually (Kennedy 1864:455). Two burr millstone companies were also reported in Montgomery County, Ohio, with $15,000 of capital invested, with $11,400 worth of raw materials, that employed 10 men, paid $3,888 in annual labor, and produced $23,000 worth of products annually (Kennedy 1864:468).

Hughes (1869:254) provided the following information on a millstone maker in Cincinnati:

> W. & E. WARD'S FRENCH BURR MILL-STONE
> MANUFACTORY, CINCINNATI, OHIO
>
> HAVING visited this establishment personally, I find many improvements in the construction of their French Burr Mill-stones, that fully warrant the assertion that they intend to keep up with the times.
>
> They are manufacturing mill-stones from both quarries, *old* and *new*, of superior workmanship; the backs are put on by a machine gotten up expressly for that purpose, which allows the stone when finished to be nearly in perfect balance; the eye of the mill-stone is put in on an entirely new plan, by which the greatest accuracy is obtained in centering the eye; and millers who design purchasing mill-stones, together with that desirable brand of bolting cloth known as "Defour's Dutch Anchor Brand Bolting Cloth," together with all other articles usually kept by mill-furnishers, such as light and heavy hoisting-screws, screen wire, Damsal's Leather Belting of all sizes, at manufacturers' prices. Cannot do better in Cincinnati, than the facilities of Messrs. Ward's furnish.

The Ohio Valley Publishing Company (1873:11) carried the following ad for millstone makers Thomas Bradford and Company of Cincinnati:

> MILL STONE MANUFACTORY
>
> BRADFORD THOS. & CO. Mnfrs of French Burr Mill Stones, Mill Castings, Mill Machinery of all kinds. This is the oldest manufactory of this kind in the Western Country. Office 61 Walnut, Factory, 135 wd.

Another Cincinnati company producing portable mills with French Burr millstones was the Straub Company (Cist 1859:244):

> Isaac Straub, warehouse No. 19 west Front Street works twenty-five hands; builds annually three hundred and fifty portable mills, running with French burr millstones; also portable saw mills and steam engines; value of products, seventy-five thousand dollars; raw materials, 35 per cent.

The April 1876 edition of *Manufacturer and Builder* included the following information about the Straub Company:

> The Straub Mill Co., of Cincinnati, O., has a very experienced foreman in Mr. R. Fuliman, who has spent several years among the quarries. The company have every facility for manu-

> facturing the mill-stone; they are very courteous to strangers visiting their novel works, and an hour can not be spent more profitably in Cincinnati than in looking through their place of business, whether as a sight-seer or a purchaser.

Pennsylvania

Dan Burnes and Son made and sold Burr millstones at their store in the late 18th century. A broadside advertisement dating to 1788 has survived and has been published by Readex Microprint in 1985 (Dan Burnes and Son 1788). No other information was encountered for the millstone manufactory of Dan Burnes and Son.

Millstone factories were opened in Franklin and Lancaster counties, Pennsylvania (Stoner 1947:424):

> It is not known that many millstones were ever quarried in this neighborhood [Chambersburg], but McCauley says three factories for fashioning stones had at various times been in operation in Franklin County. James Falkner, beginning about 1792, conducted such a factory in Chambersburg for many years. In 1820 George Walker and George Roupe carried on a "burr millstone" factory along what is now the Lincoln Highway in the neighborhood of Stoufferstown. Andrew Cleary also made millstones in Chambersburg as late as 1829. His shop was on West Market Street and he was the last person who carried on the business in Franklin County.
>
> Stones in the rough were brought by wagons to these factories, many of them from Cocalico township, near Ephrata, in Lancaster County. The stones were shaped up and dressed here and large numbers were sold in the Cumberland Valley and in other points farther west. One hundred years ago millstones constituted quite a large trade, but it is doubtful whether a set of millstones has been purchased in Franklin County within the past thirty or forty years.

In Dauphin County, Pennsylvania, the following information was provided about millstone maker William H. Kepner (Anonymous 1896:232): "KEPNER, William H., son of Samuel Kepner and Sara _____ was born in 1810, in Bern township, Berks county, Pa. His father was a millwright, came to Harrisburg in 1823, and erected the first steam flour mill in the neighborhood of Harrisburg. William H. adopted the business and trade of his father, and at the death of the latter continued the business, acquiring an extensive reputation in this and adjoining States for the superior quality of his millstones.... He died January 18, 1871, at Harrisburg, aged sixty years."

In a biographical sketch for Joseph Spang, his father Jeremiah, a millstone maker in Montgomery County, Pennsylvania, was briefly mentioned (Roberts 1904:204–205):

> Jeremiah Spang (father) was a millstone maker. He moved to York, Pennsylvania, living there a number of years, and then returned to Pottstown, where he resided until his death in 1876, in his sixty-fourth year. His wife died in 1887, aged nearly seventy-six years. He was a Lutheran in religious faith, and she was reared a Catholic.
>
> Adam Spang (grandfather) was a native of Montgomery County and also a burr-maker. He was married three times and had seventeen children in all. He died at the age of upwards of seventy years.

The U.S. Census for 1860 listed four millstone and mill furnishing companies in Philadelphia, Pennsylvania, with $61,000 of capital invested and $36,3140 worth of raw materials, that employed 49 men, paid $20,592 in annual labor, and produced $98,560 worth of products annually (Kennedy 1864:541).

Hughes (1869:259) discussed Fowler & Company's mill-furnishing establishment in Pittsburgh: "This firm are also manufacturing French Burr Mill-stones of the best quality, and are the sole agents in Pittsburg for that celebrated article of Dutch bolting cloth, known as Defoe & Co.'s, of Holland."

Another French Burr Millstone producer was Mitchell's in Philadelphia (Hughes 1869:230–231):

> *Made on an improved plan, with Kenderdine's Cast-iron Eye, Self-adjusting Irons, and Bed-stone Bush, Warranted.*
>
> The advantages of this arrangement over the old plan are, 1st. The perfect accuracy with which the stones can be built, by means of a sliding frame fitted into the groove intended for the balance-ryne, and finished with a stiff tram, II, which not only keeps the face of the stone perfectly true, but exactly at right angles to the balance-ryne.
>
> 2d. The saving of expense and trouble in setting in irons, and the ease with which these can be altered as the stone wears away, by simply shortening the blocks, EE.
>
> These irons are also *self-tramming*: the point of the spindle resting in a groove which allows the horns of the driver to come to an equal bearing, and causing the stone to wear evenly. The driver being furnished with a flange fitting over a smaller one on the cap of bed-stone bush, prevents any dust or sand getting to the neck of the spindle.
>
> The bed-stone is a simple application of wooden followers (which experience has proven to be the best,) made wedge-shape, so that they can be tightened while the stones are at work.
>
> These improvements (of which we have the exclusive right) having been in use for some time past, and in every case given satisfaction, we can safely recommend them to Millers as superior to any other plan now in use; and as all our materials are selected at the quarries in France, we can warrant our Mill-stones to be superior in every respect.
>
> N. B.—Bolting cloths of all numbers. Mill-picks, &c.
>
> J. E. Mitchell

W. W. Wallace, manufacturer of French Burr Mill-Stones and Steam-Engines, operated a factory on Liberty Street in Pittsburgh (Hughes 1869:250–251):

> The engraving [on page 251] represents O. Lull's Patent Smut Machine, made of burr-block, concave, and usually called "The Burr-stone Smut Machine." The Messrs. Wallace are the owners of this machine for the United States, and are manufacturing the same at their manufactory at Pittsburg, where millers will find every thing in their line, from a mill-stone to a tomb-stone, of the best quality.
>
> Their Smut Machines are of four sizes, and vary in price from $80 to $200. Messrs. Wallace are also extensive mill furnishers, such as steam-engines, and all other machinery wanted by millers. This firm is composed of practical men themselves, who are good judges of the construction of all descriptions of machinery.

In Northampton County, Pennsylvania, one millstone factory was once operating (Heller 1920:306).

Tennessee

The U.S. Census for 1860 listed one burr millstone manufacturer in Clairbourne County, Tennessee, with $100 of capital invested and $50 worth of raw materials. The manufacturer employed two men, paid $600 in annual labor, and produced $10,500 worth of products annually (Kennedy 1864:561).

Wisconsin

In Milwaukee, Wisconsin, the Reliance Iron Works, a successor to earlier millstone companies, was involved in millstone manufacture (Still 1948:337):

> Local industry required machines, millstones, engines, and pumps; and fast-growing inland cities duplicated the demand. Of the firms which arose to fill this need, the most outstanding

> was the Reliance Iron Works, an establishment which had grown out of the millstone factory founded by Decker and Seville in 1847 and which had come into the possession of E. P. Allis and others as a result of the financial revulsion ten years later. In 1860 Allis took over the concern himself and between that date and his death in 1889 he converted a business amounting to only $31,000 into an enterprise reporting an annual production averaging more than $3,000,000 a year.

Interestingly, the Reliance Iron Works not only manufactured millstones but also developed the technology that eventually led to the replacement of millstones (Still 1948:338):

> To meet the demands of the locality for grist and sawmill machinery, he early specialized in the production of sawmill equipment and millstones. In 1873 the millstone department of the company had fifty men engaged solely in dressing stones. In 1876 Allis enlisted the services of W. D. Gray with whose assistance he introduced to America the roller system of grinding flour. By 1878 the Allis plant had built and installed the first all-roller flour mill in America for C. C. Washburn of Minneapolis, a step which led to the introduction of the roller system in all the large milling centers with a consequent stimulation to the business of the Reliance Works.

4

The Rise and Fall of the American Millstone Industry: Producers, Annual Values, and Decline

The rise and fall of the American millstone quarrying industry can be followed to some extent through the figures published by the U.S. Bureau of Mines and Minerals. These figures were only available from 1880 to the mid–20th century. Unfortunately, the information published about millstones was not consistent throughout this period. During some years production figures were broken down by state and those localities producing millstones were mentioned. For other years, the U.S. Bureau of Mines and Minerals combined some or all the figures for the annual value of millstones sold that particular year. As the millstone industry declined, less information was published about it. In spite of the limitations, these reports are our best source of information on the annual activities of the American millstone industry.

It is the objective of this chapter to examine three aspects of the American millstone quarrying industry. First, information is presented on millstone producers, both individuals and companies. In some cases, the names of the men and companies that made millstones were recorded. To make this section more comprehensive, millstone makers mentioned in other documents are included. Second, figures are presented on the total values of millstones manufactured. For some years, these values were broken down for larger states but production figures for smaller producers were later combined into a single value. Third, information concerning the decline of the millstone industry was extracted from several annual reports. When feasible, the information on producers and the values of millstones are presented by state. General information is presented for the country as a whole.

In order to more fully understand the American millstone industry, it is important to know where most of the millstones were produced and the periods of time that the various quarries operated. To obtain this information, the author searched all the annual U.S. Bureau of Mines and Minerals reports between 1883 and 1960. The search was stopped at 1960 since the industry as a whole did not exist any more. A few granite millstones were still being quarried in Rowan County, North Carolina, after that date. The Meadows Mill Company is currently using modern technology to produce granite millstones for their portable grist mills. However, the traditional millstone industry in America ceased to exist after the mid–20th century.

A comprehensive search of reports published by the U.S. Bureau of Mines and Minerals indicated that four states produced most of the American millstones. These states were

New York, Virginia, North Carolina, and Pennsylvania. Several other states were listed as millstone producers during brief periods or just for one year. These states included Alabama, Maine, Maryland, Minnesota, New Hampshire, Ohio, and West Virginia. The U.S. Bureau of Mines noted from time to time that other states were producing a few millstones for local consumption but these quarries were not specifically identified or discussed. The first few paragraphs below discuss the value of the millstone industry as a whole as it fluctuated through time. Some of the reasons for these fluctuations are mentioned. The subsequent pages discuss those states mentioned above and the roles they played in the American millstone industry.

Millstone Producers

Between 1901 and 1963, the U.S. Bureau of Mines and Minerals frequently provided information on the number of millstone producers. Sources consulted include: Ambrose 1964; Beach and Coons 1922, 1923, 1924, 1925; Bowles 1930, 1932; Bowles and Davis 1934; Chandler and Marks 1954, 1955, 1956; Chandler and Tucker 1953, 1958a, 1958b, 1958c, 1958d, 1960, 1961; Coons and Stoddard 1929, 1930; Cooper and Tucker 1962, 1963; Hatmaker and Davis 1932, 1933a, 1933b; Johnson and Davis 1936, 1937, 1938; Johnson and Schauble 1939; Katz 1913, 1914, 1916, 1917, 1919, 1920, 1921, 1926, 1927, 1928; Metcalf 1941, 1943a, 1943b, 1949, 1950, 1951; Metcalfe and Cade 1945, 1946; Metcalf and Holleman 1947, 1948; Phalen 1908, 1909, 1910, 1911, 1912; and Pratt 1901, 1902, 1904a, 1904b, 1905, 1906. For earlier years, millstone producers were mentioned in geological reports, histories, and other sources.

The American millstone industry was very successful in 1880 ($200,000 of earnings) but had significantly dropped by 1888 ($81,000 of earnings). The industry plummeted the following year ($35,000 of earnings) and continued to decline until 1894 ($13,000 of earnings). In 1895 ($22,542 of earnings), the industry made a minor comeback and gradually increased until 1902 ($59, 808 of earnings). Between 1903 and 1911, the value of millstones fluctuated up and down between a period low of $28,217 (in 1910) to a period high of $52,552 (in 1903). The industry experienced a significant increase in 1912 ($71,414 of earnings) but gradually declined until 1917 ($43,489 of earnings). In 1918, the value of millstones increased to $92,514 but had dropped to $20,853 by 1922. The industry fluctuated up and down between 1923 and 1930 with a period low of $17,702 (in 1930) to a period high of $45,937 (in 1926). The next two years, the millstone industry hit an all time low of $5,330 in 1931 and $4,450 in 1932. Between 1933 and 1937, a gradual recovery began with annual values fluctuating from $8,305 (in 1937) to $10,609 (in 1936). The value for 1938 dropped to $3,743 but increased to $11,084 for 1939 and dropped again to $6,558 in 1940. During the final decade of the industry, prices fluctuated between $9,240 (in 1943) and $23,189 (in 1947).

The values of individual millstones were rarely reported by the U.S. Bureau of Mines and Minerals. Parker (1893a:552) gives the following figures: "The product in 1891 distributed by State was: New York, esopus stone, 353 pairs, worth $8,806; Pennsylvania, cocalico stone, 94 pairs, worth $3,801, and Virginia, Brush Mountain stone, 149 pairs, worth $3,980." Using these figures, the average cost per pair of millstones would be as follows: Esopus stone, $24.95; Cocalico stone, $40.44; and Brush Mountain stone, $26.71. However, it must be remembered that these are total production figures that reflect the range of products. For example, the New York figures may have included many small pairs of millstones that sold cheaply while the Pennsylvania quarries may have primarily produced large millstones that resulted in the average price being much higher. The point is that the millstone values are not comparable without knowing the exact price for each size of millstone.

New York

New York was the clear leader in the production of American millstones. These quarries produced a conglomerate millstone known as the "Esopus stone." Day (1886:428) stated that "in Ulster county, New York, the so-called Esopus stone has gained a definite footing as a substitute for buhrstone, for millstones for grinding chemicals and other materials except wheat." As for the name of the stone, Phalen (1908:609) stated that Esopus was "an early name for Kingston, which was formerly the main point of shipment." Day (1890:576) indicated that about 50 men were employed at the millstone quarries near Kyserike, New York. In addition to Kyserike, Parker (1898:515) mentioned that conglomerate millstones were quarried in the vicinity of the towns of Accord and Kerhonkson. Pratt (1902:793) noted, "The New York millstone quarries are located in a belt of sandstone and conglomerate on the Shawangunk Mountains, extending across the towns of Rochester, Marbletown, Wawarsing, Gardener, and New Paltz, in Ulster County. The stones vary greatly, some being very fine grained, while others are course, according to the portion of the rock from which they are manufactured. They are all, however, known as 'Esopus stone.' The largest stone made is 6 feet in diameter, and the smallest is 18 inches. The face of these stones varies from 1 to 2 feet."

Some of the Esopus millstones were distributed by middlemen that dealt in all kinds of millstones. In New York City, Munn & Company was selling "Esopus Millstones of the best material and manufacture, at the lowest prices" in 1850 (Scientific American 1850d:359). Esopus millstones were distributed by Morris & Trimble of Baltimore during 1854 along with the French Burr, Cologne (German), and Cocalico (Pennsylvania) stones (Edwards 1855:290).

A hand typed list of Ulster County millstone producers for 1918 is in the files of the Geological Survey at the New York State Museum in Albany (New York State Museum 1918). The list included these producers in the following order (Printed with permission of the New York State Museum, Albany, N.Y.):

Name	*Post Office*
Wilson C. Addis	Granite
Joseph Coddington	High Falls
Simon Coddington	St. Josen
William Countryman	Granite
Ira Davenport	Accord
Miles Decker	Granite
J. S. Depuy	Accord
Esopus Millstone Co., Benj. Schoonmaker, Jr.	Alligersville
Bruyn Hasbrouck	New Paltz
John Hendrickson	Alligersville
Harry Lawrence & B. H. Depuy	Accord
Russell Lawrence	Accord
James Lounsbury	Accord
Asa Purcell	Alligersville
David Purcell	Alligersville
W. H. Rose	Accord
Wm. D. Smith	Accord
James S. Van Etten	Kerhonkson

Hartnagel (1927:57) provided the following listing of persons and firms producing millstones in Ulster County (Printed with permission of the New York State Museum, Albany, N.Y.):

Name	*Post Office*
E. D. Coddington	Accord
Frank Coddington	Accord
Joachim Coddington	Accord
Lester Coddington	Accord
Q. D. Coddington	Accord
W. Davenport	Accord
J. S. Depuy.	Accord
Harry Laurence [sic]	Accord
Lange Brothers	Accord
Asa Purcell	Alligerville
David Purcell	Alligerville
Ross Schoonmaker	Alligerville
W. C. Addis Stone Company	Granite
Arthur Rose	Granite
Wilfred Coddington	High Falls
Esopus Millstone Company, C. Schoonmaker, Manager	High Falls
William Countryman	Kerhonkson, R. F. D.
Floyd Decker	Kerhonkson
Miles Decker	Kerhonkson
Bruyn Hasbrouck	New Paltz

New York millstone producers mentioned during 1934, 1935, 1936, 1937, and 1945 included (Printed with permission of the New York State Museum, Albany, N.Y.):

Addis Stone Co., Wilson C., Route 2, Kerhonkson, N.Y. (New York State Museum 1934; Davis 1935:1005; Johnson and Davis 1936:887; Johnson and Davis 1937:1293).

Frend Bush, High Falls, N.Y. (New York State Museum 1934).

George Coddington, Accord, N.Y. (Bowles and Davis 1934:901; New York State Museum 1934; Johnson and Davis 1936:887; Johnson and Davis 1937:1293; for 1945, Metcalf and Holleman 1947:1366; for 1946, Metcalf and Holleman 1948:100).

Oscar Coddington, Accord, N.Y. (Bowles and Davis 1934:901; New York State Museum 1934; Davis 1935:1005; Johnson and Davis 1936:887; Johnson and Davis 1937:1293).

Willfred Coddington, 507 Washington Ave., Kingston, N.Y. (New York State Museum 1934).

William Countryman, R. F. D., Kerhonkson, N.Y. (New York State Museum 1934).

Floyd Decker, R. F. D # 2, Kerhonkson, N.Y. (Bowles and Davis 1934:901; New York State Museum 1934; Davis 1935:1005; Johnson and Davis 1936:887; Johnson and Davis 1937:1293).

Esopus Millstone Co., R. F. D # 1, Box 79, High Falls, N.Y., C. Schoonmaker, Mgr. (Bowles and Davis 1934:901; New York State Museum 1934; Davis 1935:1005; Johnson and Davis 1936:887; Johnson and Davis 1937:1293). The company was advertising as early as 1875 (Howell and Keller 1977:71).

Harry Laurence [sic], Box 171, Accord, N.Y. (Bowles and Davis 1934:901; New York State Museum 1934; Davis 1935:1005; Johnson and Davis 1936:887; Johnson and Davis 1937:1293).

Henry Lawrence, Accord, New York for 1946 (Metcalf and Holleman 1948:100).

Cyrus Schoonmaker, Kerhonkson, N.Y. (Bowles and Davis 1934:901; New York State Museum 1934; Johnson and Davis 1936:887).

John Smith, Accord, N.Y. (Bowles and Davis 1934:901; New York State Museum 1934; Johnson and Davis 1936:887).

For the years 1903 to 1906, the *Mineral Resources of the United States* published the number of millstone producers per state. Pratt (1904b:999, 1905:1004, 1906:1072) provided information on millstone producers in New York during 1903, 1904, 1905, and 1906. Figures for 1912 and 1913 were provided by Katz (1913:823, 1914:258). No figures were available for 1914. Subsequent annual reports provided information for the years 1915 to 1955 (Chandler and Tucker 1953:95, 1956:124; Hatmaker and Davis 1933b:660; Johnson and Davis 1936:886; Katz 1917:68, 1919:201, 1920:217, 1921:175, 1926:329; 1927:243; Metcalf 1941:1246, Metcalf and Cade 1949:98; Metcalf and Holleman 1947:1366). Table 2 provides the number of New York millstone producers between 1903 and 1955.

Table 2. Numbers of New York Millstone Producers Between 1903 and 1955.

Year	*No. of Producers*	*Year*	*No. of Producers*
1903	17	1933	7
1904	17	1934	5
1905	14	1935	8
1906	14	1936	6
1912	14	1937	6
1913	14	1938	4
1915	12	1939	6
1916	12	1940	3
1917	11	1941	5
1918	14	1942	2
1919	7	1943	1
1920	8	1944	3*
1921	12	1945	4*
1922	12	1946	4*
1923	17	1947	4*
1924	17	1848	3*
1925	13	1949	2*
1926	12	1950	2*
1927	12	1951	1*
1928	14	1952	1*
1929	11	1953	2*
1930	7	1954	2*
1931	6	1955	1*
1932	5		

**Years for which numbers of producers were only provided for the United States as a whole.*

Conflicting figures were published for the number of millstone producers in New York for three years (Table 3). Certain editions of the *Mineral Resources of the United States* provided higher figures than those provided by other editions. However, some of the higher numbers include the names of the producers. For the sake of presenting the most comprehensive data possible, these figures are also included for 1916, 1917, and 1934 (Katz 1919:201, 1920:217; Bowles and Davis 1934:901).

Table 3. Alternate Numbers for New York Millstone Producers Between 1916 and 1934.

Year	*No. of Producers*
1916	12
1917	11
1934	7

While the conglomerate millstone industry is very well known, other types of stone were exploited in Jefferson, Oneida, and Erie counties, New York. Near the community of Antwerp in Jefferson County, a gneiss was quarried for millstones. Millstone makers in Jefferson County in 1805 included David Coffeen, James Parker, and Alexander Parker (Haddock 1895:434; Oakes 1905:1129–1130). At Utica in Oneida County, New York, several individuals were engaged in burr millstone manufacture from 1823 to ca. the 1880s. These men included Alexander B. Hart, Martin Hart, Alfred Munson, and Edmund Munson (Bragg 1892:594). Finally, John T. Noye was producing millstones at Buffalo in Erie County, New York ca. 1840s–1850s (Horton et al. 1947:229; Smith 1884:61–62).

Virginia

Virginia was usually the number two producer of American millstones and for a brief period was ranked number one. A conglomerate was quarried in Montgomery County near Blacksburg and was known as Brush Mountain stone (Parker 1894:670).

Phalen (1910:613) described the Brush Mountain millstone quarry as follows:

> The millstone industry in Virginia is confined to quarries near Price's Fork, Montgomery County, about 5 miles west of Blacksburg, the site of the Virginia Polytechnic Institute. The rock is regarded as of Mississippian (lower Carboniferous) age. The material from which the stones are quarried varies from a normal conglomerate to a fine-grained quartzitic rock. It includes pebbles, some of them as large as walnuts, though most of them are smaller. The rock has a bluish cast. Its bedding planes are very distinct, and layers only an inch thick may be observed. It is extremely hard and tough and resists erosion to a marked degree. It underlies Brush Mountain for miles, and for this reason the millstones are frequently known as Brush Mountain stones. The stone can not be quarried by blasting, and is therefore extracted by hand power, with drill and hammer, plug and feathers. Millstones and drag or rider stones are the principal products made at the Virginia quarries.

Metcalf (1941:1246) provided the following information on the Montgomery County millstones: "The diameter of the millstones produced in Virginia ranges from about 12 to 72 inches; as they are quarried by hand from underground operations, as the surface rock has been found unsatisfactory on account of its objectionable lining and lamination."

Virginia millstone producers were listed by name for 1934, 1935, 1936, and 1947. The include the following individual and companies:

R. E. Snider, Cambria, Va. for 1934, 1935, 1936, and 1937 (Bowles and Davis 1934:901; Davis 1935:1005; Johnson and Davis 1936:887; Johnson and Davis 1937:1293).

J. Fred Shealor, Route 2, Blacksburg, Va. for 1935 and 1937 (Davis 1935:1005); Johnson and Davis 1937:1293 (**Virginia Millstone Company**).

P. L. Olinger (successor to R. L. Olinger & Co.), Blacksburg, Va. for 1936 and 1937 (Johnson and Davis 1936:887); Johnson and Davis 1937:1293 (**Virginia Abrasive Company**).

Interstate Millstone Company, Christiansburg, Va. for 1945 and 1946 (Metcalf and Holleman 1947:1366; Metcalf and Holleman 1948:100).

Montgomery County, Virginia, had the largest number of known millstone makers anywhere in the United States. During the late 1700s, John Phillip Harless, Jacob Price, John Michael Price, and Michael Price made millstones at Brush Mountain (Hockensmith 1999:2; Worsham 1986a:157). Millstone makers working in Montgomery County in the mid–1800s included James R. Kent, Henry Lewis Price, and Israel Price & Company (Hockensmith 1999:2). By the late 1880s, Hugh Price, Zachariah Price, and John Snyder Surface were making Brush Mountain millstones (Hockensmith and Coy 1999:12; Hockensmith and Price

1999:65, 67, 69). Many of the Montgomery County millstone makers are known from the first four decades of the 20th century. These include Edward Cromer, Thomas Cromer, Willard Cromer, Bob Fisher, Enos Fisher, John Fisher, Donald Long, Jack Long, Ted Long, Albert Olinger, Robert Olinger, Samuel Olinger, Floyd Planke, Beck Price, John "Matt" Price, Leon Price, Leonard Price, Lester Price, Martin Price, Roy Saville, Walter C. Saville, Arthur Shealor, Byrd Shealor, Guy Shealor, J. Fred Shealor, John Shealor, Olen Shealor, Gayle Smith, Harvey Smith, Stanley Snider, Esse Snyder, John Snyder, Robert G. Surface, and Robert Huston Surface (Hockensmith 1999:2; Hockensmith and Coy 1999:9, 11–14, 16–18; Hockensmith and Price 1999:65–68).

For selected years, the *Mineral Resources of the United States* published the number of millstone producers per state. Virginia figures (Table 4) were provided for the years 1903 to 1906 (Pratt 1904b:999, 1905:1004, 1906:1072; Sterrett 1907:1045) and for 1912 to 1918 were also published (Katz 1913:823, 1914:258, 1916:553, 1917:69, 1919:201, 1920:217, 1921:1175). For several years (1919–1932), the actual number of millstone producers were only reported as two figures: New York and the rest of the country lumped together (Bowles 1930:239; Hatmaker and Davis 1932:105; Hatmaker and Davis 1932:153; Katz 1926:329). Incomplete counts of Virginia millstone makers are available for the years 1935 to 1943 (Johnson and Davis 1938:1143; Metcalf 1941:1246; Metcalf and Cade 1945:1391). For some of these years, the number of producers also included North Carolina and for one year North Carolina and West Virginia.

Table 4. Numbers of Virginia Millstone Producers Between 1903 and 1943.

Year	*No. of Producers*	*Year*	*No. of Producers*
1903	3	1933	2*
1904	3	1934	3*
1905	4	1935	3*
1906	3	1936	2
1912	5	1937	2
1913	3	1938	2
1914	3	1939	3*
1915	3	1940	2*
1916	5	1941	3**
1917	4	1942*	3*
1918	2	1943	3*

* *North Carolina and Virginia producers.*
** *North Carolina, Virginia, and West Virginia producers.*

North Carolina

Conglomerate millstones were quarried near Parkewood, North Carolina, and were known as "North Carolina grit" (Day 1890:576). Parker (1894:670) and Pratt (1902:793) both indicated that "North Carolina grit" millstones were quarried in Moore County, North Carolina. Further, Pratt (1902:793) stated: "At Faith, Rowan County, N.C., in the Dunn Mountain granite belt, a stone is being quarried and manufactured into millstones. The demand for this stone is increasing, and it is expected that the production in 1902 will be from two to three times of that of 1901. The stone are used mostly for grinding corn, oats, etc. and are sold in North Carolina, Georgia, and other Southern States."

The names of millstone producers for North Carolina were published for 1934, 1935, and 1947. These producers included:

Gardner Bros., Salisbury, N. C. for 1934, 1935, and 1936 (Bowles and Davis 1934: 901; Davis 1935:105; Johnson and Davis 1936:887).

Gardner Granite Works, Salisbury, N. C. for 1945 and 1946 (Metcalf and Holleman 1947:1366; Metcalf and Holleman 1948:100).

Millstone quarrying was conducted in Madison, Moore, and Rowan counties, North Carolina. In Madison County, George Gehagan was producing millstones in the 1860s (North Carolina Land Company 1869:106). Moore County millstone makers between the Civil War era and the 1880s included Devotion Davis, William Donnelly, Louis Grimm, Ed Taylor, George Taylor, and the North Carolina Millstone Company (Beaman 1985:18–19). The Gardner Granite Works mentioned above was the only millstone producer during the 1930s to 1970s (McGee 2001:13–16).

For selected years, the *Mineral Resources of the United States* published the number of millstone producers per state. North Carolina figures (Table 5) were provided for the years 1903 to 1906 (Pratt 1904b:999, 1905:1004, 1906:1072; Sterrett 1907:1045). The number of producers for the years 1912 to 1918 were also published (Katz 1913:823, 1914:258, 1916:553, 1917:69, 1919:201, 1920:217, 1921:1175). For several years (1919–1932), the actual numbers of millstone producers were only reported as two figures: New York, and the rest of the country lumped together (Bowles 1930:239; Hatmaker and Davis 1932:153; Katz 1926:329). Katz (1921:1175) reported that the granite millstone quarry at Salisbury, North Carolina, had moved into first place in American millstone production. Incomplete counts of North Carolina millstone makers are available for the years 1935 to 1955 (Chandler and Tucker 1958b: 124; Johnson and Davis 1938:1143; Metcalf 1941:1246; Metcalf and Holeman 1948: 105; Metcalf and Cade 1945:1391). For some of these years, the number of producers for North Carolina also included Virginia and for one year both Virginia and West Virginia. The final years of the millstone industry were reported on between 1956 and 1963 (Ambrose 1964:190; Chandler and Tucker 1958a:142, 1958b:147, 1959:133, 1960:140, 1961:149; Cooper and Tucker 1962:217, 1963:201).

Table 5. Numbers of North Carolina Millstone Producers Between 1903 and 1963.

Year	*No. of Producers*	*Year*	*No. of Producers*
1903	2	1937	***
1904	2	1938	***
1905	2	1939	3*
1906	2	1940	2*
1912	3	1941	3**
1913	3	1942	3*
1914	3	1943	3*
1915	2	1956	1
1916	2	1957	1
1917	1	1958	1
1918	8	1959	
1933	2*	1960	1
1934	3*	1961	1
1935	3*	1962	1
1936	***	1963	1

* *North Carolina and Virginia producers.*
** *North Carolina, Virginia, and West Virginia producers.*
*** *No production reported between 1936 and 1938.*

Pennsylvania

Pennsylvania was an important producer of millstones during the late 19th century. The best-known product was a conglomerate millstone known as the "Cocalico stone" quarried in Lancaster County. Day (1890:576) noted that Cocalico millstones were quarried near Durlach, Pennsylvania. Pratt (1902:793) stated that "in Pennsylvania the quarries are located in Lancaster County, and the millstones are known as 'Turkey Hill' and 'Cocalico,' the former being found on Turkey Hill, near Bowmansville, and the latter near Durlach and Lincoln."

Millstones were manufactured in several Pennsylvania counties including Berks, Bradford, Dauphin, Fayette, Franklin, Lancaster, Montgomery, Somerset, and Tioga. In Berks County, millstone makers for the early 1800s included Henry Edge, George Strohecker, Gottleib Strohecker, John Strohecker, Sr., John Strohecker, Jr., and Samuel Strohecker (Strawhacker 2004). John Northup made millstones in Bradford County ca. 1815 (Heverly 1915:363). Millstones were produced by William H. Kepner (ca. 1830–1840s) in Dauphin County (Anonymous 1896:232). Fayette County millstone makers included John Shacklett (late 1700s), Hugh Turner (early 1800s), and Samuel Jones (1825) (Maxwell 1968:482, 485; Ridenour 1977). Early millstone makers in Franklin County included James Faulkner, Jr. (1792), Andrew Cleary (1820s), George Roupe (1820s), and George Walker (1820s) (M'Cauley et al. 1878:122; Stoner 1947:424). Several millstone makers were listed in Lancaster County between 1860 to the early 20th century: Hehnly & Wike, William Konigmacher, William D. Nagel, Samuel Reifsnyder, S. P. A. Weidman, William Weinhold, Benjamin Wissler, and a Mr. Haimley (Anonymous 1962:11; Flory 1951a:76, 81; Hehnly & Wike 1880; and Zerfass 1921:25). Mid–19th century millstone producers in Montgomery County included Adam Spang and Jeremiah Spang (Roberts 1904:204–205). John Deeter of Somerset County made millstones between 1789 and 1814 (Brown 1990:22). Finally, James Hesselgessel produced millstones in Tioga County in 1836 (Berg 1986:3–6).

The *Mineral Resources of the United States* published the number of millstone producers per state for selected years. Pennsylvania figures (Table 6) were provided for the years 1903 to 1906 (Pratt 1904b:999, 1905:1004, 1906:1072; Sterrett 1907:1045). The number of producers for the years 1912 to 1918 were also published (Katz 1917:69, 1919:201, 1920:217, 1921:1175). These figures are listed below.

Table 6. Numbers of Pennsylvania Millstone Producers Between 1903 and 1918.

Year	*No. of Producers*
1903	3
1904	4
1905	5
1906	3
1914	1
1915	2
1916	2
1917	2
1918	2

Other States

Pratt (1904:879) observed that "a small number of burrstones are made in the mountain sections of North Carolina and Tennessee for local uses." Katz (1913:824, 1916:554) noted that a few Pennsylvanian age sandstone millstones were made at Dutton in Jackson County, Alabama. Conglomerate millstones (similar to the Esopus stone) were quarried near Fair Haven in Rutland County, Vermont (Katz 1916:554). Granite was quarried for millstones in Carroll County, New Hampshire (Bowles and Davis 1934: 900). During 1941, millstones were produced near Morgantown, West Virginia (Metcalf 1943a:1346). For 1945, the Jasper Stone Company of Sioux City, Iowa, quarried millstones at Rock County, Minnesota (Metcalf and Holleman 1947:1366).

Several millstone makers have been identified in records for states that were not major producers. Many of these men were involved in the millstone industry decades before the U.S. Bureau of Mines started collecting information. These states include Georgia, Kentucky, Maryland, Massachusetts, and Ohio. A number of men and one company were mentioned in Burke County, Georgia, between 1792 and 1849, including H. Hoyt, S. Hoyt, Jeremiah Miller, John Murray, George Poythress, and the LaFayette Burr Stone Company (Davis 1990:6; Hemphill 1999). Kentucky millstone makers were listed in documents between 1797 and 1821 for Clark, Franklin, Rockcastle, and Woodford counties. Clark County millstone makers mentioned in lawsuits and deeds included Spencer Adams, Peter DeWitt, Sr., Absolom Hanks, Martin Johnson, Cornelius Spry, Peter Treadway, and Moses Treadway (Clark County Court Records 1797, 1810, 1819; Fayette County Court Records 1804). In Franklin County, millstones were manufactured by Jeremiah Buckley and also by the firm of Miller, Railsback & Miller (The Argus of Western America 1821a, 1821b). Rockcastle County millstone makers included Charles Colyer, Jr., and Samuel Taylor (The Argus of Western America 1812, 1813). In Woodford County, John Tanner produced millstones (Palladium 1800).

One company and several individuals were listed as millstone producers in Maryland and Massachusetts. At Baltimore, Maryland, John M. Gorsuch, Dickinson Gorsuch, and the Maryland Millstone Manufactory were listed between 1810 and 1813 (Coxe 1814:84; McGrain 1991). Also, Herman Husbands published an ad in the *Maryland Gazette* on October 25, 1753, that he was making millstones at the mouth of the Susquehanna River (Maryland Gazette 1753). This operation would have been located in either present day Hartford County or Cecil County that flanks the mouth of the Susquehanna River. Millstone makers at Nantucket, Massachusetts, during 1670 included Edward Starbuck, Nathaniel Starbuck, John Swaine, and William Worth (Worth 1902:98).

Millstone makers were mentioned in various records for Licking, Muskingum, and Vinton counties, Ohio. In Licking County, Joseph Baird, Joshua Evans & Company, and Adam Drumm made millstones during ca. the 1820s to 1830s (Garber 1970:81; Mather 1838b:91). Millstones were manufactured in the 1830s in Muskingum County by Samuel Drumm and S. Henslee (Mather 1838b:91; Mills 1921:96–98). Individuals producing millstones in Vinton County between 1805 and 1807 included Henry Castle, Aaron Lantz, Richard McDougal, Mr. Musselman, Abraham Neisby, and Isaac Pierson (Howe 1888:427; Mather 1838a:33–34).

For many years the millstone producers were listed separately for New York and the other states were lumped together. Occasionally, producers were listed for specific states. For the years 1903 to 1906, the *Mineral Resources of the United States* published the number of millstone producers per state. Pratt (1904:999) indicated that there were 26 millstone producers in the United States during 1903 as follows: "New York, 17; Pennsylvania and Virginia, 3 each; North Carolina, 2 and Vermont, 1." For 1904, Pratt (1905:1004) mentioned 26 millstone producers in the United States as follows: "New York, 17; Pennsylvania, 4, Virginia, 3; and North Carolina, 2." During 1905, there were 25 millstone producers in the United States as follows: "New York, 14; Pennsylvania, 5; Virginia, 4; and North Carolina, 2" (Pratt 1906:1072). Finally, for 1906, Sterrett (1907:1045) listed 22 millstone producers in the United States as follows: "New York, 14; Virginia, 3; North Carolina, 2; and Pennsylvania, 3."

Table 7. Numbers of Millstone Producers in Other States Between 1903 and 1955.

Year	*No. of Producers*	*Year*	*No. of Producers*
1903	9	1934	3
1904	9	1935	3
1905	11	1936	3
1906	8	1937	2
1915	7	1938	2
1916	10	1939	3
1917	8	1940	2
1918	13	1941	3
1919	11	1942	3
1920	11	1943	3
1921	7	1944	3*
1922	4	1945	4*
1923	4	1946	4*
1924	5	1947	4*
1925	5	1948	3*
1926	7	1949	2*
1927	5	1950	2*
1928	6	1951	1*
1929	5	1952	1*
1930	5	1953	2*
1931	2	1954	2*
1932	2	1955	1*
1933	2		

* *These years also include New York.*

The *Mineral Resources of the United States* provided information on the number of millstone producers for key states (Table 7). Figures were provided for New York and all other states as one total. Thus, production figures for North Carolina (1915, 1916, 1917, 1918, 1919, 1920, 1921, 1922, 1923, 1924, 1925, 1926, 1927, 1928, 1929, 1930, 1931, 1932, 1933, 1934, 1935, 1939,1940, 1941, 1942, 1943, 1944, 1945, 1946, 1947, 1948, 1949, 1950), Pennsylvania (1915, 1916, 1917, 1918, 1919), and Virginia (1915, 1916, 1917, 1918, 1919, 1920, 1921, 1922, 1923, 1924, 1925, 1926, 1927, 1928. 1929, 1930, 1931, 1932, 1933, 1934, 1935, 1936, 1937, 1938, 1939, 1940, 1941, 1942, 1943, 1944, 1945, 1946, 1947, 1948, 1949, 1950) were combined. Minor producers such as Alabama (1916, 1917, 1919, 1920, 1921, 1922), Maine (1923), Maryland (1918, 1920), Minnesota (1923, 1945), New Hampshire (1921, 1924, 1925, 1926, 1927, 1928, 1930), Vermont (1903) and West Virginia (1941) were also occasionally included in this figure. Katz (1926:329) provided figures for 1919 to 1923, Katz (1927:243) also provided figures for 1915 to 1924; Hatmaker and Davis 1933a, 1933b: 1922 to 1931; Johnson and Davis (1936:886) published figures for 1931 to 1935; Metcalf (1941:1246) provided figures for 1936 to 1940; Metcalf and Holleman (1947:1366) provided data for 1941 to 1943; Metcalf and Cades (1949:98) dealt with 1944 to 1949; Chandler and Tucker (1953:95) for 1945 to 1950; and Chandler and Tucker (1958b:124) for 1951 to 1955.

Table 8 lists the total number of producers for America as a whole from 1903 to 1963. In the early years of record keeping by the United States government, information was not published on the specific number of millstone producers. For the years 1903 to 1906, the *Mineral Resources of the United States* published the number of millstone producers per state and for the United States (Pratt 1904:999, 1905:1004, 1906:1072; Sterrett 1907:1045). The number of producers for 1915 to 1934 were also published (Bowles and Davis 1934:901; Davis 1935:105; Hatmaker and Davis 1933a:113; Katz 1917:69, 1919:201, 1920:217, 1921:1175, 1926:329). Counts of millstone makers are available for the years 1935 to 1955 (Chandler and Tucker 1958b: 124; Johnson and Davis 1938:1143; Metcalf 1941:1246; Metcalf and Holleman 1948: 105; Metcalf and Cade 1945:1391). The final years of the millstone industry were reported on between 1956 and 1963 (Ambrose 1964:190; Chandler and Tucker 1958a:142, 1958b:147, 1959:133, 1960:140, 1961:149; Cooper and Tucker 1962:217, 1963:201).

Table 8. Total Numbers for Millstone Producers in the United States Between 1903 and 1963.

Year	*No. of Producers*	*Year*	*No. of Producers*
1903	26	1937	8
1904	26	1938	6
1905	25	1939	9
1906	22	1940	5
1912	22*	1941	8
1913	20*	1942	5
1915	19	1943	4
1916	19*	1944	3
1917	18**	1945	4
1918	26	1946	4
1919	18	1947	4
1920	19	1948	3
1921	19	1949	2
1922	16	1950	2
1923	21	1951	1
1924	22	1952	1
1925	18	1953	2
1926	19	1954	2
1927	17	1955	1
1928	20	1956	1***
1929	16	1957	1***
1930	12	1958	1***
1931	8	1959	1***
1932	7	1960	1***
1933	9	1961	1***
1934	8	1962	1***
1935	11	1963	1***
1936	9		

** Does not include figures for Pennsylvania and Alabama.*
*** Does not include figures for Alabama.*
**** North Carolina was the only state producing millstones.*

Millstone Values

Limited information has been published concerning the values of American millstones. Occasionally, the prices for specific sizes of millstones were provided. More often, the annual production figures for specific states were published. The most comprehensive figures are available for New York and Virginia and to a lesser extent North Carolina and Pennsylvania. Information was published for millstone industry in the United States as a whole for many years between 1880 and 1956.

New York

Some of the Mineral Resources of the United States provided information on the number of millstone producers for key states. The following figures were available for New York. Katz (1926:329) provided figures for 1919 to 1923; Katz (1927:243) provided figures for 1915 to 1924; Hatmaker and Davis 1933b:1922 to 1931; Johnson and Davis (1936:886) published figures for 1931 to 1935; Metcalf (1941:1246) provided figures for 1936 to 1940; Metcalf and Holleman (1947:1366) provided data for 1941 to 1943; Metcalf and Cade (1949:98) for 1944 to 1949; Chandler and Tucker (1953:95) for 1945 to 1950; and Chandler and Tucker (1958b:124) for 1951 to 1955.

Prices

A number of the New York reports provide information on the value of millstones. Nason (1894:394) stated that (Printed with permission of the New York State Museum, Albany, N.Y.), "The millstones vary in size and thus greatly in price. For milling, stones are quarried from fifteen to seven feet in diameter. The smaller stones vary in price from $5 to $15, while the seven-foot stones from $50 to $100. The millstones are used for grinding, chasers for crushing. Blocks twelve by twelve inches are used for paving the chaser floors. These chasers are used in quartz and feldspar mills, principally for preparing these materials for potters' use."

Newland (1907:44) noted that (Printed with permission of the New York State Museum, Albany, N.Y.): "The size of the stones marketed range from 15 to 90 inches. The greater demand is for the smaller and medium sizes, with diameters of 24, 30, 36, 42 and 48 inches. A pair of 30-inch millstones commonly sells for $15, while $50 may be paid for a single stone 60 inches in diameter. The largest sizes bring from $50 to $100."

For 1908, Newland (1909:39) stated that "millstones varied in value ... from $3 for 15 inch, to $45 for 54 inch stones, while chasers sold for $30 to $70 varying in size from 54 to 72 inches." The following year, Newland (1910:51) noted that "the selling prices of millstones in 1909 ranged from $3 to $4 for a 16 inch stone up to $60 for a 72 inch stone. Chasers in sizes from 54 to 72 inches sold in prices ranging from $30 to $70 each." Newland (1909:39) provided the following production figures for 1908: "the number of millstones made was 871 and chasers 182, in addition a small production of blocks and disks for use in roll crushers."

Newland (1921:155–156) made the following comments about production figures (Printed with permission of the New York State Museum, Albany, N.Y.): "Considerable difficulty is encountered in obtaining information about the production of millstones. Values of course are dependent upon the quality and size of the stones and the quarry prices range all the way from $2.50 to $3 for the smallest size, an 18-inch stone, to $75 or more for an 84-inch stone which is about the maximum size that is used. The quarrymen may sell the whole or part of their year's product to local middlemen who sometimes work quarries of their own. In the last decade the average production has been less than $20,000, although in earlier years it often exceeded $100,000."

The *Mineral Resources of the United States* published information on the value millstones produced in New York between 1883 and 1939 (Table 9). For the years 1903 to 1911, Pratt (1904b:999, 1905:1004, 1906:1072) and Phalen (1908:610, 1911:685) provided information. Production figures for 1912 and 1913 were provided by Katz (1913:823, 1914:258). No figures were available for 1914. Subsequent annual reports provided information for the years 1915 to 1955 (Beach and Coons 1923:157; Chandler and Tucker 1953:95, 1958b:124; Coons and Stoddard 1930:93; Hartnagel 1927:56; Hatmaker and Davis 1933b:660; Johnson and Davis 1936:886; Katz 1917:68, 1919:201, 1920:216–217, 1921:175, 1926:329; 1927:243; Metcalf 1941:1246,

1949:98; Metcalf and Holleman 1947:1366; Newland 1921:156; Newland and Hartnagel 1939:63). The values of millstones produced in New York for 1937, 1938, 1940, 1942, and 1943 were not available according to the note contained in these reports that the "Bureau of Mines [was] not at liberty to publish figures separately."

Table 9. Available Annual Values for Millstones Produced in New York Between 1883 and 1939 as Reported by the *Mineral Resources of the United States.*

Year	*Value of Millstones*	*Year*	*Value of Millstones*
1883	$120,000	1917	$22,103
1884	$110,000	1918	$25,488
1885	$90,000	1919	$10,155
1886	$100,000	1920	$13,331
1887	$75,000	1921	$14,672
1888	$60,000	1922	$17,025
1891	$ 8,806	1923	$10,344
1902	$39,570	1924	$18,215
1903	$35,441	1925	$14,063
1904	$24,585	1926	$23,629
1905	$25,915	1927	$26,015
1906	$28,848	1928	$26,224
1907	$23,072	1929	$18,147
1908	$18,341	1930	$6,577
1909	$13,341	1931	$2,030
1910	$13,753	1932	$1,850
1911	$13,335	1933	$5,187
1912	$34,246	1934	$3,381
1913	$21,987	1935	$4,645
1914	$16,748	1936	$5,458
1915	$16,883	1939	$2,584
1916	$10,287		

Different figures were published for the value of millstones produced in New York. The *Mineral Resources of the United States* provided higher figures than those provided by New York sources. Since it is not possible to determine which figures were in error, figures from the *Mineral Resources of the United States* were provided in Table 9. Sterrett (1907:1045) published information on the value of millstones produced in New York between 1902 and 1906. Phalen (1911:684) provided figures for 1906 to 1910 while Katz (1917:67) published data for 1910 to 1915. Newland (1921:156) compiled New York millstone values for 1904 to 1918. The New York figures which were different for 1904 to 1915 are included in Table 10.

Table 10. Alternate Annual Values for Millstones Produced in New York Between 1904 and 1915 as Reported by New York State.

Year	*Value of Millstones*	*Year*	*Value of Millstones*
1904	$21,476	1910	$6,613
1905	$22,944	1911	$13,177
1906	$22,442	1912	$15,358
1907	$21,806	1913	$13,130
1908	$18,341	1914	$12,410
1909	$19,247	1915	$10,916

Virginia

The *Mineral Resources of the United States* and the Virginia Geological Survey published information on the value millstones produced in Virginia between 1891 and 1930 (Table 11). Parker (1893a:552) published information for 1891. Watson (1907:401) and Pratt (1906:1072) provided the following information concerning the value of millstones produced in Virginia between 1902 through 1905. Bowles (1930:239) and Roberts (1942:436) listed the value of millstones produced in Virginia between 1905 and 1932. During some years, millstone values were combined with other resources, grouped under miscellaneous or were not available. The figures in Table 11 cover only those years during which separate figures were published.

North Carolina

The *Mineral Resources of the United States* published information on the value millstones produced in North Carolina between 1886 and 1952 (Table 12). Day (1888:552, 1890:576) estimated the value of buhrstones produced in the United States between 1883 and 1889 including North Carolina Grit millstones from North Carolina. Pratt (1904:999) reported the value of buhrstones sold in North Carolina and Vermont in 1902 and in 1903. Sterrett (1907:1045) published information on the value of millstones produced in North Carolina between 1904 and 1906. Phalen (1911:684) provided figures for 1906 to 1910, Katz (1913:822, 1916:552) gave figures for 1912 to 1914, Beach and Coons (1923:157) provided values for 1916 to 1920. Finally, Chandler and Marks (1956:131) reported on millstone production in the early 1950s.

Table 11. Available Annual Values for Millstones Produced in Virginia Between 1883 and 1936.

Year	*Value of Millstones*	*Year*	*Value of Millstones*
1891	$3,980	1918	$27,802*
1902	$11,435	1919	$27,792*
1903	$9,812	1920	$34,676
1904	$4,759	1921	$9,852**
1905	$8,186	1922	$3,828**
1906	$15,611	1923	$7,885***
1907	$4,684	1924	$11,910***
1908	$7,954	1925	$8,427***
1909	$22,255*	1926	$15,295
1910	$5,273	1927	$9,423
1911	$17,635	1928	$16,662
1912	$25,866	1929	$5,260
1913	$23,530	1930	$11,125
1914	$20,100	1931	$500
1915	$23,170	1932	$1,200
1916	$25,752	1933	$800
1917	$18,980	1936	$5,151

** Combined figures for North Carolina, Pennsylvania, and Virginia.*
*** Combined figures for North Carolina and Virginia*
**** All states combined except for New York.*

Pennsylvania

The *Mineral Resources of the United States* published information on the value millstones produced in Pennsylvania between 1883 and 1919 (Table 13). Day (1888:552, 1890:576) and

Parker (1893a:552) estimated the value of Cocalico millstones from Pennsylvania between 1883 and 1891. Pratt (1904b:999) reported the value of buhrstones sold in Pennsylvania in 1902 and 1903. Sterrett (1907:1045) published information on the value of Pennsylvania millstones produced between 1902 and 1906. Phalen (1911:684) provided figures for 1906 to 1910, Katz (1917:67) published data for 1910 to 1915, Beach and Coons (1923:157) published information for 1916 to 1920. The Pennsylvania millstone industry was not mentioned after 1919 and was assumed to have terminated.

Table 12. Available Annual Values for Millstones Produced in North Carolina Between 1886 and 1952.

Year	Value of Millstones	Year	Value of Millstones
1886	$30,000	1911	$9,099*
1887	$20,000	1912	$9,352
1888	$20,000	1913	$8,772
1902	$6,825**	1914	$5,164
1903	$5,902**	1915	$13,427*
1904	$6,500	1916	$8,520*
1905	$2,522	1917	$2,406*
1906	$1,507	1918	$39,224
1907	$1,969	1919	$29,025
1908	$4,052	1920	$14,226
1909	$22,255*	1951	$6,000
1910	$9,191*	1952	$9,285

** Figures combined for Virginia, North Carolina, Pennsylvania, and Alabama.*
*** Figures include North Carolina and Vermont.*

Table 13. Available Annual Values for Millstones Produced in Pennsylvania Between 1883 and 1919.

Year	Value of Millstones	Year	Value of Millstones
1883	$30,000	1908	$1,073
1884	$40,000	1909	$22,255*
1885	$19,000	1910	$9,191*
1886	$10,000	1911	$9,099*
1887	$5,000	1912	$1,950**
1888	$1,000	1913	$1,874**
1891	$3,801	1914	$1,304**
1902	$1,978	1915	$13,427*
1903	$1,397	1916	$8,520***
1904	$1,494	1917	$2,406***
1905	$1,351	1918	$27,802****
1906	$2,624	1919	$27,792****
1907	$2,016		

** Figures combined for Virginia, North Carolina, Pennsylvania, and Alabama.*
*** Figures combined for Pennsylvania, and Alabama.*
**** Figures combined for North Carolina and Pennsylvania.*
***** Figures combined for Virginia and Pennsylvania.*

Millstone Values in the United States

In addition to the production figures for specific states, annual values were published for the American millstone industry as a whole. These figures were included in the *Mineral Resources of the United States* (Table 14). Pratt (1901:793) estimated the value of buhrstones produced in the United States between 1880 and 1900. Phalen (1910:613) listed millstone values between 1880 and 1909. Katz (1920:216) listed millstone values from 1880 to 1917. Beach and Coons 1922:382) provided production figures for 1915 to 1919. For the period between 1920 and 1953, several authors provided the annual values for American millstones (Beach and Coons 1925:222; Chandler and Tucker 1953:95, 1958b:124; Hatmaker and Davis 1933a:113; Johnson and Davis 1937:1293; Metcalf 1943:1347; Metcalf and Holleman 1947:1366).

Table 14. Available Annual Values for Millstones Produced in the United States Between 1880 and 1963.

Year	*Value of Millstones*	*Year*	*Value of Millstones*
1880	$200,000	1922	$20,853
1881	$150,000	1923	$22,229
1882	$200,000	1924	$30,125
1883	$150,000	1925	$22,490
1884	$150,000	1926	$45,937
1885	$100,000	1927	$35,438
1886	$140,000	1928	$42,886
1887	$100,000	1929	$31,407
1888	$81,000	1930	$17,702
1889	$35,155	1931	$5,330
1890	$23,720	1932	$4,450
1891	$16,587	1933	$8,387
1892	$23,417	1934	$10,101
1893	$16,639	1935	$9,530
1894	$13,887	1936	$10,609
1895	$22,542	1937	$8,305
1896	$22,567	1938	$3,743
1897	$25,932	1939	$11,084
1898	$25,934	1940	$6,558
1899	$28,115	1941	$15,579
1900	$32,858	1942	$10,391
1901	$57,179	1943	$9,240
1902	$59,808	1944	$9,700
1903	$52,552	1945	$15,018
1904	$37,338	1946	$14,780
1905	$37,974	1947	$23,189
1906	$48,590	1948	$17,773
1907	$31,741	1949	$9,400
1908	$31,420	1950	$11,300
1909	$35,393	1951	$6,000
1910	$28,217	1952	$9,285
1911	$40,069	1953	$18,376
1912	$71,414	1954	*
1913	$56,163	1955	*
1914	$43,316	1956	$4,000
1915	$53,480	1957	*
1916	$44,559	1958	*
1917	$43,489	1959	*
1918	$92,514	1960	*
1919	$66,972	1961	*
1920	$63,325	1962	*
1921	$24,524	1963	*

* *Values were not release for years with only one producer.*

Millstone Industry Decline

Some of the fluctuations in the American millstone industry are readily explained while others appear inexplicable. Technological change was a major factor as was the country's economy (McGrain 1982). Gannett (1883:477) observed that "in the larger mills metallic rollers have to a large extent supplanted buhrstones. It is estimated that of the larger flouring mills from one-half to two-thirds are now using rollers." Two years later, Williams (1885:712) wrote, "The American stones are not used at all for grinding wheat, but only for the coarser cereals, and for the grinding of paints, cement, chemicals, fertilizers, charcoal, etc. The imported stones, being finer in grain and much harder, are used for grinding wheat and for all the better class of work. The use of rollers, as a substitute for buhrstones, is gaining ground with great rapidity. Indeed, at present, nearly all the large flouring mills in the country are adopting the roller system, and there is every probability that for many purposes buhrstones will be replaced by rollers or other grinding machinery in the near future."

In 1886, the *Manufacturer and Builder* carried an article entitled "Mineral Production of the United States in 1885" which noted, "*Millstones.*—The trade in millstones of all kinds has decreased markedly from the introduction of roller mills. The total value of the Esopus millstones in New York and Cocalica in Pennsylvania did not exceed $100,000 in 1885" (Manufacturer and Builder 1886:250).

Another problem for the American millstone industry was the foreign millstones. Raborg (1887:582) stated that "the foreign supply is well established and capable of supplying the best-known quality of stone at satisfactory rates."

Parker (1893b:748) stated that "the introduction of the roller process in flouring mills has practically shut out buhrstones, and the only demand at present is for grinding paints, cements, etc." The following year, Parker (1894:670) observed that "the introduction of emery rock millstones will probably cause a still further decline in those made from quartz conglomerate."

In his discussion of the American millstone industry for 1896 and 1897, Parker (1897:1219) wrote, "The industry compared with what it was when the first volumes of *Mineral Resources* were published is now very small, and there is no probability that it will regain its former importance. It has given way to modern invention. The roller process for the manufacture of flour has entirely supplanted the use of buhr in all large mills, and the use of stones is now confined to paint, cement, bone, and phosphate mills, and mills for grinding the coarser cereals."

Pratt (1901:792) stated that "what was a flourishing industry twenty years ago is now hardly worthy of that name.... The importance of buhrstones began to decline sharply in 1883, and there has been a gradual falling off since then." Phalen (1908:609) stated, "The market for millstones has been greatly curtailed of late years.... The explanation of this falling off in the millstone industry is due to the introduction of superior forms of grinding machinery, chiefly rolls, ball mills, etc. The roller process of grinding is now used almost exclusively in grinding wheat."

Katz (1916:551) presented the following comments on the decline of the American millstone industry:

> During the last 35 years for which the production of millstones is recorded the returns to the Survey from this industry have shown great fluctuations, which have been difficult to account for satisfactorily. It is natural to suppose that the market in the grain milling industry for millstones, made as they are from quartz conglomerate, would have declined in recent years, because of the introduction of other grinding machinery. The replacement of the millstones,

> it might be assumed, would be gradual and the value of millstones would, therefore, show a steady falling off. This, however, has not been the case.... Since that year [1894] the values have risen and fallen, as will be observed from the table of production, without any apparent rule.

In the *Mineral Resources of the United States for 1915*, Katz (1917:67) noted, "The setback suffered by the millstone market because of the introduction of modern grain-milling machinery has been offset to some extent in the last 20 years by the growing use of millstones for grinding mineral products such as feldspar, quartz, and pigments."

Katz (1926:328–329) offered the following comments on the decline of the American millstone industry for 1923: "Formerly the millstone manufacturing industry was much larger than at present.... The decline is due in part to the fact that the manufacture of millstones is a hand craft in which, as in many others in the United States, the old master craftsmen who are gradually disappearing are not being replaced. In part also the change is due to new processes in the grain, paint, and mineral milling industries in which the old-style burrstones and chaser mills are being supplanted by grinding equipment of an entirely different type."

Johnson and Davis (1936:886) noted that "although steel rolls, disks, tube mills, and other types of grinding equipment have superseded the ancient buhrstone, natural millstones are still preferred for certain uses, notably for grinding paint."

Soon after the U.S. Bureau of Mines began reporting on the American millstone industry, the industry was already experiencing decline. As noted in the preceding quotes, this decline was initiated when the grinding technology shifted from natural rock millstones to steel rollers. The longevity of the millstone industry was prolonged for a few years as other industries utilized millstones for grinding cement, pigments for paint, phosphate, bone, quartz, feldspar, and other substances. The American millstone industry was also affected by competition from foreign millstones and difficult economic times. Finally, the highly skilled old millstone makers gradually died off and were not replaced by younger men. By the mid–20th century, the American millstone industry had faded into history.

5

Foreign Millstones Imported to America

To meet the demands of American flour mills, large quantities of foreign millstones were imported into the United States (Webb 1933, 1935). Clark (1929:178) noted that "the first mill-stones used in the colonies were brought from Europe, but later many were supplied by local quarries ... but imported stones continued to be used; Cullen's were frequently mentioned, and French burrs were perhaps the most highly regarded of all." Beginning in 1868, records were kept on the quantity, quality, and value of these millstones. These stones were imported from France, Germany, Belgium, England, and elsewhere. The following discussion is based on data published by the U.S. Bureau of Mines and Minerals between 1883 and 1950 and other sources.

Gannett (1883:477) noted, "Most of the buhrstones used in this country are imported. The sources of supply are France, Belgium, and Germany, whence the material is brought, partially dressed, by steamer, at low rates of freight. The French and Belgian stones are a subaqueous deposit of silica mixed with shell. The stone is both hard and porous. The German stone is basaltic lava."

Two years later, Williams (1885:712) wrote briefly about foreign millstones: "Nearly all the buhrstones used in this country are imported. They come mainly from France, the principal locality of production there being the Paris basin. They are also imported in smaller amounts from Belgium, and Germany. The French and Belgium stones consist of silica mixed with calcareous material, and are both hard and porous. The German stones are said to be of basaltic lava. These stones are imported in the rough, at low rates of freight, and are finished in this country."

Wrote Parker (1894:671): "The decline in the buhrstone industry has not yet been confined to stones of domestic production, as the following table of imports will show. These show an almost steady decline from $125,072 in 1880 to $24,007 in 1887. There was then a moderate increase in 1888 and 1889, but the business again decreased in 1890 and 1891, reaching in the later year within $32 of the low-water mark of 1887. As in the case of domestic production the imports showed an increase in 1892."

Parker (1895:586) further commented: "Although classed as buhrstone, the domestic material is entirely distinct from any of the buhrs which are imported from France, Belgium, and Germany. The French buhr is considered the best. Both it and the Belgian buhr consist of small particles of silica mixed with calcareous material, and are hard and porous. The German buhr is said to be of Basaltic lava. The domestic stone is a quartz conglomerate. All of the foreign stone is quarried in small pieces, which are shipped in the rough state at cheap freight rates to this country where they are dressed to conformable shapes, fitted together, and bound into solid wheels."

Phalen (1911:686) stated, "the value of the imports of finished millstones appears to be very unsteady."

The following paragraphs briefly discuss the millstones imported from England, France, and Germany. Finally, the actual values for imported millstones are presented.

England

English settlers were accustomed to using conglomerate millstones quarried in Great Britain. Because of their familiarity with English millstones, they imported stones cut from Yorkshire and Derbyshire (Howell 1985:144; Howell and Keller 1977:69). Sass (1984:viii) noted that the English Peak and Grey millstones were used in the English colonies on the east coast of the United States. The *Virginia Gazette* contained an ad for Welch Peak stones on June 5, 1778 (Hockensmith 2006h).

On distinguishing the English millstones from the conglomerate millstone quarried in America, Flory (1951a:82) wrote, "The stones imported from England were similar to the native stones; however, the English prided themselves on the external finish of their stones, as well as the grinding surfaces. The top surface of the stones oft times had ornamental rings cut therein and were perfectly smooth; whereas the American stones were irregularly rough on the upper surface and balanced by lead weights attached to the binding irons, or else had a top coating of plaster of Paris several inches thick."

Flory also discussed the demise of the English millstone in America (Flory 1951a:75–76): "The earliest stones were imported from England, and the English had a monopoly on the millstone supply to the colonies and with the states until about 1800, at which time the French buhrs were introduced to America, and the development of native stone was begun. These two factors added to the trouble with England in the War of 1812, brought about a cessation of the use of English stones, never to be revived."

France

In 1886, Day (1986:428) stated that "French millstones are seldom imported as such, but the stone is shipped in comparatively small pieces which are then dressed to a uniform size and carefully fitted together, making one millstone of the ordinary form." A number of authors have discussed the qualities of French burr millstones (Cookson 2003; Mitchell 1855, 1869; The Miller 1878, Ward 1982a, 1982b, 1993a). Raborg (1887:581–582) reported:

> The best buhrstone is found in France, in the mineral basin of Paris and in a few adjoining districts, where it occurs in great masses. The stone has a straight fracture, and is not so brittle as flint, though its hardness is nearly the same. It has a white, gray, yellow, or bluish color, and varies in texture from the most open and porous to the closest quality possible. The stone sometimes appears to be filled with fresh water shells, and vegetable matter of inland growth. Some of the stone contains no organic forms at all. The stone is quarried in open air and sold in solid stones, blocks, quarters, panels, and half panels, and is usually imported in this condition, to be finished in this country.
>
> These foreign buhrstones, principally French, are shipped to the United States in all sizes from 16 up to 54 inches, and weighing from 500 to 5,000 pounds per pair. The average weight is 3,500 pounds per pair. The average price is $60 for the unfinished and $125 for the finished stones per pair.

Figure 24. *French Burr Millstone Imported into Kentucky. Specimen in private collection in Bourbon County, Kentucky. Photograph taken on August 9, 1988, by Charles D. Hockensmith, Kentucky Heritage Council, Frankfort.*

By 1890, the demand for the French stone was beginning to decline. Day (1892:456) noted that the "French buhr is still used in some flouring mills which have not adopted the roller process." In 1896, Parker (1896:927) wrote, "for fine flouring mills the roller process has supplanted domestic buhrstones, and to some extent French buhr also, which, while superior to domestic stone and procurable at comparatively slight expense, does not compete with the more modern roller process." Katz (1917:68) noted that "shortage in imports occasioned some inquiry for a domestic stone to supplant the French burr."

Howell and Keller (1977:73) reported that "by the 1750s French burr stones had become great favorites of colonial millers, particularly those engaged in the export business." Sass (1984:vii) noted that "French millstones are recorded as being imported into Virginia as early as 1620." The *Virginia Gazette* contained a number of ads for French burr millstones between 1770 and 1773 (Hockensmith 2006h). During the middle to late 19th century, American companies imported pieces of the French stone and assembled composite millstones (Figure 24). In Kentucky and Ohio, several companies were producing these local versions of French millstones (Hockensmith 2008a).

Germany

Dark bluish gray lava millstones quarried in the Mayen district of Rhineland in Germany were imported to America (Howell and Keller 1977:67). These were commonly known as Cullin stones and to a lesser extent Cologne stones, Blue stones, Rhine stones, and

Holland stones (Howell 1985:144; Howell and Keller 1977:67–69). Since these millstones were used in Holland, it is not surprising that they were imported to America by Dutch settlers (Howell and Keller 1977:69). Concerning the Cullin stones, Sass (1984:viii) commented that "they were once popular along the Eastern seaboard of the U.S.A. and worn out examples can be found at many old mill sites." Major (1982b:203) indicated that "in the United States blue stones were taken over by millers who emigrated from Europe from the time of the founding of the State of Pennsylvania." The *Virginia Gazette* ran ads that mentioned Cullen millstones between 1752 and 1778 (Hockensmith 2006h).

All Millstones Imported into the United States

The annual *Mineral Resources of the United States* publications provided information on the values of buhrstones and millstones imported into the United States. Between 1869 and 1893, and 1898 and 1922, these reports distinguished between the unfinished raw material for

Table 15. The Values of Buhrstones and Millstones Imported into the United States Between 1868 and 1899.

Year Ending	*Rough*	*Made into Millstones*	*Total*
June 30—			
1868	$74, 224	.....	$74,224
1869	$57,942	$2,419	$60,361
1870	$58,601	$2,297	$60,898
1871	$35,406	$3,698	$39,104
1872	$69,062	$5,967	$75,029
1873	$60,463	$8,115	$68,578
1874	$36,540	$43,170	$79,710
1875	$48,068	$66,991	$115,059
1876	$37,759	$46,328	$84,087
1877	$60,857	$23,068	$83,925
1878	$87,679	$1,928	$89,607
1879	$101,484	$5,088	$106,572
1880	$120,441	$4,631	$125,072
1881	$100,417	$3,495	$103,912
1882	$103,287	$747	$104,034
1883	$73,413	$272	$73,685
1884	$45,837	$263	$46,100
1885	$35,022	$455	$35,477
December 31			
1886	$29,273	$662	$29,935
1887	$23,816	$191	$24,007
1888	$36,523	$705	$37,228
1889	$40,432	$452	$40,884
1890	$32,892	$1,103	$33,995
1991	$23,997	$42	$24,039
1892	$33,657	$529	$34,186
1893	$29,532	$729	$30,261
1894	*	*	$18,087
1895	*	*	$20,316
1896	*	*	$26,965
1897	*	*	$22,956
1898	$22,974	$1,025	$23,999
1899	$18,368	$513	$18,881

**Not classified separately between 1894 and 1897.*

millstones and finished millstones that were imported. Information for 1868 to 1904 was extracted from Pratt (1905:1005), 1903 to 1907 from Phalen (1908:611), 1906 to 1910 from Phalen (1911:686), 1910 to 1915 from Katz (1917:68), and 1915 to 1919 (Beach and Coons 1922:383). Undoubtedly, the unfinished raw material was fashioned into millstones by American craftsmen. Table 15 provides the values of imported millstones for the last third of the 19th century. The values of imported millstones during the early 20th century are included in Table 16. Apparently, the steel roller mills gradually eroded the market for imported millstones. The *Mineral Resources of the United States* stopped reporting on imported millstones after 1922.

Table 16. The Values of Buhrstones and Millstones Imported into the United States Between 1900 and 1922.

Year Ending	*Rough*	*Made into Millstones*	*Total*
1900	$27,960	$944	$28,904
1901	$40,885	$1,302	$42,187
1902	$15,243	$915	$16,158
1903	$21,160	$8,481	$29,641
1904	$30,117	$2,269	$32,386
1905	$30,478	$938	$31,416
1906	$32,921	$277	$33,198
1907	$26,431	$877	$27,308
1908	$16,075	$2,567	$18,642
1909	$22,125	$465	$22,590
1910	$33,740	$1,023	$34,763
1911	$35,153	$875	$36,028
1912	$26,236	$1,326	$27,562
1913	$36,276	$3,922	$40,198
1914	$14,291	$709	$15,000
1915	$16,045	$982	$17,027
1916	$15,495	$4,321	$19,816
1917	$17,048	$1,179	$18,227
1918	$17,570	$2,417	$20,017
1919	$8,996	$17,360	$26,356
1920	$9,007	$11,947	$20,954
1921	$3,075	$10,481	$13,556
1922	$7,412	$7,944	$15,356

6

The Millstone Quarrying Industry Outside the United States

Millstones and querns have been manufactured for thousands of years for grinding grains and other substances. Through the centuries, many different types of stones have been used in the grinding process. In some cases, isolated boulders were exploited for this purpose. However, in most instances particular types of bedrock were quarried for millstones and querns. These quarrying activities left behind a variety of archaeological remains (abandoned millstones, quarry pits, underground mines, quarry faces, shaping debris, etc.). Querns were quarried at a very early date and continued in production until modern times. Both millstones and querns were manufactured concurrently at some of the quarries. The purpose of this overview is to briefly discuss the literature on millstone and quern quarries in Europe and other parts of the world (see Belmont and Hockensmith 2006; Hockensmith 1994b, 1994c, 1998, 2008f; Hockensmith and Ward 2007; Kling and Hockensmith 2008). Also, mentioned in this chapter are other studies that have examined millstones and querns as important artifacts worthy of detailed study. It is hoped that this overview will demonstrate the diversity and complexity of the millstone and quern industries in Europe and other areas of the world.

Industrial archaeologists and classical archaeologists have been studying millstones and querns in Europe for decades. In more recent years, millstone and quern quarry studies have been undertaken by archaeologists with increasing frequency. A number of these studies have focused on the quarries that produced millstones and querns. Other studies have looked at millstones and querns as artifacts recovered from archaeological sites. Many excellent articles have been published in English but many more studies are only available in French, German, Greek, Spanish, Swedish, and other languages. Much of the millstone literature has focused on quarries in Great Britain, France, and Germany. To a lesser extent millstone studies are available for several other countries. Undoubtedly, millstone quarries are located throughout the world, but only those quarries the author is aware of are mentioned in this summary.

Much valuable new information is available on the Millstonequarries.eu website (*http://meuliere.ish-lyon.cnrs.fr*). This website is supported by the French National Centre for Scientific Researches and the Rhone-Alps Historical Research Laboratory. The goal of the website is to provide documentation on European millstone quarries that scholars can consult. The website is setup so that new quarries can be added by interested researchers. On August 3, 2007, Alain Belmont (personal communication, 2007), who established the website, noted that it contained information on millstone quarries for the Czech Republic (n = 1), France (n = 133), Germany (n = 8), Greece (n = 6), Italy (n = 12), Luxemburg (n = 1), Norway (n = 6), Spain (n = 41), Sweden (n = 5), and Switzerland (n = 4). Since that time there has been a steady increase of new quarries added to the website. By April 8, 2008, there were the following frequencies of millstone quarries recorded: Czech Republic (n = 1), France (n = 197), Germany (n = 14), Greece (n = 6), Italy (n = 14), Luxemburg (n = 1), Norway (n = 6), Slovenia (= 3), Spain (n = 111), Sweden (n = 5), and Switzerland (n = 5). Nearly every week new quarries are added to the website. The website also contains information for Libya (n = 1), Scotland (n = 1), and Turkey (n = 1). The website also has some outstanding color photographs illustrating the quarries. Some of these quarries are briefly discussed under the countries listed below. Most of the quarry descriptions are currently available in French, German or Spanish. A few entries are available in English. Plans are underway to have English and German translations in the future.

Millstones have been an essential component of mills used for grinding grains since antiquity. Different cultures have utilized different stone types and different technologies for manufacturing millstones. While the millstone literature is very extensive, language barriers prevent a detailed discussion of the known publications in this overview. To adequately synthesize the available millstone quarry literature in the world would require a lengthy book and fluency in many languages. It is the intent of this chapter to briefly expose the reader to this larger body of literature. More in-depth quarry discussions are provided for those geographical areas with publications in English. The discussions are arranged alphabetically by country in the following pages.

Albania

A recent study deals with determining the sources of millstones found at Greek and Roman period sites in Albania (Gerke, Stocker, Davis, Maynard, and Dietsch 2006). Using a sample of 31 millstones, the authors analyzed the stone (vesicular volcanic rocks) to determine the origin of the raw material (Gerke et al. 2006:137). The millstones in the sample were obtained from two sites. The first site, Apollonia in west central Albania, was originally founded in the 7th century B.C. as a Greek colony and later occupied by the Romans in the middle of the 2nd century B.C. (Gerke et al. 2006:137). The second site, Butrint in southern Albania, was established by the end of the 8th century B.C. and later became part of the Roman Empire (Gerke et al. 2006:139). Samples obtained from the millstones were subjected to three types of analysis: 1) hand, 2) X-ray Fluorescence, and 3) Loss-On-Ignition (Gerke et al. 2006:140). The authors compared the chemical compositions of the millstones from Apollonia and Butrint to that of eleven millstone sources located within 800 km of Albania by sea (Gerke et al. 2006:140). This provenience study (Gerke et al. 2006:144) revealed that "[a]ll of the millstones from Apollonia derive from Sicilian sources (the Iblean Fields and Mt. Etna). Butrint millstones were quarried from Sicilian (the Iblean Fields and Mt. Etna) and Melian sources. Quarrying of millstones on Melos in Roman and Early Byzantine times was previously unknown."

Austria

Millstones were quarried at Perg in Austria. Grimshaw (1882:288) in his book *The Miller, Millwright, and Millfurnisher* stated, "The quartz sandstone of Perg, in Upper Austria, is much used for rye milling. It works better on dry than on moistened grain. Can be used for either breaking or flouring.... The Perg quartz granite is very hard and durable, but has not the sharpness of pores peculiar to the French burr-stone."

Additional literature is available on the Austrian millstone quarries. Eibensteiner (1933) wrote about the millstone industry near Perg. In an article entitled "A Survey of Millstones from Morgania," White (1963) discussed lava millstones dating between the Greek and Roman periods. Two 1950s era studies of Austrian millstones were conducted by Zirkl (1954, 1955). Zirkl (1963) later conducted a petrological study of Roman millstones from Magdalensburg in Carinthia, Austria. Of these studies, only White's 1963 article was published in English. Undoubtedly, other non–English publications exist for the millstone literature in Austria.

Canada

Little is known about the millstone quarries of Canada. Granite millstones were made in Nova Scotia in the early 19th century (Haliburton 1829:417). In her book, *Grist and Flour Mills in Ontario: From Millstones to Rollers, 1780s–1880s*, Leung (1997) provided some interesting comments on potential sources for Canadian millstones. It is suspected that the early geological literature for Canada may be a productive source for future millstone research. Millstones were imported into Canada from France, England, and the United States (Leung 1997:122–123). Concerning local millstones, Leung's (1997:122–123) summary is our best source (her note numbers have been omitted here, and her sources noted in brackets):

> Canada's mineral sources were reported to yield "millstones of an inferior quality" [Taché 1855] though useful. In 1804 Lord Selkirk observed in Oxford County "loose stones of red granite" similar to rock seen on Matchedash River (between Georgian Bay and Lake Simcoe) which were reported to make good millstones [Selkirk 1958]. In 1851, the Canadian geologist W. E. Logan wrote that "aside from the numerous and accidential granitic and syenitic boulders strewed about the country," sites for good millstones were found in the Eastern townships at Bolton, Knowlton, Stanstead, Barnston, Barford, Hereford, Ditton and Marston [Brown 1851]. One highly esteemed site of granite was at the "Vaudreuil Beauce Seigniory, near the band of serpentine" (possibly Rigaud area) [Brown 1851]. Pseudo-granite without quartz strains was found in mountains north and south of the St. Lawrence river at Sainte-Thérèse, Beloeil, Rougemont, Yamaska, Shefford and Brome. Silicious conglomerate rocks in situ were located at the Vaudeuil seigniory, the Cascades, and Point du Grand Détroit, as well as in the Gaspé at Port-Daniel and L'Ance à la Vieille. Taché's *Sketch of Canada* stated that the best rock for millstones in Canada was found in the district of Gaspé.

Czech Republic

Only three references were encountered for quern and millstone quarrying in the Czech Republic. Salac (1993) published an article entitled "Production and Exchange During the La Tene Period in Bohemia." His study focused on the manufacture and trade of quern-stones during the La Tene period (Salac 1993). Beranová (1988) published an article entitled "Manual Rotation Grain Mills on Czechoslavak Territory Up to the Incipient 2nd Millenium AD." The Oparno millstone quarry was reported in the Czech Republic on the Millstonequarries.eu website (*http://meuliere.ish-lyon.cnrs.fr*) by Stefanie Wefers (April 2007). This quarry is located at Oparno near Bezirk Litom in the Ústecký Kraj region of the Czech Republic. A volcanic rock, a rhyolite, was quarried at this location during the Middle Ages. The description of this quarry is only available in German currently.

France

The French burrstones were considered by many millers to be the best millstones ever produced. As early as the fifteenth century, the French millstone quarries were renowned and

Figure 25. *Four Millstone Extraction Depressions at the Quaix-en-Chatreuse Millstone Quarry North of Grenoble, Southeast France. Photograph taken on September 23, 2005, by Charles D. Hockensmith, Kentucky Heritage Council, Frankfort.*

the quarries continued in operation until the mid–twentieth century (Ward 1982a:205). After the mid–1800s the burr stones were removed from the quarries and taken to larger workshops where the burrs were carefully shaped to form composite millstones (Ward 1982a:207). Most of these stones were quarried at La Ferté-sous-Jouarre east of Paris (Ward 1982a).

Ward (1982b:36) described the French burrstone as "a chalcedonic hornstone or fresh water quartz found among beds of fresh water limestone that lie above the chalk" and filled with "irregularly large and small cavities which ... forms a kind of network or skeleton." Apling (1984:14) described the stone as "a decalcified siliceo-calcareous rock, a freshwater flinty limestone now a freshwater form of secondary silica, but not just plain 'freshwater quartz.'" Prior to 1836, monolithic French burrstones were common but were rapidly replaced by composite millstones comprised of shaped blocks cemented together (Ward 1982a). The author observed many of the monolithic French burr millstones that were used in a retaining wall along the La Marne River at La Ferté-sous-Jouarre (Hockensmith 2003a:17–18).

The most comprehensive book on the millstone quarries at La Ferté-sous-Jouarre is Owen Ward's (1993a) *French Millstones: Notes on the Millstone Industry at La Ferté-sous-Jouarre.* This excellent book discusses the early references to the millstones and then discusses the quarries and their locations. Next, the monolithic millstones and the composite burr stones are described in detail. Ward also discusses the extraction and the steps in constructing the burrstones. Finally, the last part of the book deals with the evolution of the industry through time. The magnitude of the industry is demonstrated by Ward's (1993a:18–23) listing of millstone

Figure 26. *Abandoned Millstone at the Second Quarry Area at the Quaix-en-Chatreuse Millstone Quarry North of Grenoble, Southeast France. A depression has been created around the millstone by cutting the rock away. Photograph taken on September 23, 2005, by Charles D. Hockensmith, Kentucky Heritage Council, Frankfort.*

Figure 27. *One of the Many Millstone Extraction Areas at the Mount Vouan Millstone Quarry in the Savoie Region of France Near Border with Switzerland. Three levels of millstone extraction depressions are visible on the cliff face. Photograph taken on September 25, 2005, by Charles D. Hockensmith, Kentucky Heritage Council, Frankfort.*

quarries which included 27 quarries to the southeast of La Ferté-sous-Jouarre, seven quarries north of La Ferté-sous-Jouarre, and 14 quarries south and west of La Ferté-sous- Jouarre. Another important book for the La Ferté-sous-Jouarre area is *Les Meuliers: Meules et Pierres Meulières* (The Millstone Industry in the Paris Basin) (Agapain 2002).

A number of studies have focused on the millstone quarries at La Ferté-sous-Jouarre. Studies published in English include Mitchell 1855; Peacock 1860; Tucker 1982b; Ward 1982a, 1982b, 1984a, 1984b, 1986a, 1986b, 1988, 1990b, 1992a, 1993a, 1995a, 1996a, 1996b, 1997, 1998, 2000, 2002a, 2002b, 2003, 2007a. Many additional millstone studies are available only in French including Agapain 2002; Anonymous 2002a; 2002b; 2002c; 2002d; 2002e; Aris 1963; Barboff 2002; Beauvois 1978, 1980, 1982, 2002a, 2002b, 2002c; Belmont 2001, 2003a, 2003c, 2006a, 2006b; Bonneff 2002; Boyer 2003; Dufrénoy 1834, 2002; Erpelding 1982; Fassbind-Bacq 2002; Gaucheron 1981, 1982, 1985, 1991a, 1991b, 2001, 2002; Geist 1991, 1995, 2003; Prevot 1975; Siguat 2002; and Spiteri 2002. The literature on the millstone quarries at La Ferté-sous-Jouarre is quite impressive and has extensively documented this world-famous millstone production center.

Millstones were also quarried at Epernon which is 67 km west of Paris (Tucker 1982b, Ward 1986a:12). These millstones were very similar to those from La Ferté-sous-Jouarre. The quarries at Epernon and the millstones produced there have been discussed by several other

Figure 28. *Millstone Extraction Area at the Mount Vouan Millstone Quarry in the Savoie Region of France Near Border with Switzerland. Note that multiple millstones were removed in a straight line as the stone was cut back into the cliff. Photograph taken on September 25, 2005, by Charles D. Hockensmith, Kentucky Heritage Council, Frankfort.*

Figure 29. *One of the Many Millstone Extraction Areas at the Montmirail Millstone Quarry in Southeast France Near Beaumes de Venise. Note the partially shaped millstone above the arch. Photograph taken on September 26, 2005, by Charles D. Hockensmith, Kentucky Heritage Council, Frankfort.*

Figure 30. *Millstone Within an Extraction Area at the Montmirail Millstone Quarry in Southeast France Near Beaumes de Venise. Photograph taken on September 26, 2005, by Charles D. Hockensmith, Kentucky Heritage Council, Frankfort.*

Figure 31. *The Cap d'Ail Millstone Quarry Located on the Mediterranean Coast at the French Riviera (Département des Alpes-Maritimes) in Southeast France. Photograph taken during March 2006, by Alain Belmont, University of Grenoble, LARHRA, Grenoble, France.*

Figure 32. *Rejected Millstone at the Cap d'Ail Millstone Quarry Located on the Mediterranean Coast at the French Riviera (Département des Alpes-Maritimes) in Southeast France. Photograph taken during March 2006 by Alain Belmont, University of Grenoble, LARHRA, Grenoble, France.*

Figure 33. *Hendaye Millstone Quarry Located in Southwest France on the Atlantic Coast (Département des Pyrénées-Atlantiques). View of quarry in the Atlantic Ocean. Photograph taken during February 2007 by Alain Belmont, University of Grenoble, LARHRA, Grenoble, France.*

Figure 34. *Hendaye Millstone Quarry Located in Southwest France on the Atlantic Coast (Département des Pyrénées-Atlantiques). Photograph taken during February 2007 by Alain Belmont, University of Grenoble, LARHRA, Grenoble, France.*

authors (Beauvois 1980, 1982; Duc 2003, 2005; Tucker 1982b). In 2005, an excellent book, *Carriers et Meuliers de Région d'Epernon*, was produced by Duc (2005) which describes the quarries at Epernon and includes many wonderful historic photographic of the millstone industry. In his review of Duc's book, Ward (2005:43) remarked, "The *pierre meulière* was of the same nature as around La Ferté-sous-Jouarre and of such good quality that some considered it to be superior to that of La Ferté. But none of the millstones ever bore the name

Figure 35. *Rejected Millstone at the Hendaye Millstone Quarry Located in Southwest France on the Atlantic Coast (Département des Pyrénées-Atlantiques). Photograph taken during February 2007 by Alain Belmont, University of Grenoble, LARHRA, Grenoble, France.*

Epernon; either the burrs were sent to La Ferté or else the complete stones were built at Epernon by firms that came under the aegis of older establishments in La Ferté and were sold under their name."

Apling (1984:14) provided the following comments about Epernon:

> The finest burrs for millstones — the hardest and least porous — came from a bed called "Calcaire de Brie," which stretches from Vernon, between Paris and Rouen, to Rheims. The principal quarries were those of Bois de la Barre and La Justice du Nord at La Ferté sous Jouarre on the River Marne about 60 km. east of Paris. There were also quarries further east along the river at Épernay.
>
> Other sources of burr stone were in the Departments of Vienne and Sarthe, at Tours and at Bergerac in the Dordogne east of Boreaux. However, stones from La Ferté were regarded as being of the finest quality for millstones, with the others being softer and more porous, requiring to be dressed more frequently and not lasting so long.
>
> On extraction from the quarries, the blocks of stone were dressed on one side by *Epaneures* to discover their qualities for grading purposes. The blocks known as *panneaux*, became "panes" in English. The medium sized panes would be some 14–16 ins. long, by 12 ins. broad and 6 ins. thick. The best quality close hard stones were known in France as "English," the large open porous stones as "French" with the medium quality as "Semi-English." The panes were imported here as separate pieces and were made up into millstones by local millwrights and later by specialist firms. The plaster of Paris on which they were bedded, imported with the stones, was calcined gypsum from nearby deposits of the same geological age. All French burr stones were fitted with iron bands which had to be knocked down as the stone wore away. These were put on cold — not hot and shrunk on.

In addition to the research on La Ferté-sous-Jouarre and Epernon, the millstone quarries of southeast France in the Alps near Greenoble have been studied in detail. Alain Belmont (2006a) recently published a two volume study entitled *La Pierre à Pain: Les Carrières de Meules de Moulins en France du Moyen Age à la Revolution Industrielle*. This book deals with the millstone quarries all over France dating from the Middle Ages to the Industrial Revolution. Professor Belmont (personal communication, 2004) indicated that there were possibly 500 millstone quarries within only 100 kilometers of Grenoble. He noted that only the largest quarries were included in his book (Belmont 2006a) while most of the smaller quarries are not well documented (Belmont personal communication, 2004). Harverson (2006:43) published an excellent English review of Belmont's book, calling it "a magnificent contribution to mill studies, so far unmatched in any other country."

A recent publication by Belmont (2006b:90) provided an overview of the Medieval millstone quarries in the Dauphiné, a former principality, in southeast France. Belmont's (2006b:90) research revealed that several hundred millstone quarries were located in this area of the French Alps by the end of the Middle Ages. However, he pointed out that only a small number of these quarries produced good quality millstones that were exported on a regional scale (Belmont 2006b:90). Other studies on the millstone quarries near Grenoble, France, in the Alps include the following publications: Belmont 2001, 2002a, 2002b, 2002c, 2002d, 2003a, 2003b, 2004, 2006b; Belmont and Bois 1998; and Martel 1973.

Many studies have been published on the millstone quarries in other regions of France. Sandstone millstones were quarried in different areas during various time periods. The millstone quarries at Ardèche, Coux, and Aubenas, France (which includes sandstone millstones), were discussed by Azéma, Meucci and Naud (2003). Bulliott (1888) discussed the sandstone millstones from St. Emilon near Autun, Saône-et-Loire, in east central France dating from the Gallo-Roman period to the Middle Ages. Sandstone millstones were also made by the Romans in the L'Oise Valley on the France/Belgium border (Chambon 1954; Jottrand 1895).

Figure 36. *Montagne-Saint-Emilion Millstone Quarry Located in Southwestern France in the Bordeaux Vineyard (Département de Gironde). View of the entrance of underground quarry. Photograph taken during June 2007 by Alain Belmont, University of Grenoble, LARHRA, Grenoble, France.*

Figure 37. *Montagne-Saint-Emilion Millstone Quarry Located in Southwestern France in the Bordeaux Vineyard (Département de Gironde). Closeup view of quarry face and abandoned millstones. Photograph taken during June 2007 by Alain Belmont, University of Grenoble, LARHRA, Grenoble, France.*

Laville (1963) discussed the production centers of Gallo-Roman millstones within sedimentary rocks (sandstone) in France. A large modern sandstone millstone quarry was located in central France (Cabezuelo, Connier, and Gauthier 2001).

Several studies have examined millstones and querns made from flint, granite, and limestone. Coutard (1998) reported on the modern flint millstones quarries located near Villaines-la-Gonais. The extensive millstone quarry at Périgord, France, produced flint millstones as early as the Middle Ages to the mid–20th century (Gibert 1987). Flint millstones were also quarried at Campagne near Reims during modern times (Guérin 1985, 1990, 1991). Granite millstones and querns were manufactured at Normandy, France (San Juan, Gasnier, and Savary 1999). Durham (2006:3–4) discussed granite querns recovered from the excavations at Le Yaudet at Brittany in northern France. Limestone millstone quarries have been studied at Claix and Angoulème (Agard 2003, 2005).

Longepierre (2006) recently reported on the discovery of significant millstone quarries at Saint-Quentin-la-Poterie in the Gard area of southern France which date to Roman times. These quarries exploited a "micro conglomerate" deposit and the unfinished millstones were transported to workshops located between rural settlements where they were completed (Longepierre 2006:54). Archaeological excavations have been undertaken at the workshops associated with these quarries (Longepierre 2006:54). Longepierre (2006:54) reported that the quarries of Saint-Quentin-la-Poterie began following the decline of the millstone quarries at Agde (Hérault) which exploited basalt. Longepierre (2004) also published an earlier study of Saint-Quentin-la-Poterie.

Another recent publication provides information on the millstone quarries of Marèze including Saint-Martin-Laguépie and Riols in southern France at Tarn (Servelle 2006:69). Beginning in the Neolithic era and continuing until the end of the prehistoric period, sandstones and micro conglomerates were utilized for manufacturing grinding stones and rotary querns. Servelle examined the geological and geomorphological restrictions on extracting blocks of stone for producing performs. Finally, he looked at extraction areas, structures, and stone mason workshops found over an area greater than 15 hectares (Servelle 2006:69).

Cabezuelo (2006:114) presented the results of a survey conducted in the locality of "Les Meules" at Vic-le-Comte (Puy-de-Dôme) in central France in the general area of Clermont-Ferrand. Cabezuelo located two quarrymen's shelters and many unfinished abandoned millstones. He noted that millstone quarrying was one of the principal industries in this region during the 18th century (Cabezuelo 2006:114).

Recent studies have been conducted for underground millstone mines in southern France. Azéma (2006:161) has undertaken an inventory of the millstone mines in Tarn (n = 9) and Aveyron (n = 2) in southern France. Azéma notes that the most impressive example of these mines is the site of Clairac. The scientific study of these mines and the development of millstone production is yet to be undertaken (Azéma 2006:161). Bailly-Maitre (2006:170) examined millstone quarrying and mining techniques in southern France. When looking at the types of stone deposits, she found that "the techniques for organizing the underground space are often the same" (Bailly-Maitre 2006:170).

During a six day tour of mills and related sites during June of 2006, 40 members of TIMS visited many mills, three museums, and a millstone quarry in southwest France as reported by Hawksley (2006:2). The millstone quarry at Clairac (see Azéma 2006 above) produced millstones from a silicaceous limestone. According to Hawksley's report, the quarry began operation in the 13th century and continued until 1836. Quarrymen created a cave by extracting rock from an area 60 m wide, 60 m deep, and 12 m high with pillars left to support the roof. Hawksley indicates that the workers created cylinders that were split into individual millstones. The quarry contains millstones abandoned in different stages completeness and a nearly complete millstone measured over 1.5 m in diameter (Hawksley 2006:9).

Gibbings (2006:32–33) shared the results of his archival research which provided new insight into the trade of millstones in Côtes d'Armour. The records that he found dated from approximately A.D. 1400 to 1800. He noted that millstones were being imported from the Brie region, east of Paris to northern Brittany, a distance of about 600 km in northwest France. Gibbings found that millstones were primarily shipped to the ports of Saint-Malo and Le Légué but also to the ports of Vannes, Dinan, Port à la Duc, Saint-Cast, and Dahouet. These early records also contained information on the prices of millstones, transportation, and financial risks (Gibbings 2006:32–33).

Herrscher et al. (2006:108) conducted a study looking at dental pathologies and the use of millstones. The authors selected 125 individuals who had lived in the Grenoble district between the 13th and 18th centuries. The wear on the teeth was examined for different periods. It was assumed that dental wear would decrease through time as better millstones became available (Herrscher et al. 2006:108).

Many other diverse millstone and quern studies have been undertaken at French quarries and archaeological sites. Bruggeman (2003) studied the millstone quarries in Flanders (border of northwest France and Belgium) dating between the Middle Ages and the Renaissance. Several authors (Deffontaines 2002, 2003; Durand-Vaugaron 1969; Kleinman 1984) described the production of millstones at Cinq-Mars-la-Pile, France. Millstones were also made at Savoie, France (Deschamps 2003). Eck (1901) studied the Gallo-Roman millstones between Paris and Rheims, Seine-et-Marne, France. Millstones were also quarried on the Born Plain at Dordogne, France (Lacombe 2003). Belmont (2005) has also studied the French quarries near the German border and the quarries near Carcassonne, France, near the border with Spain (Belmont, Allabert, Marty and Zanca 2003). Recently, Boyer et al. (2006:13) published an article dealing with the production and diffusion of querns in France during the Neolithic to Antiquity. Several studies have focused on querns from Roman sites (Boyer and Buchsenshutz 2000; Cantet 1995; Lacroix 1963; Py 1992) and Celtic querns (Boyer and Buchsenshutz 1998). One recent study focused on early quern quarries in four areas of France: Vosges, Franche-Comté, Picardy, and Burgundy (Boyer et al. 2006:13). Another study was conducted by geologists to determine the petrographic qualities of the different stones used in France for millstones (Fabre et al. 2006:97). The goal of this study was to understand the selection of certain stones for millstone from the Middle Ages until the 19th century (Fabre et al. 2006:97). Ward (2007a:52–54) provided commentary and additional information in response to two TIMS publications. He briefly comments on a millstone quarry at Clairac in southwest France and the millstone trade in the Côtes d'Armor area in northwestern France (Ward 2007a:52–54). Lorenz and Wolfram (2007) published an article entitled "The Millstones of Barbegal" which discussed millstones from Roman ruins in southern France.

Many additional studies on millstone quarries and millstones are available only in French. These include: Allabert, Marty, and Zanca 2000; Amouric 1990, 1991 (at Provence); Andrieu 1990 (at Langeac); Anonymous 2002a, 2002b, 2002c, 2002d, 2002e; Aris 1974; Bajulaz 1987; Beauvois 1978, 2002c; Bottin 1905; Buchsenschutz and Pommepuy 2002; Carcauzon 1986 (at Saint-Crépin-de-Richemont); Chevillot et al 2005 (at Saint-Crépin-de-Richemont); Claracq 1994; Clochard 2004; Coquebert-Montbret 1796; Cordier 1807; Couston 1982; Coustet, Valette, and Alos 2001; Coutard 1998; Delmas 1975; Dendaletche and Saint-Lebe 1997; Desirat 1985; Dubech and Gaborit 2006; Ducassé 1995; Duhard 1996; Gandilhon 1986;

Geist 1995, 2003; Guettard 1758, 2002; Jean 1989; Kleinmann 1984; Lacombe 2000, 2003; Lechevin 1975; Longepierre 2004; Lorenz, Turfland, and Boitel 1992; Mallet 1995; Mary 2003; Menillet 1985; Moreau and Ruel 2003; Oliva, Béziat, Domergue, Jarrier, Martin, Pieraggi, and Tollon 1999; Ortuno 1999; Palausi 1965; Passemard, Laoust, and Bourrilly 1923; Peacock 1860; Piot 1871; Prevot 1975; Sauldubois 2003; Triboulet, Langouet and Bizien-Jaglin 1996; and Valero 1983, 1984. See Belmont and Hockensmith (2006) for an extensive annotated bibliography of the literature on French millstone quarries and millstone studies that includes several studies not cited in this overview.

At this writing, 197 French millstone quarries can be viewed on the Millstonequarries.eu website (*http://meuliere.ish-lyon.cnrs.fr*). While most of the regions within France have at least one recorded millstone quarry, the highest number of quarries are located within the southern half of the country. The distribution of these quarries by region, from north to south, are as follows: Haute-Normandie (n = 1), Picardie (n = 2), Champagne-Ardenne (n = 6), Lorraine (n = 2), Alsace (n = 2), Pays de la Loire (n = 2), Centre (n = 14), Ile de France (n = 8), Bourgogne (n = 4), Franche-Come (n = 8), Poitou-Charentes (n = 12), Auvergne (n = 3), Rhone-Alpes (n = 64), Aquitaine (n = 25), Midi-Pyrénées (n = 7), Lanqueduc-Roussion (n = 20), and Provence-Alpes Côte d'Azun (n = 17). Most of these quarries were reported by Alain Belmont while a few were reported by other French scholars. Since there is so much literature published on the French millstone quarries and the website includes many previously published quarries, no attempt will be made to comment on these quarries here.

The reader is directed to the Millstonequarries.eu website (*http://meuliere.ish-lyon.cnrs.fr*) to review specific information for each quarry and to view photographs of the quarries. The author has personally visited the Quaix-en-Chatreuse Millstone Quarry north of Grenoble (Figures 25–26), the Mount Vouan Millstone Quarry in the Savoie Region of France near the border with Switzerland (Figures 27–28), and the Montmirail Millstone Quarry in southeast France near Beaumes de Venise (Figures 29–30). Professor Alan Belmont graciously shared photographs of several French quarries: the Cap d'Ail Millstone Quarry (Figures 31–32) on the Mediterranean coast at the French Riviera (Département des Alpes-Maritimes) in southeast France; the Hendaye Millstone Quarry (Figures 33–35) in southwest France on the Atlantic coast (Département des Pyrénées-Atlantiques); and the Montagne-Saint-Emilion Millstone Quarry (Figures 36–37) in southwestern France in the Bordeaux vineyard (Département de Gironde).

Germany

Millstones were quarried in different areas of Germany. Perhaps the best known area was the Eifel region of West Germany that began producing millstones as early as Roman times and continued until the modern era. Millstone quarries were located near the communities of Mayen (Figure 38), Mendig, Kottenheim, Gerolstein, and Daun (Major 1982b:194). These underground quarries or mines exploited areas containing volcanic deposits. Very porous dark blue basalt was used for millstones (Howell 1985; Major 1982b). These quarries were exploited as early as 1200 B.C. for saddle querns (Major 1982a:343). During the Roman period, double-coned millstones for donkey mills were quarried (Major 1982a:343). The modern millstones were commonly known as Cullen stones but were also known as Holland stones, Blue stones, Rhine stones, and Cologne stones (Howell 1985:144).

Over a century ago, Raborg (1887:581–582) stated that: "German buhrstone comes from a district on the Rhine, near Cologne. It is a basaltic lava, and is found near old craters of extinct volcanoes, at a depth of from 100 to 150 feet under ground. A shaft is sunk and the

Figure 38. *Millstone Quarries of Mayen, Germany (Rhineland-Palatinate), Towards the End of the 19th Century. Hundreds of years of production resulted in huge piles of debris. The photograph shows millstones, workers, and working shelters constructed with the production debris. To the right are two old style wooden cranes used to lift the millstones from the mines. In the foreground a more modern quarry is partially filled with debris. From the Archives of the Geschichts- und Altertumsverein Mayen, Germany. Photograph provided by Fritz Mangartz, the Römisch-Germanisches Zentralmuseum, Mainz, Germany.*

stone quarried out so as to leave natural columns to support the earth above. The stones increase in size with the depth. The structure is very uniform and of a dark blue color. It is too soft for grinding wheat."

A recent paper by Mangartz (2006) provided an overview of millstone production in the eastern Eifel region of West Germany. Mangartz (2006:34) noted that the volcanic fields of Eifel were exploited for querns and millstone from prehistory to the Middle Ages. More than one hundred places have been identified where millstone and quern were quarried (Mangartz 2006:34). Mangartz and his colleagues have previously documented archaeological remains associated with the millstone industry (Harms and Mangartz 2002; Mangartz 1993, 1998, 1999, 2003a, 2003b, 2005, 2006a, 2006b, 2006c; Mangartz and Pung 2002). Joern Kling shared several photographs of millstone quarries in the Mayen area (Figures 39–43).

Millstone quarries were also located in eastern Germany at Jonsdorf near Dresden (Bost 2002:30–32; Bauer; Müller and Lorenz 2002; Ruben 1954, 1961, 1967, 1973; Sitte 1954). Situated on the border with the Czech Republic, Jonsdorf was home to the millstone industry between 1580 and 1918 (Bost 2002:31). The quarries exploited sandstone that had been changed by contact with glowing magma (Bost 2002:31). This sandstone was both very hard and porous (Bost 2002:31). Millstones manufactured at Jonsdorf included both monolithic and composite types (Bost 2002:31). Bost (2002:31) illustrated a composite millstone which was made from seven shaped segments. Based on the hardness and purity of the stone, eight dif-

Figure 39. *Entrance of a 19th Century Millstone Quarry at Mayen, Germany. This passage was exposed by modern quarrying in the 1950s. The height of the opening is about 9 meters. Since the late 19th century, the abandoned quarry was used as stocking place for a brewery. Photograph taken during 2007 by Joern Kling, Bonn, Germany.*

Figure 40. *A Small Pillar Supporting the Roof of an Early 19th Century Mayen, Germany, Millstone Quarry. The roof height is about 3 meters. In the background is a larger chamber where millstones were removed. Photograph taken during 1996 by Joern Kling, Bonn, Germany.*

Figure 41. *An Old 18th Century Millstone Quarry at Mayen, Germany. As typical for the old quarries, the discarded stone fragments have been piled up to the roof. The height of the passage is about 6 meters. Photograph taken during 2005 by Joern Kling, Bonn, Germany.*

Figure 42. *Narrow Steps Providing Access to the Underground Millstone Mines at Mayen, Germany. Such stairs are rare in the Mayen area and were installed just before the end of the 18th century. At the Niedermendig field, an old plan shows over 40 of such mine entries. They were used until the end of the mining, about 1960. Photograph taken during 1996 by Joern Kling, Bonn, Germany.*

ferent qualities of millstones were produced at Jonsdorf (Bost 2002:32). They were shipped as far as Russia and England and were used in corn mills as well as edge runners in bone mills, oil mills, tanbark mills, and color mills (Bost 2002:31–32). The Jonsdorf quarries were in direct competition with those in the Eifel (Bost 2002:32).

Falkenstein (1983, 1988a, 1988b:19) reported on a millstone mine located in the German town of Waldshut-Tiengen on the Rhine River close to the border with Switzerland. He felt that there may be a number of these mines in the valley in the Liederbachtal (Falkenstein 1988b:19). It is thought that the millstone mines were worked as early as the 15th century and

Figure 43. *An Open Shaft into a Millstone Quarry at Mayen, Germany. The height is about 20 meters. Over 450 such shafts were excavated to reach the basaltic lava. At Mayen, where the lava has been covered by only 8 meters of volcanic ashes, shafts were excavated every 15–20 meters. Photograph taken during 2006 by Joern Kling, Bonn, Germany.*

Figure 44. *Reconstruction of an "Göpelwerk" at Niedermendig, Germany, by a Private Association. Horse powered machines of this type were used to lift millstones from the underground quarries until the introduction of electric machines in the early 20th century. At Niedermendig the shafts are 30 meters deep. Photograph taken during 2004 by Joern Kling, Bonn, Germany.*

Figure 45. *Collection of Millstones at Niedermendig, Germany, Dating to the Mid–19th Century. The specimens on the right side are small handmills (querns), the center specimen is a millstone, and the specimen on the left is the base for an oil mill. Photograph taken during 2005 by Joern Kling, Bonn, Germany.*

Figure 46. *Pillar Constructed from Abandoned Millstones in an Old Brewery at Niedermendig, Germany. The height is about 7 meters. Unsafe areas in old millstone mines (later used as brewery stocking places) were reinforced with additional pillars. The walls in the background were constructed to protect the brewery stock from robbery. Photograph taken during 2005 by Joern Kling, Bonn, Germany.*

that there was competition from other mines (Falkenstein 1988b:19). The raw material used is not specified but one photographic caption mentions sandstone (Falkenstein 1988b:22). Of particular interest is an English translation of an edict published by the city of Waldshut in 1531 that specified some obligations and responsibilities for millstone makers (Falkenstein 1983:1–2, 1988b:19). The article does not discuss the longevity of the millstone mine or the sizes of millstones produced. Falkenstein (1986, 1987, 1989, 1997, 2001; Falkenstein and Koerner 1989) has published many additional details concerning the millstone mines at Waldshut in his studies written in German. His 2001 article included information on both the millstone quarries and mines in the general area around Waldshut (Falkenstein 2001:227–232).

Grimshaw (1882:288) provided the following information on several types of German millstones:

> The Münden stone is a white, sharp, fine porous sandstone, used in north Germany on rye. By reason of its grindstone-like action, it is generally used working against Crawinkler stone.
>
> The Jonsdorfer stone is like the Münden. It is found near Zittan, and used in Bohemia in large runs for hulling and in small ones for rye grinding.
>
> Rhine stone (lava or basalt) comes from Andernach, is of a dark, grayish blue color, very porous, and was once much demanded for wheat milling. This stone is not too hard, and dresses easily and sharply. Somewhat like it is the stone from Volvic, in Auverge.
>
> The Crawinkler and Ohrduffer quartz porphyries, come in very hard but porous masses. The harder are used for wheat, and the softer on rye. The color is gray, sprinkled with feldspar crystals.

Several recent studies of millstones and querns have been published in German but with English abstracts. Wefers (2006) compiled a catalogue of rotary querns from the La Tène period in Germany from Hessen to north of the River Main. She identified quarries, established a typology for the querns, and looked at transportation routes (Wefers 2006:24). Kling (2006) inventoried the millstone mines of Niedermendig, Germany. These mines exploited a basaltic lava between the early Middle Ages and the 20th century. Maps suggested that the millstone quarries moved from west to east over time. Kling documented and mapped two areas as a sample. His work "allowed a reconstruction of the underground extraction fields and of the dynamics of extraction as well as a determination of the associated extraction shafts" (Kling 2006:144). Kling shared several photographs of the millstone quarries at Niedermendig, Germany (Figures 44–48). Another inventory, undertaken for the millstone mines of Eifel and Luxembourg, has documented about 110 millstone mines to date and is still ongoing (Kremer 2006:154).

Some major publications are available for the millstone industry in Germany. An excellent overview of the millstone industry in the Eifel region of West Germany is presented in Fridolin Hörter's (1994) book entitled *Getreidereiben und Mühlsteine aus der Eifel*. Another major publication on the German millstone industry is Eduard Harms and Fritz Mangartz's (2002) book entitled *Vom Magma zum Mühlstein: Eine Zeitreise Durch die Lavaströme des Bellerberg-Vulkans*, which deals with the basalt millstone industry in the volcanic area at Mayen, in northwestern Germany. Also, Dankmar Leffler's (2001) book entitled *Das Crawinkler Mühlsteingewerbe: Zur Geschichte eines der ältesten Gewerbe im Thüringer Wald* deals with the millstone industry at Crawinkel, Germany. Leffler illustrated this book with many excellent photographs including a number of views showing millstone workers in 1928.

Fourteen German millstone quarries are described in French on the Millstonequarries.eu website (*http://meuliere.ish-lyon.cnrs.fr*). The Barsinghausen quarry is located in the Niedersachsen region near Barsinghausen where millstones were made from sedimentary rocks

Figure 47. *Millstone Quarry Passage at Niedermendig, Germany, Exposed by Modern Mining in the 1970s. Note the five/six-sided basaltic lava pillars. At Niedermendig, these pillars have heights ranging up to 15 meters. The tops of the pillars support the roof which is covered by 20 meters of ashes from the "Laacher See Volcan." The height of the passage is about 5 meters. Photograph taken during 1995 by Joern Kling, Bonn, Germany.*

Figure 48. *Typical Steps Used to Access a Millstone Quarry at Niedermendig, Germany, During the Middle Ages. In the early Middle Ages, millstone manufacture was accomplished in open quarries. Today the old quarry pits are exposed by modern quarrying activities. Photograph taken during 1996 by Joern Kling, Bonn, Germany.*

between the 12th and 20th centuries (information provided by Alain Belmont). Millstones were made during antiquity from sedimentary rocks at the Schweigmatt quarry (information provided by Timothy Anderson and Cornel Doswald) in the Baden-Württemberg region near Schopfheim (Anderson, Agustoni, Duvauchelle, Serneels, and Castella 2003; Joos 1975). The remaining quarries are located in the Rheinland-Pfalz region of Germany. The Hochstein quarry (information provided by Alain Belmont) is located near Obermendig and produced millstones from volcanic rock during the Middle Ages (Hörter 1994; Mangartz 1993). Basalt millstones were manufactured at the Niedermendig quarry (information provided by Alain Belmont) near Mendig between the 15th and 20th centuries (Harms and Mangartz 2002; Hörter 1994; Major 1982a, 1982b). The Mayen quarries (information provided by Alain Belmont) are famous and produced basalt millstones from the Neolithic period to the 20th century (Harms and Mangartz 2002; Hörter 1994; Major 1982a, 1982b). Millstones were made in the 18th and 19th centuries from volcanic rock at the Birresborn quarry (information provided by Alain Belmont) near Birreson at Eishöhlen (Hörter 1994). The Andernach, Hohen Buche quarry (information provided by Alain Belmont) exploited basalt for millstones during antiquity, the Middle Ages, and 18th to 19th centuries near Andernach-Namedy (Hörter 1994; Mangartz 1999). The Hohenfels-Hensigen, Mühlenberg quarry (information provided by Alain Belmont) near Hohenfels-Hensigen produced basalt millstones from the Middle Ages to the 19th century (Hörter 1994). Finally, information was provided on the Borken quarry at Stadtwald in the Hessen region of Germany on the Millstonequarries.eu website (*http://meuliere.ish-lyon.cnrs.fr*) by Stefanie Wefers (September 2007). A volcanic rock was quarried at the Borken quarry for rotary handmills during the La Tène period.

The last five German quarries on the Millstonequarries.eu website (*http://meuliere.ish-lyon.cnrs.fr*) were reported by Fritz Mangartz in November 2007. The first three quarries are located in the Rheinland-Pfaltz region. The Andernach-Eich, Sattelberg, quarry was located near the community of Andernach-Eich. During the 18th and 19th centuries, a volcanic rock from the Sattelberg volcano was exploited for millstones (Hörter 1994). About 100 millstones were made at this quarry which is no longer visible. The second quarry, Bad-Bertrich, Seesenflürchen, was used during the 19th century. A volcanic rock from the Seesenflürchen lava flow was exploited for a small number of millstones (Hörter 1994). The Bad-Bertrich-Kennfuß, An den Dachslöchern, quarry exploited the lava flow of the Dachskopf volcano. The quarry was located at An den Dachslöchern near the community of Bad-Bertrich-Kennfuß (Hörter 1994). The Bad Godesberg-Mehlem, Rodderberg, quarry was visible in the scoria cone of the Rodderberg volcano until the 19th century. Situated in the Nordrhein-Westfalen region near Bad Godesberg-Mehlem, it contained a leucite quarried for millstones. This quarry was destroyed by more recent quarrying (Hörter 1994). Finally, the Jonsdorf quarry mentioned above was added to the website by Carmen Schaupp in 2008.

The literature on the German millstone quarries is quite extensive. Several studies have been published in English as previously noted including: Bost 2002 (quarries at Jonsdorf); Crawford 1953, Crawford and Röder 1955 (querns and quern quarries at Mayen in the Eifel); Falkenstein 1983, 1988a, 1988b (the millstone Mines of Waldshut); Major 1982a, 1982b (millstones in the Eifel Region); Mangartz 2006a (quernstone production from the Prehistoric to Medieval periods near Mayen); Parkhouse 1976, 1977, 1991, 1997 (German querns found in England and elsewhere in Europe); and Smith 1978 (German found in the Graveney Boat). However, the majority of the millstone studies have been published in German or other languages including Dohm 1950; Falkenstein 1983, 1986, 1987, 1988a, 1988b, 1989, 1997, 2001; Falkenstein and Koerner 1989; Faujas de Saint-Fond 1802; Harms and Mangartz 2002; Holtmeyer-Wild 2000; Hörmann and Richter 1983; F. Hörter 1979, 1990, 1994, 2003, F. Hörter, Jr. 1950, P. Hörter 1914, 1917, 1922; Hörter, S. 1942; Hörter, Michel and Röder 1950–1951, 1954–1955; Kars 1983; Konschak 1996; Krämer 1948; Laumanns 1987; Leffler 2001, 2002, 2003; Mangartz 1993, 1998, 2003a, 2003b, 2005; Mangartz and Pung 2002; Mason 1950; Moog 1989; Müller and Lorenz 2002; Reichstein 1987; Röder 1955, 1956, 1970, 1972; Ruben 1961; Schaaff 2006; Schmandt 1930; Schön 1989, 1995; Schüller 2006; Schulze 1828; and Siegert 1921. Joern Kling and the author (Kling and Hockensmith 2008) are currently preparing an annotated bibliography of the millstone and quern literature for Germany.

Great Britain

Many studies have been published on the querns, millstones and their quarries in Great Britain. Because of this excellent coverage, it is possible to discuss these studies by the individual countries comprising the United Kingdom: England, Ireland, Scotland, and Wales. The following paragraphs highlight this research.

England

In England, querns and millstones were manufactured during different periods in history. The earliest rotary querns were manufactured during the Iron Age. A quarry located near Lodsworth at West Sussex produced some Iron Age querns (Peacock 1987). Peacock (1987:70) noted that Iron Age querns were about 115 mm thick which made them thicker than Roman querns (about 75 mm thick). Querns were also studied from an Iron Age and Medieval site at Brooklands at Weybridge (Tomalin 1977). At Gravelly Guy in Oxfordshire, Iron Age querns were recovered during the excavations conducted at this site (Bradley, Roe, and Wait 2004). Iron Age querns were also documented in the Tees Valley of York (Gwilt and Heslop 1995). Wisdom (1981) conducted a typological and spatial study of querns made from the quartz conglomerate found within the Forest of Dean. The Iron Age hamlet of Ashton Keynes in Wiltshire produced querns (Saunders 1997). Querns and mortars dating to the Iron Age were recovered from Cornwall (Watts 2003a). Finally, Spain (1986) dealt with millstones recovered from Barton Court Farm at Abingdon, Oxon, dating to the Iron Age and other periods.

A number of studies have focused on the beehive style querns found in the United Kingdom. Watts (2002:27) has noted the beehive quern came into existence during the Iron Age. These thick rotary querns represented a significant technological development in milling (Watts 2002:28). Hayes, Hemingway, and Spratt (1980) studied the distribution and lithology of beehive querns in northeast Yorkshire. Ingle (1987) examined the production and distribution of beehive querns in Cumbria and two years later submitted her Ph.D. thesis entitled “Characterisation and Distribution of Beehive Querns in Eastern England” (Ingle 1989). The manufacture of beehive querns in the southeast Pennines was reported on by Wright (1988). Heslop (1988) and Chapman (1990) also studied beehive querns. Two types of beehive querns made from Old Red Sandstone were discussed by Shaffrey (2006a:37).

Many archaeological studies have dealt with querns produced during the Roman occupation of Britain. As previously noted, Peacock (1987:62) documented a quern quarry located near Lodsworth at West Sussex that exploited a sandstone from the Greensand formation that was used to produce Roman rotary querns. At the east end of the Pitts Copse area, quern fragments were found among the rubble, and chipping debris was uncovered when a nearby quarry mound was opened (Peacock 1987:63). Peacock’s (1987) study sampled different geo-

Figure 49. *Small Unfinished Millstone (Scale 10 cm) Lying Under a Wall by Access Track Near Fell End Farm, Quernmore, Lancashire. Photograph taken during May 1986 by Phil Hudson, Procter House, Settle, Yorkshire, United Kingdom.*

Figure 50. *View of a Roughed Out and Partly Worked Millstone Blank (size 76 inches in diameter and 16 inches thick). It has fallen off its props, and is possibly too damaged to warrant further work. Seen as evidence of early, possibly late mediaeval, working of the Namurian Grits to manufacture millstones at Baines Cragg in Littledale, Lancashire. Photograph taken on December 12, 1986, by Phil Hudson, Procter House, Settle, Yorkshire, United Kingdom.*

Figure 51. *View from top of Baines Cragg, Littledale, Lancashire. Shows knapping areas, abandoned partly worked blanks and other debris/remains of millstone working and quarrying. Photograph taken during December 1986 by Phil Hudson, Procter House, Settle, Yorkshire, United Kingdom.*

Figure 52. *Very Weathered Chisel Marks on a Large Block of Gritstone Taken out of the Quarry Face on Baines Cragg, Littledale, Lancashire. Photograph taken during November 1986 by Phil Hudson, Procter House, Settle, Yorkshire, United Kingdom.*

Figure 53. *Small Unfinished Millstone (scale 10 cm) Lying in Knapping Debris Under Baines Cragg, Littledale, Lancashire. Photograph taken during November 1986 by Phil Hudson, Procter House, Settle, Yorkshire, United Kingdom.*

logical outcrops of the Greensand formation in an attempt to match the material used for querns. His study stressed petrology and typology to examine the distribution of the querns in southern England (Peacock 1987). Shaffrey (2006a) studied 1,200 rotary querns from 180 sites in the United Kingdom which were manufactured from the Old Red Sandstone. She collected reference samples from various rock outcrops and then examined querns curated from earlier archaeological investigations. Petrological study of these querns allowed the identification of the geological formations from which they were made (Shaffrey 2006a). Shaffrey's (2006a) research represents a major study that contributes much new information on the Romano-British rotary querns and millstones manufactured from the Old Red Sandstone.

A few studies deal with querns recovered from Roman period archaeological sites. These studies include querns recovered from Alfoldean at Slinfold (Aldsworth 1984), Gravelly Guy in Oxfordshire (Bradley, Roe, and Wait 2004), the Roman hamlet of Ashton Keynes in Wiltshire (Saunders 1997), Trethurgy Round in Cornwall (Quinnell 2004), Roman Alcester (Evans 1994), the Roman Fort of Vindolanda at Chesterholm in Northumberland (Welfare 1985), a Romano-British roadside settlement in Somerset (Roe 2001), a Romano-British villa estate at Mantles Green at Amersham in Buckinghamshire (Stewart 1994), a Romano-British site at West Blatchington, Hove (Curwen 1950), the Cataractonium at the Roman Catterick (Wright 2002), a Roman gold mining complex near Dolaucothi (Burnham and Burnham 2004), Wanborough (Buckley 2001), and Roman Cornwall (Watts 2003a). A Roman quern made from Hertfordshire Puddingstone (a conglomerate) was found at a quarry in Hertfordshire (Lovell and Tubb 2006). Since these studies were not readily available to the author, no further details can be provided.

Figure 54. *Broken and Abandoned Millstone on Birk Bank Fell, Caton, Lancashire. Photograph taken during August 1986 by Phil Hudson, Procter House, Settle, Yorkshire, United Kingdom.*

Figure 55. *View of Quarry Face on Clougha Pike, Quernmore, Lancashire. This is where attempts were being made to chisel and prise out the bedded gritstone, to be moved down the hillside and propped up for knapping and finishing (scale 10cm). Old and weathered chisel marks can be seen on the front edge of the block, and debris from previous workings in the background. Photograph taken during May 1987 by Phil Hudson, Procter House, Settle, Yorkshire, United Kingdom.*

Figure 56. *Broken Millstones, on Black Fell, Quernmore, Lancashire (scale 10 cm). Lying abandoned, by the side of the "rake" (the route way down) they are thought to have been damaged during transportation down the fell side. These are good examples of stone cut from the finer gritstone found on the top of the fell. Photograph taken during May 1986 by Phil Hudson, Procter House, Settle, Yorkshire, United Kingdom.*

Several regional studies include information on Roman period querns. These include querns from Tees Valley of York (Gwilt and Heslop 1995), querns made from the Old Red Sandstone (Saunders 1998; Shaffrey 2006a), donkey mills in Roman Britain (Williams-Thorpe and Thorpe 1988), and the raw materials used for Roman querns (Buckman 1866). A rotary quern from the Totternhoe Roman villa was reported by Williams-Thorpe, Watson, and Webb (1992). Addison (1995) studied the trade in querns during the Roman and Anglo-Saxon periods in Britain. Roman era querns have also been recovered from ship wrecks (Renouf 1993).

Querns made from lava have been recovered from urban and military sites in Roman Britain as well as other site types (Watts 2002:33). These lava querns were imported from the Eifel region of Germany (Watts 2002:33). During the middle of the Saxon period, lava querns were once again imported into Great Britain (Watts 2002:38). Freshwater (1996) reported on a lava quern workshop in late Saxon London. Parkhouse (1977) studied early Medieval basalt lava quernstones and how they were traded. He also wrote about the lava querns from the Thames Exchange Site in London (Parkhouse 1991). Parkhouse (1997) expanded his research on Mayen lava quernstones, used during the early Medieval period, to include their distribution in northwestern Europe.

While Roman period querns have received considerable attention, little information is available on Roman millstones. Shaffrey (2006a:30) reported that 63 rotary querns in her sample may have actually been millstones on the basis of their larger diameter. She suggested that millstones manufactured from the Old Red Sandstone may be restricted to the late Roman period (Shaffrey 2006a:30). A Roman nether millstone at Bramley was reported on by Rawnsley (1925). Addison (1995) studied the trade in millstones during the Roman and Anglo-Saxon periods in Britain. Millstones were also studied at the Romano-British town of Wanborough (Buckley 2001). Spain (1986) dealt with Romano-British millstones from Barton Court Farm at Abingdon, Oxon.

Querns and millstones have been discussed for the Medieval period. Biddle and Smith (1990) reported on the querns from Winchester. Fox (1994) published information on the millstone makers at Dartmoor. Graham and Farmer (1990) reported on the Medieval and post–Medieval millstone from the old malthouse at Abbotsbury in Dorset. Kars (1980) wrote about the Tephrite querns from early Medieval Dorestad and (Stoyel 1992) also discussed Medieval millstones.

Crawford (1953) mentioned four quarries that produced querns or millstones in his book *Archaeology in the Field*. The Pen Pits quarry was located near the point where the counties of Wiltshire, Somerset, and Dorset converge (Crawford 1953:100; Skinner 1996). This large site includes pits where the Greensand formation was quarried for querns (Crawford 1953:100–101). Crawford (1953:102) suggested that querns were produced at Pen Pits during the Saxon period and possibly earlier. The second site, Cole's Pits, was reported as being located in North Berkshire, south of Faringdon and east of Little Coxwell (Crawford 1953:102). The pits at the Cole's Pits site were deeper (up to 30 feet deep) than the pits at Pen Pits and abandoned specimens indicate that millstones rather than querns were manufactured there (Crawford 1953:102). The age of this quarry (exploiting the Greensand formation) was not known but Crawford (1953:102) thought it might date to the Medieval period. A third site thought to be a quern quarry was on Worm's Heath at Chelsham at Surrey (Crawford 1953:105). This quarry (exploiting puddingstone conglomerate) was covered with pits but was smaller than the two previous quarries (Crawford 1953:105). The fourth quarry was located at Waverly in Surrey (Crawford 1953:105). Exploiting the Folkestone formation, this poorly known quarry contained many pits (Crawford 1953:105).

Figure 57. *Partly Worked Millstone on Black Bank, Quernmore. It is still on its working props, but has been abandoned. Photograph taken during May 1984 by Phil Hudson, Procter House, Settle, Yorkshire, United Kingdom.*

Figure 58. *New Cut, Possibly Tooled as an Edge Runner, Cut from Grit Outcrop Near Outhwaite, Roebandale, Lancashire. Photograph taken during May 1989 by Phil Hudson, Procter House, Settle, Yorkshire, United Kingdom.*

Other publications also discuss querns and quern quarries in the United Kingdom. Querns and millstones were recovered from the Glastonbury Lake Village (Bulleid 1917). Barford (1984) discussed some possible quern quarries in the Bristol area. Butcher (1970) described the Wharncliffe quern workings. A study entitled "The Lithology, Origin and Disposal of Some Quernstones from Yorkshire" was published by Briggs (1988). Pearson (2000) also reported on the quern manufacturing at Wharncliffe Rocks in Sheffield in South Yorkshire. Philips (1950) wrote "A Survey of the Distribution of Querns of Hunsbury or Allied Types," studying querns recovered from excavations at Breedon-on-the-Hill.

Several studies that deal with millstone quarries have titles that don't mention the quarries' exact age. Beadle (n.d) discussed a quarry at Carr Crags in Newbiggin-in-Teasdale. Clarke (2003) reported on the millstone quarries at Whittle Hills in Lancashire. General millstone studies were written by Crawford (1953) and Gleisberg (1977). The millstone quarries in Northumberland were reported on by Jobey (1981, 1986) while Linsley (1990) studied the Brockholm quarry in Northumberland. A millstone found at Bishop's Castle, Shropshire, was described by Gifford (1998). Meredith (1981a) wrote about millstone making in Yarncliff. Finally, the millstone quarries in the Forest of Dean described by Mullin (1988, 1990).

A number of books on mills have included discussions on millstones. Some examples include Aikin's (1838) study on corn mills, Apling's (1984) book on Norfolk corn windmills, Bennett and Elton's (1898a, 1898b, 1900) books on the history of corn milling, Elton's (1904) volume on corn milling, Jones (1983) and, Rahtz and Meeson's (1992) study of an Anglo-Saxon watermill at Tamworth. Russell's (1949) commented on millstones in wind and water mills. Watts' (2002) book *The Archaeology of Mills and Milling* discussed millstones. Williams-Thorpe and Thorpe's (1988) studied donkey mills from Roman Britain

Concerning millstone quarries during the Domesday Book period of England (A.D. 1086), Bennett and Elton (1898b:128) provided the following information: "Millstone quarries, let at rentals, are occasionally mentioned. At Watone (Notts), is molaria where millstones are dug, molaria ubi molae fodiunt, yielding four silver marks per annum; at Bigenevre (Sussex), is a molaria paying 4s. a year."

Some interesting information on British millstone quarries was documented by Bennett and Elton (1900:95–96):

> Various British localities long supplied quernstones and millstones. Among the earliest known are the quarries of Notts and Sussex, mentioned in Domesday, which, however, do not seem to have endured till more recent times. At Everton, near Liverpool, the compotus of the king's receiver in 1444 mentions a millstone quarry, of which also at the present day nothing is known. A compotus of the receiver of Henry de Lacy, lord of Clithero, Lancashire, in 1296, records that "the rock of millstones" in that honour was let for quarrying at £1 6s. 8d. per annum; while another entry in the same accounts records the purchase of a stone, probably from this quarry, at the price of 7s. 7d. The quarry seems to have been at Quernmoor, near Lancaster, where, as Baines states, the stones are "full of those hard flinty particles which constitute what is called 'hunger-stone,' similar to that of ancient Roman querns, and of which small millstones were made in England; and from the aptitude of this stone for the purpose, it is probable that Quernmoor derived its name." ...There were various well-known millstone quarries in England during the Middle Ages; and, in fact, till they were superseded by the quarries of the Continent, which produced harder and finer stones. Millstone rock was found, for example, in the Bucks conglomerate in the neighbourhood of Winslow. Mention of these stones occur as early as 1312. ... A third variety was obtained from the Trillek quarries in Monmouthshire, which, on the spot in the thirteenth century, cost no more than a shilling each. A fourth variety was found in North of England, at Barkby and Kibworth in Leicestershire, and Finchale in Durham; while quarries at Bastlow, Derbyshire

> have been worked for centuries; and perhaps there was a fifth class, as the stones used in Wiltshire.

Elton (1904:130) provided an interesting footnote pertaining to millstones at the Everton manor near Liverpool:

> At Everton was conserved, till towards 1516, as a source of manorial income, the licensing of millstone makers. In the early days of difficult and costly carriage, local stones were of course used for hand or power mills wherever obtainable; and for digging or quarrying them lords of manor ordinarily exacted a rent. At Everton in the time of John of Gaunt mention occurs of a licence for the millstone quarry; and in 1516 Sir R. Molyneux, Dutchy Receiver for that manor, includes in his accounts the same old licence quaerendi petres molars in quaerra ibidem; though "nihil" for the same. The stones available at Everton could but have been boulder-stones from the drift overlying the red sandstone formation of the district, useful probably only for querns.

The modern millstone industry has been well documented. An excellent summary of the British industry was published in Gordon Tucker's (1977) article entitled "Millstones, Quarries, and Millstone-Makers." Eleven years later, Tucker (1987) wrote a more comprehensive overview entitled "Millstone Making in England." Tucker's article provided a general discussion of the millstone industry with an emphasis on the last three centuries. It should be noted that two major types of millstones were being produced. The first type consists of monolithic millstones made from indigenous rock (Tucker 1987:169–172). The second type, composite millstones, were assembled from imported French burr stone (Tucker 1987:172–174). In his overview of British millstone making Tucker (1977) presented an excellent summary of known quarries and millstone firms. He listed 30 millstone quarries in England, Wales, and Scotland, most of which exploited a conglomerate but a few quarries exploited other materials such as sandstone, limestone, granite, and basalt (Tucker 1977:13–16). Tucker (1977:16–20) also discussed British firms that imported French burr blocks and assembled them into millstones in larger English cities. In a 1987 follow-up article, Tucker provided information on more recently discovered millstone quarries and the urban French burr stone manufacturers. Bonson (1999) published an excellent bibliography of Gordon Tucker's writings including his many publications on millstones.

Most of the English millstones were monolithic stones manufactured from conglomerate. Tucker (1987:170) described "suitable rock" as "almost always ... to be some form of millstone grit which would have been a coarse texture with small quartz pebbles perhaps up to 1 cm across; or alternatively a sandstone/quartz conglomerate which was basically similar but with larger pebbles which in many places gave the rock the local name of 'pudding stone.'" Millstones were quarried in two major areas of England. The best known stones were quarried in the Peak District of Derbyshire and these quarries produced more millstones than all the other English quarries combined (Tucker 1985:42). The Peak stones were made from a conglomerate and three different types of millstones were produced at various times (Tucker 1985:46–47). Cossons (1975:64–65) briefly discussed the Peak millstones. Concerning these quarries, Apling (1984:16) noted:

> Peak Millstones were quarried from the Millstone Grit beds in the Peak District of Derbyshire dating from the middle of the Carboniferous Period, some 300/250 million years ago. The rock consist mainly of cemented grains of white quartz and pink feldspar, the chief constituents of granite and is a deltaic deposit resulting from the denudation of a granite landscape. Variation in grain size indicates how the carrying power of the river or rivers concerned varied from time to time, stronger currents bring down coarser material.

> The chief quarries were at Hathersage, where there is still a 'Millstone Edge,' at Grindleford Bridge and Stanton Moor. ...The Peak millstone is all in one piece with a central hole, round and about 8 ins. in diameter in a runner stone, but square in a bedstone.

Several millstone quarries have been documented in Lancashire in northwestern England. Phil Hudson (1989, 1995, 1998) conducted a study of the economic development in Quernmore Forest. He noted that the Namurian Gritstone was used for the manufacture of millstones (Hudson 1995:232–233). In this area, beehive querns were used during the Roman period, larger querns during the Anglo-Norse period, and larger millstones were utilized during the Medieval period (Hudson 1989:2, 1995:233). For the Medieval period and later periods, Hudson (1995:234) noted that abandoned millstones in various stages of completion have been found across the landscape. He has documented many millstone quarrying areas and has assembled an impressive series of early references concerning the millstone industry. His fieldwork at Quernmore Forest has identified several sites where gritstone was quarried for millstones including Fell End Beck (Figure 49), Baines Cragg (Figures 50–53), North East End of Birk Bank (Figure 54), Clougha east of the Pike (Figure 55), Black Fell (Figure 56), Clougha Scar, Trough Brook, Windy Clough Quarry, Windy Clough, Pike Foot, Cragg Wood, and Rowden (Ashpotts) Wood (Hudson 1989:5, 1995:234). Another area with millstone quarry remains is at Fox's Clough (Hudson 1995:234). He goes on to provide additional information on some of the millstone quarry remains (Hudson 1989:3, 1995:235–237). Hudson (1989:6) in his article "Old Mills and Millstone Quarries in the Forest of Lancaster" illustrates a four-stage sequence for working millstones. Hudson (1996) also published on the millstone quarries in Lancaster area. Additional quarry photographs by Phil Hudson include Black Bank at Quernmore (Figure 57) and Outhwaite at Roeburndale (Figure 58).

Raistrick (1979:56) noted that "on the Millstone Grit moors of the Pennines and on other fells where suitable sandstone is found, there are very many places where a large boulder has been split to get a lump which could be dressed into a millstone." He then went on to describe the techniques employed for accomplishing this work (Raistrick 1979:56–57).

Quarries elsewhere in England have also been described. In the Derbyshire District, millstone quarries have received considerable attention from several authors (Polak 1987; Radley 1966; Tomlinson 1981; and Tucker 1985). In northern England, millstones were quarried in Northumberland (Jobey 1986). The Fell Sandstones were exploited in some areas while Millstone Grits were quarried elsewhere in Northumberland (Jobey 1986:56–59). Telford (1938:208) reported that "millstones are made from stones got in Monmouthshire, but they are hard and flinty, they soon wear smooth and glaze, and require much picking." Further, Telford (1938:208) noted that "there are likewise many millstones made from a stone which is got on Mowcop-Hill in Cheshire; it is an open, gritty stone and answers for coarse work." The millstones produced at Mowcop have also been discussed by Bonson (2001) and Browne (2005). Millstone quarries have also become points of interest on public walking trails (Cleare 1988).

A number of other English millstone and quern studies have been published. These include: Brown 1992 (querns from Suddern Farm); Brown and Laws 1991 (querns from Bury Hill); Butcher 1970 (Wharncliffe quern workings); Cruse 2006 (Yorkshire quern survey); Cumming 1984 (John MacCulloch's millstone survey); Gleisberg 1977 (millstone quarries); Goffin 2003 (quernstones dating to the Middle Saxon period recovered during the excavations at the Royal Opera House in London); Henderson 1959 (millstones in Sheffield District); Higham 1907 (millstones of Muscovy); Hume 1851 (querns); Ingle 1982 (querns from Bristol region); Ingle 1984 (querns from Bristol region); Ingle 1993 (quernstones from Hunsbury Hillfort); Jecock 1981 (rotary querns in Wessex); Jecock 1985 (querns from Winnall

Down, Winchester); Jobey 1981 (excavations on Millstone Hill); Keller 1989a (Folkestone querns); Keller 1989b (Folkestone querns, southeast, Kent); King 1986 (querns and millstones in Bedfordshire); Laws 1987 (quernstones from Hengistbury Head Dorset); Linsley 1990 (millstone from Brockholm Quarry); Lorimer 1932 (millstones); Lovell and Tubb 2006 (querns quarries in Hertfordshire Puddingstone); McWhirr, Viner, Wells, Witt et al. 1982 (querns from the Cirencester Romano-British cemeteries); Mayer 1993 (quern from Horwich, Bolton); Meredith 1981a (millstone making at Yarncliff); Meredith 1981b (millstones at Hathersage); Moritz and Jones 1950 (grinding with a Romano-British quern); Parkhouse 1976 (Dorestad quernstones); Parkhouse 1998 (a quern from Buckinghamshire); Pearson 2000 (quern manufacture at Wharncliffe); Pitt-Rivers 1884 (excavations at Pen Pits); Radley 1964 (millstone makers smithy at Gardom's Edge); Rawlinson 1954 (millstones at Hathersage); Roberts 1994 (Lutyens' and Jekyll's Garden millstones); Roe 1991 (a quern from Alvaston); Roe 2004 (Saxon querns); Rogers 1993 (querns from 46–54 Fishergate at York); Rudge 1966 (the distribution of the puddingstone quern); Russell 1949 (millstones in mills); Shaffrey 2003 (querns from excavations at Silchester); Shaffrey 2006b (querns from southern England); Stoyel 1967 (millstones in the northeast); Stoyel 1992 (medieval millstones); Tucker 1980a (millstone making at Anglesey); Tucker 1980b (millstones); Tucker 1984a (millstone dressing); Tucker 1988a (14th century millstone transaction); Wallis 1988 (petrology study of 19th century millstones in the Wells Area, Somerset); Walton 1997 (millstone production in Yorkshire); Ward 1985 (British burrstones); Ward 1990b (English millstones and Seven Years War); Ward 1992b (millstones from Sacrewell); Ward 1992c (millstone makers in London); Ward 1993b (millstones at Willsbridge Mill); Ward 1994 (British millstone makers at Paris); Ward 1995b (balancing burrstones); Watts 2003b, 2006 (Longis querns from Alderney, the Channel Island in the English Channel); Williams-Thorpe 1995 (stone objects from Tattenhoe and Westbury); Williams-Thorpe and Peacock 1995 (quernstones from the Biferno Valley Survey); Williams-Thorpe and Thorpe 1988 (Roman millstones); Wilson 2003 (unusual millstone at Heckington Windmill); Winwood 1884 (excavations at Pen Pits); and Wisdom 1981 (conglomerate quernstones from Forest of Dean). Hockensmith and Ward (2007) have published a detailed bibliography of the millstone and quern literature for the United Kingdom for those desiring a detailed listing of available sources.

Several authors have focused on the dressing or sharpening of millstones. Some studies have focused on the tools used to dress the millstones. Studies on millstone dressing include Gibbons (1994) and Roberts (1993). Tool studies focusing on millstones include Major (1985), Tebbutt (1978), and Watts (1986).

Ireland

Several studies have examined querns and millstones in Ireland. A study of millstones from Carrig-na-m Brónta was published by the Mallow Archaeological and Historical Society of Ireland (Maume 1991). Mallow is located in southwestern Ireland between two mountain ranges. Unfortunately, a copy of the Carrig-na-m Brónta publication was not available. In 1966, John Caufield (1966) wrote his M.A. thesis, University College in Dublin, on "The Rotary Quern in Ireland." S. Caulfield (1977) published an article on "The Beehive Quern in Ireland." Two early studies appearing in the *Journal of the Royal Society of Antiquaries of Ireland* include Crawford's (1909) study "Some Types of Quern, or Hand-Mill" and Curran's (1908) article on "An Ancient Quern, or Millstone, County Kerry." Two brief articles were also published in the *Old Kilkenny Review* that dealt with "Rotary Querns" (Ryan 1972) and "A Rotary Quern Still in Use" (Anonymous. 1971). Power (1939) examined the religious significance of querns in an article entitled "A Decorated Quern-Stone and its Symbolism."

Since most of these studies were not available to the author, it not known whether the querns were quarried in Ireland or imported from other areas.

Scotland

Millstone making in Scotland occurred in two major forms. At Kaim Hill, Ayrshire, a quartz conglomerate was quarried for millstones (Tucker 1982a, 1984b). Three types of millstones were produced at Kaim Hill: monolithic, nine-piece, and four-piece (Tucker 1982a:191, 1984b:547). Millstones were being produced at Kaim Hill in the 18th century and production continued into the early 20th century (Tucker 1982a:192). Tucker (1984b:546–553) listed 26 other millstone quarries in Scotland. These primarily exploited conglomerate and sandstone while other stones such as granite, basalt, and schists were rarely used (Tucker 1984b:546–553). Butt (1967:95) noted that "there are several quarries which produced millstone in Ayrshire, but remains are extensive at Craigmaddie Muir in Stirlingshire." He also noted that the millstone quarries at Craigmaddie Muir contained millstones representing different manufacturing stages (Butt 1967:309). The second form of millstone manufacture combined local and imported stones. These composite millstones combined a central piece made from a local stone which was surrounded by segments of imported French burr stones (Tucker 1982a:186). Granite was quarried for millstones at Glenstocken in the Urr Valley of southwest Scotland between 1780 and the early 1900s (Dalbeattie Town website site: *http://www.dalbeattie.com/history/dbtquar.htm*).

John Pickin provided information on the Glenstocken millstone quarry in Scotland on the Millstonequarries.eu website (*http://meuliere.ish-lyon.cnrs.fr*). According to Pickin, the quarry is located south of the town of Dalbeattie. The millstones were made from a conglomerate sandstone (Rascarrel Formation) along the coast during low tide. The quarry dates from the 1790s to the end of the 19th century. Pickin noted that the quarry extended over 400 meters and contained some extraction pits as well as many unfinished and broken millstones.

Querns were also quarried in Scotland. Bennett and Elton (1898a:158–159) provided the following information including an 1868 quote from Alex. Carmichael:

> An ancient Scotch quern quarry evidences the manner in which the early Briton achieved the same end. At Heisgeir, North Uist, is a sea-beaten, rocky promontory in a small creek, where scores of the native querns, known as abrachs, have been quarried. "The original surface of the rock is cut away, and the size of each quern cut, and the marks of the tools used, are visible. Querns are there in all stages of progress: some had been just begun, and the marking on the rock could only be faintly traced: others had been half-cut and then abandoned: while in not a few cases the stones had broken as they were being separated from the rock, and had been left. Yet these quern quarries cannot have been worked for many long years." The abrach querns in question were never grooved, but merely dressed somewhat smooth; and as the tradition is, were provided with a good grinding surface by the simple expedient of placing them for a time in a stream of water.... The peculiar term "abrach," applied to these kind of stones, has reference not to their nature or quality, but to their supposed place of origin, the term "abrach," or, more correctly, "aberach," indicating an origin in Lochaber; though, curiously enough, the stone from which abrach querns were made is not found at Lochaber. As a rule the abrach was smaller than the ordinary quernstone.

Several other sources are available on querns found in Scotland. Buchanan (1912) published an article entitled "Notice of a Pair of Quernstones Found at Highland Dykes Near Falkirk." Feachem (1958) briefly reported on "A Quern from Mullochard, Duthil, Inverness-shire." This 16-inch diameter quern, made from a micaceous schist, was found near the River Dulnan two miles below Carrbridge (Feachem 1958:189). Another quern was described by

Stirling (1958) in an article entitled "An Upper Quernstone from Perthshire, Near Bridge of Allan." This Roman type quern (15½ inches in diameter), made from the German Niedermendig lava, was found about one half mile from Bridge of Allan Station (Stirling 1958:187). Campbell (1987) reported on "A Cross-Marked Quern from Dunadd and Other Evidence for Relations Between Dunadd and Iona." Close-Brooks (1983) prepared an article on "Some Early Querns." MacKie wrote three articles on querns including "Some New Quernstones from Brochs and Duns" (MacKie 1972), "Three Iron Age Rotary Querns From Southern Scotland" (MacKie 1995), and "Two Querns From Appin" (MacKie 2002). Finally, Spence (1988) wrote about millstone making in "The Quarryman in Sandwick."

Concerning quern manufacture in the Shetland Islands off the coast of Scotland, Bennett and Elton (1898a:157–158) observed, "The conglomerate 'pudding stone' of which in modern times Shetland querns are occasionally made, is believed to have been used in the Roman period: some specimens of the kind having been discovered at a reputed camp of the Caesars at Abbeville and one or two other places in France."

A more recent study for the Shetland Islands is the Hansen and Larsen (2000) article "Miniature Quern- and Millstones from Shetland's Scandinavian Past." The author has not seen the article but the title suggests that it deals with millstones and querns manufactured in Scandinavia and imported to the Shetland Islands.

Wales

In Wales, millstone quarries have been reported in different areas of the country. Tucker (1971:231) in his article "Millstone Making at Penallt, Monmouthshire" reported five main quarries at Penallt. These quarries produced monolithic millstones made from a quartz conglomerate (Tucker 1971). After field investigations were conducted, Tucker (1971:231) noted that the millstone industry was "very widespread over the parish, it was a sizeable industry, and ... it made stones for both cider mills and corn mills." Three years later, Tucker (1973) published an article entitled "Millstone Making in Gloucestershire: Wm. Gardner's Gloucester Millstone Manufactory with a Note on Hudsons of Penallt and Redbrook." Tucker (1973:7–11) discussed millstone makers John G. Francillon, William Gardner, the Youngs, and the Hudsons. He reported that the "Welsh stones" were made in eastern Monmouthshire in the parish of Penallt (Tucker 1973:11). A quartz conglomerate which outcrops over a large area in the parish was quarried for millstones (Tucker 1973:11). The quarry remains included numerous abandoned millstones, many of which were unfinished (Tucker 1973:12).

On the island of Anglesey, three millstone quarries (Pen' rallt, Cors Goch, and Bwlch Gwyn) have been documented (Tucker 1980a:18). These millstones were monolithic and made from a conglomerate (Tucker 1980a). Bennett and Elton (1900:96) also mentioned "the Welch millstones, obtained chiefly from Anglesea." Off the coast of Wales, Telford (1938:208) reported that "millstones are also procured from the Isle of Anglesa; they are grey, soft and gritty and do not last." Davies (1997:104–105), in his book *Watermill: Life Story of a Welsh Cornmill Being the History of Felin Lyn, Dyffryn Ceiriog, Denbighshire, North Wales*, shared the following information about the millstone quarry at Anglesey:

> There were several millstone pits in Anglesey, but the best-known quarry was at Enys, about two miles inland from Red Wharf Bay on the east coast of the island. It produced millstones and farm rollers for some 200 years until closure in 1939. Geologically, the stone was defined as a carboniferous millstone grit, a form of sandstone. Only a small part of the quarry's output went to Welsh mills, there being a large trade within northern Europe — in particular Norway and Sweden.

The diameters of the millstones at Felin Lyn, at 55 inches and 57 inches, were slightly unusual but were not a rarity. Millstones in Britain ranged from 32 inches to 64 inches, with 48 inches and 54 inches being the favoured diameters. A discarded but surviving millstone at Tregeiriog, six miles upstream from Felin Lyn, has a related diameter of 55 inches. It is of interest to learn that of the few millstones surviving at the small, abandoned quarries in Anglesey, in partial or finished state, some are 55 and 56 inch diameter. All Anglesey stones had iron bands affixed to them and those at Felin Lyn were similarly treated.

Several other studies have focused on millstones in Wales. Millstones were quarried at Conway, Wales (Malaws 1990). Ward (1990a) wrote an overview of Welch millstones. Burnham and Burnham (2004) wrote about the "Early Millstones Recovered from the Dolaucothi Area" which were recovered during archaeological investigations at Roman Gold-Mining Complex.

Several studies have focused on querns. Beehive querns produced during the Iron Age were studied by Chapman (1990). Stanley (1975) reported on querns from South Cadbury. The rotary querns in Wales have been examined by Watts (1996, 1997). Finally, Welfare (1981) wrote about "The Milling Stones" recovered from a farmstead in South Glamorgan occupied during the Iron Age and Roman periods.

Greece

Several important studies have been made for millstones in Greece. Curtis Runnels and his colleagues have made many major contributions. Runnels and Murray (1983:62) in their article "Milling in Ancient Greece" noted:

> We set out ... in one area of Greece where we studied stratified and well-dated millstones spanning the last 10,000 years from the Mesolithic period to modern times. We discovered that millstones, even those from the Neolithic, were used for many other purposes besides grinding grain, and that contrary to Childe's opinion they were quite variable in shape, use and material through time. These mundane tools are, in fact, valuable artifacts for the archaeologist because they reveal a great deal about ancient technology and economy and can easily be used, like pottery, to date archaeological deposits.
>
> Study of millstones from archaeological excavations as well as from modern context in the Argolid, Korinthia and Attica, districts located in southern Greece, revealed continuous change in their form and function through time.

Concerning the stone types exploited for millstones in Greece, Runnels and Murray (1983:62) summarized as follows:

> The materials from which millstones were made changed through time just as the shapes and uses of the tools did. In the Mesolithic and Neolithic periods, two basic kinds of stone were employed: sandstone and andesite. Several different kinds of andesite, a rough volcanic rock, were utilized during these early periods. With the coming of the Bronze Age there was a major increase in the amount of this material employed. Andesite continued to gain in popularity after the Bronze Age, but during the Classical period the range of sources narrowed to two. Beginning perhaps in the later Mediaeval period and continuing to the present, a white metamororphosed volcanic rock replaced andesite as the most common material from which millstones were made.

Curtis Runnels' (1981) Ph.D. dissertation focused on millstones and was entitled "A Diachronic Study and Economic Analysis of Millstones from the Argold, Greece." Geographically, the study included the districts of Argold, Korinthia, and Attica and provided information on over 400 unpublished millstones (Runnels 1981:4–5). Runnels' study is very broad and comprehensive in scope. The first three chapters define millstones and describe the study area, the chronology and previous research; establish a theoretical context by discussing uniformitarianism in archaeology, economic theory and choice, and behavioral theory; and provide information on the sites, sampling problems, and classification. The fourth chapter describes the raw materials and their properties, sources of raw materials, previous research on ancient millstone quarrying, manufacturing methods, fieldwork at sources of materials, and evidence for quarrying at material sources. Chapter five provides an overview of the cultural periods (Palaeolithic, Mesolithic, Neolithic, Bronze Age, Greek, and Roman) and a discussion of millstones (shape, methods of manufacture, raw materials, use, and use life). The sixth chapter deals with economic analysis, microeconomics, ethnographic supply and demand, supply and demand in antiquity, prehistoric millstone production, Greek millstone production, Roman millstone production, and the scale of ancient millstone production. Chapter seven provides the conclusions of the study. Appendix one provides descriptions of the millstones examined in the study while appendix two is an ethnoarchaeological study of millstone production and modern quarries. The study includes a glossary and impressive bibliography.

The millstones quarries on the Greek islands of Aegina and Melos were briefly described by Runnels and Murray (1983:63):

> In an outcrop adjacent to the ancient sanctuary of Zeus on Mt. Oros, the highest peak on the island of Aegina, we found abundant traces of quarrying. Wedge holes, cut channels and chisel marks indicate that andesite had been extracted there. At least some of the stone had gone into building the nearby ancient sanctuary, but samples from this quarry also matched the rocks which had been fashioned into implements in the Classical period and shipped to the mainland.
>
> Most other ancient quarrying traces had probably been obliterated by subsequent extraction of stone. On the island of Melos, for example, there is a deposit of white volcanic rock matching that fashioned into millstones and widely used in Mediaeval and later times. This deposit was definitely the source of modern millstones, for there we also found a mine shaft, railroad tracks, bins, to carry the quarried material, and fragments of the millstones themselves — all abandoned only recently. Mediaeval millstone works were certainly located at this same spot because no other source of this unique rock exists, but few traces could be found of those earlier activities.

Homans (1859) in *A Cyclopedia of Commerce and Commercial Navigation* mentioned millstones that were quarries on the island of Milo. The island of Milo is the same as Melos or Mílos. Homans (1859:1359) stated that "Milo, in the Archipelago, furnishes mill-stones of a very excellent quality. They are exported to Greece, Italy, and other countries on the Mediterranean, where they are employed in grinding the hard wheat, or *grano duro*, used in the manufacture of macaroni, vermicelli, etc. The quarries are wrought on account of government, and the stones sold at moderate prices fixed by a tariff, which, however, leaves a handsome profit to the State."

Millstones have been studied in detail in Greece. Kardulias and Runnels (1995) defined the different types of milling tools in Southern Argolid area of Greece. These include saddle querns, handstones, hopper querns, rotary, and rotary olive mills or crushers (Kardulias and Runnels 1995:110). Curtis Runnels and his associates have focused much of their research on the study of querns recovered from surveys and excavations (Kardulias and Runnels 1995; Runnels 1981, 1990, 1994). Some of the studies focused on the trade and sources of Greek millstones (Cohen and Runnels 1981; Runnels 1985, 1988; Runnels and Murray 1993). General research on Greek millstones and querns included the following studies: Grypari and

Karapidakis 2003; Kardulias and Runnels 1995; Runnels 1981, 1988, 1990, 1992, 1994, 2004; and Cohen and Runnels 1981; Williams-Thorpe and Thorpe 1990. Chalkousaki (2003) discussed millstone quarrying in Milos, Greece.

As part of his Ph.D. research, Runnels (1981:225–232) gathered information on modern millstone quarries in Greece for comparative purposes. He noted that the areas containing the most important quarries were in Aigina, Poros, and Melos (Runnels 1981:225). Limestone millstones were also quarried at one time at Patras (Runnels 1981:225). At Aigina, basaltic andesites were shaped into olive oil millstones (Runnels 1981:225–226). Quarries were operating in 1935 and one quarry was still in operation in 1976 (Runnels 1981:226–227). Runnels (1981:228–230) described nine steps from clearing the surface to the finished millstone. At Polos, olive oil millstones were produced from andesite in the 19th century until 1960 (Runnels 1981:233–234). The quarries ranged from small to large. On the eastern coast of Melos, millstones were quarried from medieval times to about 1957 (Runnels 1981:235–238). The quarries exploited a hard white igneous rock for millstone for wind and water mills as well as some querns (Runnels 1981:235). Because of the limited size of the stone, only composite millstones were produced (Runnels 1981:236).

Greek archaeologist Margarita Vrettou-Souli has studied the millstone industry at Milos, Greece. She has written an excellent book in Greek entitled *H Milopetra tis Milou: Apo tin exorixi stin emporiki a diakinisi* (Vrettou-Souli 2002). The title translates into English as *The Millstone of Milos. From Mining to Commercial Circulation*. She graciously prepared a detailed summary of her research in English for this author (Vrettou-Souli 2007) and granted permission for it to be quoted:

> Milos is an island of the Cyclades, in the Aegean Sea, which has been characterized by its very long mining history until today. Owning to the mining of the obsidian, Milos has been a commercial center in the Aegean Sea since the Neolithic times. Mining continues uninteruptedly until nowadays.
>
> The millstone of Milos has been famous and used in the Aegean Sea as well as in the Mediterranean Sea generally. This kind of stone was so famous that several researchers believe that the name of the island has stemmed from its millstone. It is a quartz-trachite stone, a stone with great holes (0,5–1 cm.), which is suitable mainly of barley. That is why it is called "barley-stone."
>
> There are different types of millstone, such as the "crassato" (= of the wine), which was expensive and hard, particularly suited for the grinding of wheat; another type was the "tyflo" (= blind), and was carved the same way as the French millstone; another type was the "routhounato" (= of the nostrils), which was the most common type suitable to the grinding of barley.
>
> The greatest site Age of mineral wealth is located in the area of Rema, a beach on the East Coast of the island, where the excavation produced millstones from the Middle Ages until the mid–20th century. The mining was intensive and a great part of the population worked there. For this reason, it is possible that during the Byzantine era the capital of Milos was transferred from Klima to Zephyria, a part of the island nearby the quarry. From time to time, immigrants arrived to work in Rema. Very many of these immigrants arrived from Tsakonia (mountainous Arcadia, Peloponnesus); as a result, we still have first names and surnames on Milos nowadays originating from Tsakonia. Next, during the Turkish occupation there have been mining excavations on the island of Kimolos, in the site of Brovarma (= Observatory), and also at Stis Hiromili (= At the Hand mills); finally, in the neighboring uninhabited island of Poliegos. These latter minerals were of inferior quality to Milos, consequently cheaper. Still, their mining and transport to the boats was easier.
>
> The quarry of Rema is near the sea; however this beach is attacked by strong winds. As a consequence, boats had great difficulty in approaching the coast, and always for a short time.

> Workmen had to speed with the millstones to the cargo boat, using a rough sack on their shoulders to avoid possible injuries. Labor was hard and unhealthy but profitable.

The millstone workers of Milos had specialized job duties and had steps that they followed for making different types of millstones (Vrettou-Souli 2007):

> There were four specializations of working people: the "minadoroi" (= the explosive-setters), the "ypourgoi" (= pullers), who were pulling the millstones out of the tunnel; the "mastores" (= craftsmen), who were working out the millstone there and then; next the non-qualified workers; finally there was a foreman. Safety measures were taken much later, in the 20th century. They created columns to support the tunnel, and they also installed a ventilation system. There was also a small rail-wagon to transport the minerals.
>
> Initially, the elaboration was done on the spot. They would choose the biggest stones which they would call "pelekites" (= chopped, hack). They would need 5–6 of these to create a whole millstone. Then, the smaller and cheaper ones, which they would call "karykou." The smaller fragments, almost square in shape, were called "Poleos" (deriving from Constantinople), as they were sold in Constantinople; the fragments of "mastorou" (= master craftsman), which were smaller and cheaper; also the simpler fragments were necessary for the construction of the millstone. At the beginning of the decade of the 1950s there was a new type: "2A," an idea which had been inspired by the last master craftsman of the quarry in Rema. This type was much bigger and more expensive; with only 2–3 fragments one could construct a whole millstone. The master craftsman would receive a bonus for the construction of this last type. When they found fragments suitable for the hand mills, they would put them apart and destine them for gifts. These fragments were very expensive and hard to find. Even today, in several houses on Milos one can find hand mills.

Of the working conditions Vrettou-Souli wrote (2007):

> The salaries generally were high, much higher — almost double — than the usual salaries and this is why this kind of job was preferable, although they had to work under unhealthy and dangerous conditions. Accidents were not rare. A local tradition describes an accident which caused the death of 40 workers from Tsakonia, Peloponnesus. The workers generally were working five days a week, because the distance from their villages was very long and they had to go to the quarry on foot from home. On their way back, they used to work together on their own farming fields. These working conditions were comparatively forward for this era: five working days a week and work solidarity. In the decade of the 1930s an association of workers was founded on Milos.

The millstones of Milos were widely distributed but had to be sharpened often since the stone was soft (Vrettou-Souli 2007):

> The commercial circulation was great. The millstones were sold in the entire country, as well as it being exported abroad: to Constantinople, Egypt, Italy (Naples, Trieste), Serbia etc. In the international commercial exposition in London in 1851, the millstone of Milos was among the exhibits. In 1905, for example, 10,477 units of millstones were sold for 29,000 drachmas.
>
> The millstone of Milos is soft enough, compared to the French millstone and one had to carve it often, almost every thirty days. The carvings were close to one another (almost 0,5 cm. and the depth was between 1–1,5 cm.) like a hair comb. There are many legends, proverbs and traditions regarding the millstone and generally to the function of the mills and the profession of mill-keepers, who confirm in this way the importance of the millstone in the traditional society.

In the mid–20th century, the millstone industry ceased, but the archaeological remains are a reminder of its former significance (Vrettou-Souli 2007):

> In 1956 the mining excavations in the quarry of Rema were stopped for good. The modern technology made the millstone redundant. The ruins of this quarry are evidences of the glory

of the past and are a part of the living symbol of this long history and importance of the mining history of Milos.

Six Greek millstone quarries have been reported on the Millstonequarries.eu website (*http://meuliere.ish-lyon.cnrs.fr*). The Nisyros Island quarry was reported on the website by Fritz Mangartz (February 2007) and is in the Dodekanes Islands in the South Aegean. Millstones were quarried during the Roman period from a basaltic andesite. The exact location of the quarry is not known but these millstones are discussed in the archaeological literature by Williams-Thorpe 1988; Williams-Thorpe and Thorpe 1990; Williams-Thorpe, Thorpe, Elliot, and Xemophontos 1991. Fritz Mangartz (February 2007) also added the Aegina Island quarry on the Aegina Island to the website. Andesite millstones were produced during the Roman period and earlier periods but the quarries have not been located. Archaeological literature mentioning these millstones includes Runnels 1981; Williams-Thorpe 1988; Williams-Thorpe, Thorpe, Elliot, and Xemophontos 1991. The third quarry is Milos, Rema 1, which is located at Rema near Milos (reported on the website by Alain Belmont, April 2007). The quarry exploited a volcanic rock during antiquity to the modern era. Several archaeologists have mentioned this quarry (Peacock and Williams 2006; Williams-Thorpe 1988; Vrettou-Souli 2002). The Milos, Rema 2, quarry is also located at Rema near Milos (reported on the website by Alain Belmont, April 2007). Volcanic rock was exploited at this quarry between antiquity and the Middle Ages as well as between the 18th century and 1956. This quarry is also discussed in the archaeological literature (Williams and Peacock 2006; Williams-Thorpe 1988; Vrettou-Souli 2002). The fifth quarry, Kimolos, Mprovarma, is located at Mprovarma near Kimolos (reported on the website by Alain Belmont, April 2007). Millstones were made at this quarry during antiquity, the Middle Ages, and the modern era from a volcanic rock. Previous investigators have mentioned this quarry (Williams and Peacock 2006; Williams-Thorpe 1988; Vrettou-Souli 2002). Finally, the Olympie Temple de Zeus quarry at the ancient temple of Zeus site near Olympie (reported on the website by Alain Belmont, April 2007) produced millstones from a sedimentary rock during antiquity.

Hungary

Some nineteenth century authors commented on the millstone industry in Hungary. Lyell (1853:421) noted that "the Hungarian lavas are chiefly felspathic, consisting of different varieties of trachyte; many are cellular, and used as millstones." Dana (1857:129) made the following comments on alum stone: "the variety found in Hungary is so hard as to admit of being used for millstones." Finally, Egleston (1872:111) reported that "Jaspery Alunite is so hard that it is used in Hungary for millstones."

Grimshaw (1882:288) provided the following comments on some Hungarian millstone quarries: "The millstone of Tilleda, in Kyffhauser, is a sharp, open red sandstone, suitable for either breaking or flouring.... The trachyte quarries of Hungary have been worked for 800 years. The ancient Sarosptaker quarries yield good burr for low middlings. Sardinia and Germany also have quarries."

Attila Selmeczi Kovács has published three studies dealing with hand mills and querns. The first study focused on the hand mills in the Carpathian Basin (Selmeczi Kovács 1989). The following year, Selmeczi Kovács (1990) published a study of querns. Finally, Selmeczi Kovács (1999) published a study of millstones, mills, and mill work in Carpathian Mountains Region of Hungary with a summary in English.

Italy

Italy played a major role in the development of circular grinding of grain by stones. Bennett and Elton (1898a:128) provided the following information on the Italian quern:

> The quern, an Italian contrivance apparently not known till about 2,000 years ago, constituted the earliest complete grinding machine; the first in which the parts were combined into a perfect mechanism; and from its advent corn milling was no longer conducted by mere loose stones. The quern is distinguished from the more primitive corn stones primarily by its circular motion: the upper stone revolving upon or with a pin upon the lower.

Concerning the raw materials used for Roman querns, Bennett and Elton (1898a:155) noted:

> Originally the Romans, like other people, seem to have used all varieties of native rock. Ovid speaks of millstones "rough as pumice," testifying to the use of the local porous volcanic tufa; and Pliny records the use of the same "porous" variety. According to Pliny, the hard fine grained quartzite of Volsinii (where the quern was invented) would be the material of which querns were made....

Bennett and Elton (1898a:157) further noted:

> Quarries, mentioned by Pausonius at the Island of Nesis near Naples, "where abundant millstones are quarried the workers selling large numbers to the neighbouring countries," are found to have been quarries of basalt.

A number of publications deal with millstones studies in Italy. Antonelli, Nappi, and Lazzarini (2001) focused on Roman millstones from Orvieto, Italy. Millstones from Pompeii in southern Italy were studied by Buffone, Lorenzoni, Pallara, and Zanettin (2003). Some of these publications have focused on ancient exchange and trade networks. Galetti (2006:80) published a paper that dealt with "Mills, Millstones, Millstone Rocks and Millstone Quarries in Medieval Italy." Her paper pulled together information on mills, millstones, and millstone quarries from archaeological excavations and early written sources (7th to 15th centuries) from north and central Italy (Galetti 2006:80). Galetti (2006:80) noted that many millstones have been recovered from archaeological investigations and were transported from quarries (such as those exploiting garnet slate) along road systems and water courses. Available studies include: Antonelli, Bernardini, Capedri, Lazzarini, and Kokelj 2004 (prothistoric grinding tools of volcanic rocks in the karsts of Italy); Antonelli, Nappi, and Lazzarini 2000; Capedri, Venturelli, and Grandi 2000; Cattani, Lazzarini, and Falcone 1997; Ferla, Alaimo, Falsone, and Spatafora 1984 (saddle querns from five sites in Sicily); Lorenzoni, Pallara, Venturo, and Zanetti 1996 (volcanic millstones from Neolithic-Roman archaeological sites of the Altamura area, southern Italy); Lorenzoni, Pallara, and Zanetti 2000a, 2000b (Bronze Age millstones made of volcanic rock of Apulia, Southern Italy); Peacock 1980a (the Roman Millstone trade), Peacock 1980b (the mills of Pompeii), Peacock 1986 (Roman millstones near Orvieto); Renzulli, Santi, Nappi, Luni and Vitali 2002 (volcanic millstones from Roman archaeological sites in central Italy); Tucci, Azzaro, Morbidelli, Agostini, and Misiti n.d. (lava millstones from Pompeii, Naples, Italy); Volterra and Hancock 1984 (millstones from two Roman villa sites in southern, Italy); and Williams-Thorpe and Thorpe 1989 (Roman millstones from Sardinia, Italy).

Fourteen Italian millstone quarries are described in French on the Millstonequarries.eu website (*http://meuliere.ish-lyon.cnrs.fr*). Seven of these quarries were reported to the website by Alain Belmont (during October and December of 2006, April and November of 2007). The Valmeriana quarry is located at Alpe de Valmeriana near Pontey in the Valle-d'Aosta region. Millstones were carved out of vertical faces of metamorphic rock at this quarry from between the 13th and 19th centuries. Second, the Borgone Susa, Roccafurà, quarry is located at Roccafurà near Borgone Susa in the Piemonte region. A metamorphic rock was exploited at this quarry between the 17th and 19th centuries to produce an estimated 2,000 to 3,000 millstones (Grillo 1993). Quarry photographs indicate that millstones (123–125 cm in diameter and ca. 30 cm thick) were produced by cutting deep grooves around the millstones and splitting the isolated millstone blank from the rock face (vertical to steeply sloping). The third quarry, Borgone Susa, Arco di Maometto, is located at Maometto near Borgone-Susa in the Piemonte region (Grillo 1993). Between the Middle Ages and the 19th century, a metamorphic rock was manufactured into millstones. The Susa quarry in the Piemonte region is located at Arco di Augusto near Susa. Millstones (1.10 to 1.16 m in diameter) were quarried during the Middle Ages from a metamorphic rock. The fifth quarry, Borgone Susa, Molere, is located at Molere near Borgone Susa in the Piemonte region. Millstones were quarried from a metamorphic rock during the Middle Ages and modern era (Grillo 1993). The Lambrugo quarry is located at Parco di Valle Lambro, Ca' di Lader, near Lambrugo in the Lombardia region. During the 18th and 19th centuries, millstones were produced from a conglomerate. The seventh quarry, Recoaro Terme, Sentiero delle Mole, is located at Rotolon, Campo Grosso, La Guarida, near Recoaro Terme in the Veneto region. Millstones were made from a sedimentary rock between the 16th to 19th centuries.

The remaining seven Italian millstone quarries are described in English on the Millstonequarries.eu website (*http://meuliere.ish-lyon.cnrs.fr*). These quarries were reported to the website by Fritz Mangartz (during January and February of 2007). The Orvieto quarry is located southwest of Orvieto in the Umbria region. Millstones were manufactured from a volcanic leucitite rock from the Bronze Age to Roman times. Several researchers have described these quarries (Antonelli, Nappi, and Lazzarini 2001; Lorenzoni, Pallara, and Zanetti 2000b; Peacock 1986). Second, the Monte Vulture quarry, used from the Bronze Age to Roman times, is known only through mineralogical analysis and the actual quarries have not been located. Located somewhere in the Basilicata region, this quarry produced millstones that have been discussed in earlier studies (Lorenzoni, Pallara, and Zanetti 2000b; Williams-Thorpe 1988). The Island of Ustica quarry is also expected to be located in Sicily but has not yet been identified. Roman millstones from this area made from a basalt have been previously discussed by Williams-Thorpe (1988). Fourth, the Mulargia quarry consists of several shallow extraction pits near the village of Mulargia in the Sardegna region. Unfinished Roman millstones made from volcanic rhyothic stone are used as decorative objects in the village and have been discussed by several authors (Peacock 1980; Williams-Thorpe 1988; Williams-Thorpe and Thorpe 1989, 1991). The Monti Iblei quarry is known only from mineralogical analysis and is thought to be somewhere in Sicily. A basaltic andesite was exploited in this area from the Bronze Age to Roman times. These millstones have been studied by Lorenzoni, Pallara, and Zanetti (2000b). Sixth, the Pantelleria Island quarry in Sicily is another potential millstone quarry known only from mineralogical analysis. These basalt millstones were produced during Roman times (Williams-Thorpe 1988; Williams-Thorpe and Thorpe 1990). Finally, the Etna quarry, known only from mineralogical analysis, is thought to be located near Catania in Sicily. The volcanic rock used for these millstones was used from the Bronze Age to Roman times (Lorenzoni, Pallara, and Zanetti 2000b; Williams-Thorpe 1988).

Luxembourg

The Berdorf, Hohllay millstone quarry is located at Hohllay near Luxembourg, in the Grevenmacher region near Berdorf. A summary of this quarry was placed on the Millstonequarries.eu website (*http://meuliere.ish-lyon.cnrs.fr*) by Alain Belmont in May of 2007. A conglomerate was exploited there to produce millstones during the Middle Ages to the 19th century (Hörter 1994). Monolithic millstones were manufactured at the quarry that ranged in diameter from 113 to 152 cm and 25 to 35 cm in thickness. The general photograph of the quarry indicates that it is a subterrian quarry with a cave-like appearance. The quarry was originally reported in German (Hörter 1994) and is summarized on the website in French. Interested readers should consult these sources for more details.

Norway

Norway has a long history of millstone quarrying. Quarries have been recorded at Selbu, Tydal, Hyllestad, and Kvennberget. These quarries are briefly discussed below. An interesting book entitled *Kvernfjellet* included information on quarries, quarrying, and millstones near Selbu and Tydal, Norway (Rolseth 1947). The book appears to be published in the Norwegian language. It has many excellent photographs of the millstone quarries on the west coast of Norway near Trondheim. The millstone quarries at Selbu exploited a staurolite-schist (Kresten, Elfwendahl, and Pettersson 1996:80). The Selbu and Lalm millstone quarries in Norway are included on the Millstonequarries.eu website (*http://meuliere.ish-lyon.cnrs.fr*). The

Figure 59. *Nearly Complete Quern Still Attached to the Bedrock at Hyllestad, Norway. Note the drill holes at the base of the quern for the purposes of wedging it away from the bedrock. Photograph taken by Astrid Waage of Hyllestad, Norway.*

information for these quarries was added to the website in French by Alain Belmont. The Selbu quarry is located at Høgfjellet near Selbu in the Sør-Trøndelag region. At this quarry, a metamorphic rock was quarried for millstones between the 16th century and 1940 (Alsvik, Sognnes, and Stalsberg 1978; Rolseth and Alsvik 2000). The Lalm quarry is located at Kvennberget near Vågå in the Oppland region. Between the 15th century and 1890, millstones were quarried at this locality from a metamorphic rock.

A major center of the millstone industry once existed near Hyllestad, Norway which is north of Bergen (Carelli and Kresten 1997). Baug (2004:3) noted that "in Hyllestad, you can visit one of Norway's largest historic monuments in terms of area. Over 27 km2 of the landscape is scattered with hundreds of mill stone quarries. During the Viking era and the Middle Ages, Hyllestad was a centre for mining and was an important site both locally and internationally." Peter Kresten visited the millstone quarries at Hyllestad in 1996 and noted a goal of documenting the quarries and a desire to protect them (Carelli and Kresten 1997:117). Carelli and Kresten (1997:117) reported that 25 larger quarries had been recorded in the vicinity of Åfjorden but there were undoubtedly many additional small quarries. Baug (personal communication, 2008) shared that many additional quarries have been discovered since the work of Carelli and Kresten (1997). An excellent four page booklet has been published to serve as a guide through the millstone quarries at Hyllestad, Norway (Anonymous 2005).

Figure 60. *View of Quarry Face at Hyllestad, Norway Showing the Depression where Querns Were Extracted from the Quarry. Note the drill hole scars visible inside the depressions where the querns were removed. Photograph taken by Astrid Waage of Hyllestad, Norway.*

Figure 61. *View Looking Down into a Millstone Quarry at Hyllestad, Norway. Abandoned millstones and querns are visible on the quarry floor. Photograph taken by Astrid Waage of Hyllestad, Norway.*

The millstone quarries of Hyllestad, Norway have a very long history (Baug 2004:8):

> Archaeological surveys have indicated that mining in this area dates back to the late Iron Age, maybe as early as the 700's A.D. and mill stones were extracted in Hyllestad in large amounts during the Viking era. The period from the 11th century until the 14th century saw the highest level of activity in the oldest stone quarries, even though production continued right up until the 16th–17th century in the oldest quarry types. In more recent times, we now know that explosives were used in production and that this method was utilized until around 1930. We can therefore establish that mill stones have been produced in Hyllestad for well over 1000 years.

The technology used in making millstones at Hyllestad varied through time (Baug 2004:8):

> Hyllestad has seen two different types of techniques used to cut the mill stones. The oldest technique involved cutting the shape of the mill stone directly into the rock wall, then breaking the piece loose. In Hyllestad, this method has been used to produce both manually operated mill stones and mill stones for watermills. The second cutting method came into use after the introduction of explosives in the millstone quarries, probably in the early 18th century. The mill stones were cut out of blocks of stone which had already been blasted out of the rock.

Baug (2004:5–7) provided the following information on the rock utilized for the millstones at Hyllestad:

Figure 62. *View of Two Shaped Querns on a Cliff Face at Hyllestad, Norway. Photograph taken by Astrid Waage of Hyllestad, Norway.*

The natural source for the mill stone quarries in this area was the kyanite-garnet-muscovite-mica schist often referred to as garnet mica schist.

The materials required for mill stones had to be neither too hard nor too soft, and the combination of the soft mica schist and the hard garnet produced a perfect type of rock for grinding corn. The hard garnet took longer to wear down than the mica schist, and this meant that the surface of the mill stone remained rough. It also meant that the mill stone lasted longer.

The combination of garnet and mica schist can also be found in other mill stone quarries in Norway, but with local variations.

> The mill stone quarries in Hyllestad is the only one in Norway to contain all minerals; garnet, muscovite, staurolite and kyanite. And this allows us to identify Hyllestad as the source of products found in many different parts of Norway and Europe.

Baug (2006:55–56) has noted that the unique mineral content of the stone at Hyllestad made it possible for researchers to identify trading routes for these millstones. However, she says, variations in the stone do not permit researchers to attribute millstones to particular quarries. Also, Baug explains that the schist content of the stone affected the way that it fractured. This variation in the schist was responsible for millstones being removed from vertical faces at some quarries and from horizontal beds at other quarries. In general, querns were manufactured at the earliest quarries but the archaeological evidence indicated that both querns and millstones were produced at some of the quarries (Baug 2006:56).

The Hyllestad millstone industry appears to have been forced out of business by the competing millstone quarries at Selbu, Norway, that were located to the northeast (Baug 2006:57–58). Baug (2004:11) indicated that "[w]ith modern times, production at Hyllestad saw a decline. Why is this? There are signs that competition with the mill stone quarries in Selbu, Sør-Trøndelag could be the cause. By the start of the 20th century, the Selbu stone had completely wiped the Hyllestad stone off the market nationwide."

In order to obtain a better understanding of the quarries at Hyllestad, Baug (2006:57)

Figure 63. *View of Quarry Base Showing Millstone and Quern at Hyllestad, Norway. Millstones were previously cut from the from quarry creating the cliff face. Photograph taken by Astrid Waage of Hyllestad, Norway.*

conducted archaeological investigations. She selected four areas containing different types of quarries and located different distances from the coast (Baug 2002, 2006:57). The trenches that she excavated provided evidence for quarrying activities as early as the 8th century (the Merovingian period). Baug uncovered evidence for the Viking period as well as the period of transition from the Viking period to the Middle Ages. She discovered that the greatest period of production was between A.D. 1000 and A.D. 1300. After A.D. 1500, the industry experienced decline but continued to produce stones until the early 20th century (Baug 2006:57).

There are some other millstone studies for Norway (Baug 2002, 2005; Ekroll 1997; Rønneseth 1968, 1977) and Scandinavia in general (Carelli and Kresten 1997; Hansen 1989; Kresten 1996a, 1996c, 1996e; Liebgott 1989; Madsen 1967; Müller 1907). The reader is directed to these sources for additional information.

Four millstone quarries at Hyllestad, Norway, have been reported on the Millstonequarries.eu website (*http://meuliere.ish-lyon.cnrs.fr*). The information for these quarries was added to the website in French by Alain Belmont between April 2006 and July 2007. First, the Skor quarry is located at the Skor farm, Krivollen, in Hyllestad in the Sogn og Fjordane region. A metamorphic rock was exploited at this quarry for querns during the Middle Ages. The Skor farm is just one of several farms in Hyllestad containing quarries (Baug personal communication, 2008). The second quarry is located at Kvernsteinsparken in Hyllestad. A metamorphic rock was used for millstones from the 8th century to the beginning of the 20th century. This quarry has been previously described by archaeologists (Baug 2006; Baug, Heldal,

Figure 64. *Two Partially Shaped Millstones on a Cliff Face at Hyllestad, Norway. Photograph taken by Astrid Waage of Hyllestad, Norway.*

Figure 65. *Closeup View of a Millstone in the Millstone Park at Hyllestad, Norway. Note holes at the base of the millstone for the purposes of wedging it away from the bedrock. Photograph taken by Astrid Waage of Hyllestad, Norway.*

Englert, Marøy Hansen, Gullbekk, and Kilger 2006; Thue 2000). There are several quarries located at Kvernsteinsparken (Baug personal communication, 2008). Third, quarries are located at Otringsneset in Hyllestad. Baug (personal communication, 2008) shared that Otringsneset is a small place located on the Rønnset farm. Metamorphic rock was quarried at this location for millstones between the 8th century and the beginning of the 20th century (Baug 2006; Baug, Heldal, Englert, Marøy Hansen, Gullbekk, and Kilger 2006; Thue 2000). Finally, at the farm Rønnset in Hyllestad several millstone quarries have been located. During the Middle Ages (8th to 13th centuries) querns (40 to 50 cm in diameter) and millstones were cut from a metamorphic rock at this quarry (Baug 2006; Baug, Heldal, Englert, Marøy Hansen, Gullbekk, and Kilger 2006; Thue 2000).

Through the generosity of Astrid Waage, several of her photographs of the quarries of Hyllestad are included in this book. These include a photograph of a nearly complete quern still attached to the bedrock (Figure 59), a view of quarry face showing the depression where querns were extracted (Figure 60), a view looking down into a millstone quarry showing abandoned millstones and querns on the quarry floor (Figure 61), a view of two shaped querns on a cliff face (Figure 62), a view of quarry base showing a millstone and a quern (Figure 63), two partially shaped millstones on a cliff face (Figure 64), and a closeup view of a millstone in the millstone park at Hyllestad (Figure 65).

Slovenia

Limited research has been conducted on querns and millstones in Slovenia. Smerdel (2003) published an article in French that dealt with the making, selling and using of hand querns in rural areas of Slovenia. Her research provided an overview of ethnological and archaeological research into hand mills in Slovenia (Smerdel 2003:468). She also mentioned the manufacture of hand mills at the quarries of Donack Gora and Rifnik (Smerdel 2003:468). The raw material for querns and the black millstones from the Rifnik quarry are referred to as "'conglomerat calcaire du miocene moyen' with 'grains de silice'" in French (Smerdel 2007, personal communication). Inja Smerdel graciously shared photographs of the Rifnik quarry (Figures 66–68). Smerdel (2003:468) also explores the role of hand mills in the lives of women. An English translation of another article by Smerdel (2006:127) is entitled "Stones, People and Oxen. Memories on the Work in Millstone Quarries in Slovenia." She discusses both well-known and recently discovered millstone quarries (Smerdel 2006:127). Further, Smerdel (2006:127) compiled ethnological information from accounts of individuals that worked in the Slovenian millstone quarries. Črnilec (2003) has written about the manufacture of conglomerate millstones near the community of Naklo. Finally, Horvat and Župančič (1987) have published the results of their petrographic analysis of prehistoric and Roman querns from western Slovenia.

Figure 66. *A Close-up View of an Unfinished Millstone on the Quarry Face at the Rifnik Quarry Near Šentjur by Celje, Slovenia. Note the depressions where other millstones and querns were removed at this ancient quarry, existing "from times of yore" in the local oral history. (There are indeed some late Roman querns from Rifnik archaeological sites in the collection of the Celje Regional Museum, that seem to be manufactured from the same stone.) Photograph taken during the early spring of 2002 by Inja Smerdel, Slovene Ethnographic Museum, Ljubljana, Slovenia.*

Figure 67. *A Close-up View of an Unfinished Quern on the Quarry Face at the Rifnik Quarry Near Šentjur by Celje, Slovenia. Photograph taken during the early spring of 2002 by Inja Smerdel, Slovene Ethnographic Museum, Ljubljana, Slovenia.*

Figure 68. *Stanko Zupanc (born 1927), the Present Owner of the Rifnik Quarry and an Excellent Informant. He is here telling the story of millstones and querns manufactured in their quarry until the 1960s. Photograph taken during the early spring of 2002 by Inja Smerdel, Slovene Ethnographic Museum, Ljubljana, Slovenia.*

Three Slovenian millstone quarries are described in French on the Millstonequarries.eu website (*http://meuliere.ish-lyon.cnrs.fr*). The information was reported to the website by Alain Belmont with the cooperation of Inja Smerdel (during September of 2007). The Jama quarry is located at Jama near the community of Mavčiče in the Gorenjska region. Between 50 and 100 conglomerate millstones (120 to 130 cm in diameter) were estimated to have been pro-

Figure 69. *A Close-up View of the Quarry Face at the Polica by Naklo Quarry Near Kranj, Slovenia. This part of the quarry was exploited until 1970s. It's surface clearly shows the conglomerate structure of the so-called labora stone ("conlomerat quaternaire," or even "tertiaire," with "des cailloux gris," mostly "calcaires"), in the area for centuries used to manufacture the so-called white millstones. Photograph taken during the summer of 2005 by Inja Smerdel, Slovene Ethnographic Museum, Ljubljana, Slovenia.*

duced at this quarry during the 19th and 20th centuries (Smerdel 2006). The second quarry, Poliča, is also in the Gorenjska region but is located at Poliča near Naklo (Figure 69). An estimated 5,000 to 7,000 conglomerate millstones (ca. 180 cm for earlier stones and 80 to 92 cm in diameter for the later millstones) were removed from this quarry during the 17th century to 1974 (Smerdel 2006). The Podgrad quarry is the third site which is located at Podgrad, Starigrad, near Ljubljana in the Osrednjeslovenska region. Conglomerate millstones (80 to 110 cm in diameter and 20 cm thick) were quarried from the 16th to 18th centuries at this quarry (Smerdel 2006). It is estimated that perhaps 10,000 millstones were produced at the Podgrad quarry. One photograph of the quarry shows round extraction depressions on a nearly vertical cliff face. Another photograph illustrates a completely shaped millstone that is still attached to the bedrock.

Spain

The Millstonequarries.eu website (*http://meuliere.ish-lyon.cnrs.fr*) contains information on 111 millstone quarries in Spain. The information available on the website was provided by several individuals and is primarily written in Spanish and French. Quarries were reported for several regions with the following numbers: Aragón (n = 1), Cataluna (n = 2), Comunidad de Andaucía (n = 8), Comunidad Autónoma de Castilla y León (n = 31), Euskadi/ País Vasco (n = 3), Comunidad Floral de Navarra (n = 10), Comunidad de La Rioja (n = 20), Islas Canarias (n = 4), and Isla de Menorca (n = 32). The author of this book is not aware of any articles in English concerning the Spanish millstone quarries. A very brief summary of the Spanish millstone quarries is provided below, although specific information is provided for those quarries for which Pilar Pascual Mayoral and Pedro García Ruiz shared photographs. Researchers fluent in Spanish and French are directed to the above website for details about these quarries. Articles published in Spanish and French are available for some of these quarries. Apparently, many of these quarries are not yet included in the archaeological literature.

Two regions of Spain have a total of three recorded quarries. In the region of Aragón, a single millstone quarry has been recorded. This quarry is the Prov. de Zaragoza, Cerro Redondo (Pardos Abanto), quarry which is near Abanto (*http://meuliere.ish-lyon.cnrs.fr*). Querns and millstones were made from a porphyry granite during the Roman and Medieval periods. María Pilar Lapuente Mercadal and Tim Anderson placed this information on the website during October 2007. A reference for this quarry is the publication by Cisneros Cunchillos, Lapuente Mercadal, Magallón Botaya, and Ortiga Castillo (1985). Two millstone quarries have been reported (*http://meuliere.ish-lyon.cnrs.fr*) for the Cataluna region. The Olot quarry was located somewhere near Olot, Spain, but the precise location is not known. A volcanic basalt was used for Roman millstones. Fritz Mangartz provided the information to the website during February of 2007. The Olot quarry is mentioned by Williams-Thorpe (1988). The second quarry, Tarragona, is located near Tarragona, Spain, at Pedrera de El Mèdol. Millstones were produced at this quarry during Roman times from an unknown type of stone. Natalià Alonso provided information to the website concerning this quarry in February 2007.

Eight millstone quarries have been reported in the Comunidad de Andaucía region of Spain (*http://meuliere.ish-lyon.cnrs.fr*). Most of these quarries produced millstones during the modern era, but one dated to Roman times and two were used the Medieval period. Five of the six quarries exploited limestone and one produced millstones from a sandstone. Timothy Anderson reported most of these quarries but was assisted by Angel Serrano García for two quarries. Reyes Mesa (2006) is cited as a reference to the Prov. de Granada–Moclín quarry, the Prov. de Granada–Loja–Camino del Calvario, and the Prov. de Granada–Loja–Cerro de la Fuente Santa.

The Comunidad Autónoma de Castilla y León region contains 31 known millstone quarries (*http://meuliere.ish-lyon.cnrs.fr*). All these quarries were reported to the website by Pilar Pascual Mayoral and Pedro García Ruiz between February 2007 and March 2008. Nearly all the quarries exploited conglomerate deposits but some may have quarried limestone. Most of the quarries in this region date to the modern era. Four quarries were used during the Medieval period and two during the Roman period. Three of these quarries are illustrated here due to the kindness of the recorders. First, the Prov. de Soria–Fuentelárbol quarry is located at Fuentelárbol near Quintana León. Approximately 300 millstones were at this locality. A photograph shows millstones that have been placed end to end to form a long fence (Figure 70). Second, the Prov. de Soria–Fuentelárbol 2 quarry is located at Las Canteras near Fuentelárbol (Figure 71). This modern era quarry exploited a sedimentary rock to produce about 1,000 millstones. Third, the Prov. de Soria–Canos–La Cuerda quarry was a source of conglomerate millstones (1 to 1.3 m in diameter) during the Medieval period and the modern era (Figure 72). This quarry is located at Canos (La Cuerda) near Aldehuela de Periáñez.

Three millstone quarries have been reported from the Euskadi/ País Vasco region of Spain (*http://meuliere.ish-lyon.cnrs.fr*). Pilar Pascual Mayoral and Pedro García Ruiz reported all three of these quarries to the website in November 2007. The Prov. Álava (Araba)–Barambio-Garrastatxu quarry is located at Barambio (Santuario de Barambio) near Amurrio. A sedimentary rock (conglomerate?) was quarried at this locality. Millstones produced at the quarry ranged between 1 and 1.3 m in diameter. The Prov. de Vizcaya–Arbaitza–Barrio Arbaitzarte quarry is located at Arbaitza (Barrio Arbaitzarte) near Orozko. During the 18th to 20th centuries, more than 50 millstones were made from a sedimentary rock at this quarry. The third quarry, Prov. de Vizcaya-Manzarraga, is also near Orozko. This modern quarry exploited a sedimentary rock during the modern era.

A total of ten millstone quarries have been reported in the Comunidad Floral de Navarra region of Spain (*http://meuliere.ish-lyon.cnrs.fr*). Most of these quarries were used during the modern era and three date to the 20th century. At least one quarry was also used during the Medieval period. Various sedimentary rocks were quarried including sandstone and conglomerate. Some of the quarries produced millstones between 1 and 1.2 m in diameter. Pilar Pascual Mayoral and Pedro García Ruiz documented all of these quarries.

Twenty millstone quarries have been documented in the Comunidad de La Rioja region of Spain (*http://meuliere.ish-lyon.cnrs.fr*). All of these quarries were reported to the website by Pilar Pascual Mayoral and Pedro García Ruiz between February 2007 and January 2008. Sedimentary rocks, most conglomerates, were quarries for millstones and querns in this region. The majority of the quarries were used during the modern era. One Roman era quarry, one Medieval, and one "Celtibero (Hierro II)" were also reported. The researchers shared photographs for three of these quarries. The first quarry is the San Vicente de Robres (Robres del Castillo) quarry near Robres del Castillo was used during the modern period (Figure 73). Conglomerate millstones between 1 and 1.7 m in diameter were produced at this quarry. A second quarry, Robres del Castillo, is located at Los Molares. Large millstones (1.4 to 1.8 m in diameter) were made from a sedimentary rock (Figure 74) at this quarry during the modern era. Finally, the San Vicente de Robres 2 (Robres del Castillo) quarry is another modern era quarry located near San Vicente de Robres (Figure 75). A conglomerate was quarried at this location.

Publications are available for several millstone quarries in the Comunidad de La Rioja

Figure 70. *Millstones Placed in a Vertical Position (End to End) to Form a Fence at the Fuentelárbol Millstone Quarry, C. A. Castilla y León, Soria, Spain. Photograph taken during October 2004 by Pilar Pascual Mayoral and Pedro García Ruiz, Logroño, Spain.*

Figure 71. *General View of the Fuentelárbol 2 Millstone Quarry, C. A. Castilla y León, Soria, Spain. Photograph taken during March 2007 by Pilar Pascual Mayoral and Pedro García Ruiz, Logroño, Spain.*

Figure 72. *General View of the Canos–La Cuerda Millstone Quarry, C. A. Castilla y León, Soria, Spain. Note the partially finished millstone in the foreground. Photograph taken during August 2007 by Pilar Pascual Mayoral and Pedro García Ruiz, Logroño, Spain.*

Figure 73. *Close-up View of a Partially Completed Millstone at the San Vicente de Robres Quarry, C. A. La Rioja, Spain. Photograph taken during February 2007 by Pilar Pascual Mayoral and Pedro García Ruiz, Logroño, Spain.*

Figure 74. *Close up View of a Partially Completed Millstone, Robres del Castillo Quarry, C. A. La Rioja, Spain. Photograph taken during February 2007 by Pilar Pascual Mayoral and Pedro García Ruiz, Logroño, Spain.*

Figure 75. *Close-up View of a Partially Completed Conglomerate Millstone, San Vicente de Robres Quarry, C. A. La Rioja, Spain. Note the circular groove cut into the bedrock around the millstone. Photograph taken during February 2007 by Pilar Pascual Mayoral and Pedro García Ruiz, Logroño, Spain.*

region including San Vicente de Robres (Robres del Castillo) quarry (Pascual Mayoral and García Ruiz 2001, 2002, 2003c), Jubera (Santa Engracia de Jubera) quarry (Pascual Mayoral and García Ruiz 2001, 2002, 2003c), the Torrecilla en Camerous quarry (Pascual Mayoral and García Ruiz 2003b), the Arnedillo quarry (Pascual Mayoral and García Ruiz 2002), Robres del Castillo quarry (Pascual Mayoral and García Ruiz 2003c), Jubera 2 (Santa Engracia de Jubera) quarry (Pascual Mayoral and García Ruiz 2003c), the Robres del Castillo 2 quarry (Pascual Mayoral and García Ruiz 2003c), the Igea quarry (Pascual Mayoral and García Ruiz 2003a; Pascual Mayoral and Moreno Arrastio 1980), the Villaroya quarry (Pascual Mayoral and García Ruiz 2003a; Pascual Mayoral and Moreno Arrastio 1980), the Grávalos quarry (Pascual Mayoral and García Ruiz 2003a; Pascual Mayoral and Moreno Arrastio 1980), the Muro de Aguas quarry Pascual Mayoral and García Ruiz 2003a), the Robres del Castillo 3 quarry (Pascual Mayoral and García Ruiz 2003c), the Luezas quarry (Pascual Mayoral and García Ruiz 2003c), and the Almarza de Cameros quarry (Pascual Mayoral and García Ruiz 2003b).

Four millstone quarries have been reported on Spain's Islas Canarias (Canary Islands) located in the North Atlantic Ocean near southern Morocco (*http://meuliere.ish-lyon.cnrs.fr*). The Cantera de La Calera (La Suerte-Las Piletas) quarry located near Agaete on the island of Gran Canaria was reported by Amelia Rodríguez Rodríguez in March of 2007. Volcanic stone was exploited for millstones 35 50 cm in diameter during the prehistoric period. A quarry photograph shows several millstone preforms isolated on a nearly vertical rock face. The Cantera de El Queso quarry is located near Santa Lucía de Tirajana on Islas Canarias. At this location, millstones were manufactured from volcanic deposits. The quarry was reported by Amelia Rodríguez Rodríguez in March of 2007. A quarry photograph shows scars on a nearly vertical rock face where millstone had been cut. The third quarry on the Canary Islands is Cantera de Riquiánez quarry located near Las Palas de Gran Canaria. It was reported by Amelia Rodríguez Rodríguez in March of 2007. A volcanic basalt was exploited for millstones. The final quarry reported on the Canary Islands is Cantera de Cuatro Puertas near Telde. It was reported by Amelia Rodríguez Rodríguez in March of 2007. Millstones were made from a basalt. A photograph of the quarry shows circular millstone extraction areas on a nearly vertical rock face. References cited for all four quarries were Rodríguez Rodríguez and Barrosa Cruz (2001) and Rodríguez Rodríguez, Martín Rodríguez, Mangas Viñuela, González Marrero and Buxeda-Garrigós (2006).

A total of 32 quarries have been reported for Spain's Isla de Menorca (*http://meuliere.ish-lyon.cnrs.fr*). The island of Menorca (or Minorca) is one of the Balearic Islands off the east coast of Spain. Joaquin Sanchez Navarro has personally recorded all these quarries. The period of use for all these quarries was between the ninth and thirteenth centuries A.D. It is assumed that people returned to these quarries repeatedly over the centuries when querns were needed. The quarries are located along the coast of the island and exploited a sedimentary rock. Quarries ranged from small extraction areas to quarries that produced large quantities of querns. Mr. Sanchez Navarro graciously shared some photographs for this book. The first quarry in these photographs is Punta de Sa Mioca (Figure 76) which is located on the coast; it exploited a sedimentary rock for querns. It is estimated that 16,000 to 25,000 querns were produced at the quarry. Photographs of the quarry show an expanse of horizontal bedrock with round extraction holes. Second, the Morro Llevant-ses Anglades-Cap d'en Font quarry (Figure 77) is located on the coast at Morro Llevant-ses Anglades-Cap d'en Font near Sant Lluís. This quarry (Figure 78) is in a sedimentary rock where an estimated 750 to 2,250 millstones were produced. An example of a completed millstone from S' Aranjif on Menorca is illustrated in Figure 79.

Figure 76. *General View of the "Es Corral fals–Punta de sa Miloca" Millstone Quarry on the Island of Minorca, Spain. Photograph taken by Joaquín Sánchez Navarro, Ciutadella de Menorca, Illes Balears, Spain.*

Figure 77. *Partial View of the "Morro Llevant–Ses Anglades" Millstone Quarry on the Island of Minorca, Spain. Photograph taken by Joaquín Sánchez Navarro, Ciutadella de Menorca, Illes Balears, Spain.*

Figure 78. *Detail of a millstone extracted at the "Morro Llevant–Ses Anglades" Millstone Quarry on the Island of Minorca, Spain. Photograph taken by Joaquín Sánchez Navarro, Ciutadella de Menorca, Illes Balears, Spain.*

Figure 79. *Millstone of S'Aranjí on the Island of Minorca, Spain. Photograph taken by Joaquín Sánchez Navarro, Ciutadella de Menorca, Illes Balears, Spain.*

Several studies have provided descriptions of the millstone and quern quarries of Spain. Most of these sources have been cited in the discussions above, but for the reader's benefit they are here listed together, with their Spanish titles approximately translated into English: Cisneros Cunchillos, Pilar Lapuente Mercadal, Magallón Botaya, and Ortiga Castillo (1985) published "Archaeological-Geological Study of Round Hill (Pardos, Zaragoza)" which includes millstones. Rodríguez Rodríguez and Barrosa Cruz (2001) produced a study entitled "Working the Stone to Grind the Grain: The Prehistoric Exploitation of the Mill Quarries of tufa in the Grand Canary Islands." Pascal Mayoral and Garcia Ruiz (2001) wrote about the "Quarries and Technology of Millstones in the Jubera River (the Rioja)." The following year, Pascal Mayoral and Garcia Ruiz (2002) published "New Quarries of Millstones (and Presses), Valley of Cidacos." In 2003 Pascal Mayoral and Garcia Ruiz produced three studies: (2003a) "Millstone Quarries and Presses, Basin of Linares River: Wall of Water, Villaroya and Gravalos"; (2003b) "New Discoveries of Millstone Quarries, Iregua River"; (2003c) "The Millstone Quarries: An Unknown Industry of Rioja." An earlier study by Pascal Mayoral and Moreno Arrastio (1980) was "Roman Oil Presses in La Rioja." Reyes Mesa (2006) dealt with millstone quarries in "The Hydraulic Mills: Flour Mills of the Province of Granada." Rodríguez Rodríguez, Martín Rodríguez, Mangas Viñuela, González Marrero and Buxeda-Garrigós (2006) wrote a study entitled "The Exploitation of the Lithic Resources in the Grand Canary Islands. Towards the Reconstruction of the Social Relationships of Production in the Pre-European and Colonial Epochs." Three studies dealing with the quern industry on the Island of Menorca are published in French (Sanchez Navarro 2001, 2005, 2006). Two other studies for Spain include Barbera Miralles' (2003) study of Catalan millstones and Harrison's (2002) article entitled "Quernstones from a Survey Around Nájera, Spain."

Sweden

Extensive millstone quarries are present in several areas of Sweden. A very useful bibliography entitled "Millstones-literature Scandinavia" was prepared by Holger Buentke (2003) which listed many sources on the millstone industry in Sweden (largely the Lugnås area where Mr. Buentke lives) and to a lesser extent other Scandinavian countries. Most of this literature is published in Scandinavian languages and is thus not accessible to English readers. Kresten, Elfwendahl, and Pettersson's (1996:80) study of quernstones, grindstones, and hones from Sweden mentioned several locations where quernstones (small millstones) were quarried. Areas producing quernstones included Storsjö area (sandstone), Roslagen (sandstone), Malung (metamorphosed sandstone), Yttermalung (metamorphosed granite), Lugnås (kaolinized gneiss), and Frostviken (garnet-schist) (Kresten, Elfwendahl, and Pettersson 1996:80). Unfortunately, as Kresten, Elfwendahl, and Pettersson (1996:78) observe: "millstones and quernstones have attained little interest only in Swedish archaeological reports."

Among the best-known Swedish millstone quarries are those located near Lugnås. These quarries are southwest of Stockholm between lakes Vanern and Vattern. An excellent booklet was prepared by Berit Hange-Persson (1982b) entitled *Kvarnstensbrotten i Lugnås* that discusses the millstone quarries of Lugnås and includes many historic photographs. An useful website (*www.qvarnstensgruvan.se*) in Sweden provides information on "The Millstone of Lugnås, The Abbey of Varr and The Abbey Ruin of Gudhem." The website provides the following information about the quarries:

> Millstones have been made here for 800 years and man's industry has left traces all around the hills. Dense deciduous forest hides much of the mile-long heaps of discarded stone, but

> all around you can see the heritage left behind by old inhabitants of Lugnås in the form of around 600 opencasts and 55 mines one of which is open to the public....
>
> According to spoken tradition mining was begun by monks in the 12th century and it grew into an industry which in the 1850's became larger than all the other industries in Mariestad together. The Lugnås dominance over the market can be explained by the good supply of suitable gneiss rich in feldspar and kaolin. The quarrying was done using simple tools and methods in opencasts or in mines. The work was very hard and many quarrymen died young.

While many publications mention the millstone quarries at Lugnås, a few examples mentioned by Buentke (2003) and others will suffice. Carelli and Kresten (1997) studied Late Viking Age and Medieval querns from the southern part of Scandinavia. Medieval querns from the Uppsala area of Sweden were subjected to geoarchaeological study (Elfwendahl and Kresten 1993). Hange-Persson (1979, 1982a, 1982b) published on the millstone quarries at Lugnås. Kresten (1996b, 1996d) and Pettersson (1977a) also published on the quarries of Lugnås. Buentke and Gustafsson (2006) published a paper presenting the results of their experiments on the quarrying and shaping millstones at a mine at Lugnås. A recent overview of the millstone mines at Lugnås was prepared by Beiron (2006). Situated 15 km from Mariestad, the quarries at Lugnås were extensively worked between A.D. 1145 and 1915 (Beiron 2006:203). Only querns were manufactured in the early centuries but a variety of stone sizes were produced beginning in the late 18th century (Beiron 2006:203). Beiron's article provides some important details about the about the millstone industry at Lugnås and includes

Figure 80. *A Mine Opening at the Lugnås Millstone Quarry, located at Lugnås, Near Mariestad in the Västra Götalands Län Region of Sweden. Photograph taken by Ingemar Beiron, Mariestad, Sweden.*

Figure 81. *Rejected Millstone at the Lugnås Millstone Quarry, located at Lugnås, Near Mariestad in the Västra Götalands Län Region of Sweden. Photograph taken by Ingemar Beiron, Mariestad, Sweden.*

Figure 82. *Display of Millstones Produced at the Lugnås Millstone Quarry, located at Lugnås, near Mariestad in the Västra Götalands Län Region of Sweden. Photograph taken by Ingemar Beiron, Mariestad, Sweden.*

some interesting historic photographs with millstone quarry workers. Beiron, of Mariestad, Sweden, graciously shared three photographs for this book: a mine opening at the Lugnås Millstone Quarry (Figure 80), a rejected millstone at the Lugnås Millstone Quarry (Figure 81), and a display of millstones produced at the Lugnås Millstone Quarry (Figure 82).

Another important millstone quarry area was at Malung which is northwest of Stockholm in western Dalecarlia. High quality quernstones were manufactured at Malung during the Middle Ages (Kresten, Elfwendahl, and Pettersson 1996:80). Among the many publications that mention the millstone quarries at Malung are Buentke 2003; Carelli and Kresten 1997; Jacobson 1998; Kresten and Elfwendahl 1994; and Pettersson 1973, 1977b, 1981. Kresten (1998) examined the millstone quarries at Malung from the Viking Age to modern times. Jacobson's (1998:12) produced an excellent booklet entitled *Upptäck Kvarnstensbrottet i Östra Utsjö, Malung*. Jacobson's booklet explains that over the span of several hundred years, at least 30,000 pairs of millstones were likely quarried at Malung. This huge quarry is approximately 1,700 meters long with a width varying between 50 and 170 meters. Oral tradition suggests that the quarry can be divided into the "Old Mountain" and the "New Mountain" areas. Millstones were quarried from the Dalarna sandstone which was considered to be self-sharpening (Jacobson 1998).

In the vicinity of Malung, several millstone quarrying areas were mentioned by Kresten, Elfwendahl, and Pettersson (1996:86), who stated that south of Malung at Kvarnberget (near Östra Utsjö), extensive quarry remains were distributed over 1 km. They noted that the Kvarnberget quarries could be separated by age into the earlier "Gamla Kvarnberget" portion and the later "Nya Kvarnberget" portion. Other quarrying remains in the vicinity included "small prospecting pits and quarries ... at Byråsen and Årstjärnen, west of Malung, and still smaller ones near Gärdås and Västra Fors, northwest of Malung.... Some minor quarrying occurred at Äfsålberget, about 7 km SSW of Yttermalung" (Kresten, Elfwendahl, and Pettersson 1996:86).

Glacial boulders were also exploited for millstones in Sweden. Kresten, Elfwendahl, and Pettersson (1996:81) noted that "[a]nother major rock type represented among the quernstones from Uppsala and in the surrounding province of Uppland is jotnian sandstone, quarried from glacial boulders in the Storsjö area of Gästrikland and the coast of Roslangen in Uppland."

Five Swedish millstone quarries are described in French on the Millstonequarries.eu website (*http://meuliere.ish-lyon.cnrs.fr*). The information was reported to the website by Alain Belmont during March and August 2006. The first quarry is Östra Utsjö at Kvarnstensbrott near Malung in the Dalarnas Län. Approximately 30,000 millstones (60 cm to 1.2 m in diameter) were produced at this at this quarry. The quarry was in use between the 9th century and 1880 (Jacobson 1998). The Lugnås quarry is located at Lugnås, Kvarnstensgruva, near Mariestad in the Västra Götalands Län region. A gneiss rock was exploited at Lugnås between the 13th century and 1919. The third quarry is Gislovshammar in the Skåne Län region near Simrishamn. Millstones were made from a sedimentary rock (limestone?) during the 18th century at this quarry. Fourth, the Suède quarry is located at Steinskogen near Höör in the Län region. Between the years A.D. 1100 and 1920, millstones (87 cm to 1.8 m in diameter) were manufactured from a conglomerate at this quarry. The final quarry is Dals-Rostock located at Grönhult near Mellerud in the Västra Götalands Län region. A sedimentary rock was quarried for millstones (40 cm to 2 m but generally 1 to 1.2 m in diameter) from the Middle Ages to the beginning of the 20th century.

Switzerland

Several articles have been published on the quern quarries of Switerland. Most of these studies were published in French. These studies include: Anderson 2005; Anderson, Castella, Doswald, and Villet 2004; Anderson, Duvauchelle, Serneels, and Agustoni 2000; Anderson, Villet, and Doswald 2002; Anderson, Villet, and Serneels 1999; Anderson, Duvauchelle, and Agustoni 2001; Castella and Anderson 2004. The work by Timothy Anderson and his colleagues has primarily focused on a Roman era quarry in eastern Switzerland near Bern. A comprehensive book on the Châbles quern quarry in eastern Switzerland was published in 2003 and was entitled *Des Artisans à la Campagne: Carrière de Meules, Forge et voie Gallo-Romaines à Châbles (FR.)* (Anderson, Agustoni, Duvauchelle, Serneels, and Castella 2003). In his review of this monograph, Peacock (2004:650) noted that the book "presents detailed information on rural industries, giving a vivid picture of quarrying for querns and building stones and iron-smithing between the 1st and 3rd c. A. D." Concerning the preservation of the Châbles quern quarry he stated that "the marks in the rock face are remarkably clear, and it is possible to use them to work out in detail the precise way in which the querns were marked out, extracted and dressed" (Peacock 2004:650).

Anderson (2005) presented a paper in English at the second international millstone conference entitled "The Three Roman Rotary Handmill Quarries in Switzerland." The published version of Anderson's (2006) paper "Three Roman Quern Quarries in Switzerland" provided interesting details about the quarries. All three of the quarries exploited the same raw material, a stone called *grès coquillier* in French, which could be called "shell limestone" in English (Anderson 2006:42). The excavations at the Châbles quarry site in the Canton of Fribourg in western Switzerland were conducted in advance of highway construction (Anderson 2006:42). Anderson (2006:42–43) estimated that as many as 450 querns were potentially extracted at the 70 m^2 quarry site located near a major Roman road. The author was able to reconstruct the methods employed for quarrying these ca. 45 cm diameter querns (Anderson 2006:41–42). After a central point was established, the outline of the quern was traced onto the top of the stone (Anderson 2006:42). Anderson (2006:42) described the shaping process as follows: "To insure the visibility of the tracing, the circumference was pecked. Then a circular trench was dug around the roughout with a pick. This work methodically followed three vertical planes of work ... producing the multiple diagonal tool marks typical of quarry fronts. The cylinder was split from the bedrock by multiple strokes placed at regular intervals (approximately 10) at its base. This activity produces concentric marks at the base."

Following the archaeological excavations, stone blocks from the quern quarry were carefully transported by truck to a large shelter for storage (Anderson 2006:44). It is hoped that these blocks, with the quern extraction areas, can be displayed in a museum at a future date (Anderson 2006:44).

Two other quern quarries have been documented in Switzerland including Châvannes-le-Chêne and Würenlos. Châvannes-le-Chêne is a Roman era quern quarry at Vaud about 4 kilometers from Châbles (Anderson 2006:44). This quarry was previously reported by Bosset 1943 and Weidmann 2002. The Châvannes-le-Chêne quarry is slightly larger than Châbles and has nearly identical extraction areas (Anderson 2006:44). It was initially unearthed in 1943 and largely subjected to 50 years of weathering which erased most of the tool marks (Anderson 2006:44). When the quarry was cleared in 2002 the portions that had been covered with soil had survived the weathering (Anderson 2006:44). Klausener (2001) also wrote about the Châvannes-le-Chêne quern quarry. Alain Belmont shared two photographs of the

Figure 83. *The Roman Era Chavannes-le-Chêne Millstone Quarry (Canton de Vaud) Located in Northwestern Switzerland. Photograph taken during August 2006 by Alain Belmont, University of Grenoble, LARHRA, Grenoble, France.*

Figure 84. *The Roman Era Chavannes-le-Chêne Millstone Quarry (Canton de Vaud) Located in Northwestern Switzerland. Close-up view of stack of abandoned millstones. Photograph taken during August 2006 by Alain Belmont, University of Grenoble, LARHRA, Grenoble, France.*

Châvannes-le-Chêne quern quarry (Figures 83–84). The other quern quarry, Würenlos, is located in the canton of Aargau in northern Switzerland (Anderson 2006:45). The quarry was mentioned in earlier studies by Haberbosch 1938 and Doswald 1994. Anderson (2006:45) indicated that the Würenlos quarry was used for the manufacture of small rotary querns like those found at Châbles and Châvannes-le-Chêne. In 1938 when the quarry was first found, it was already impacted by modern construction and then became overgrown (Anderson 2006:45). The quarry was cleared of vegetation in 1994 and studied but no actions were taken to protect this site (Anderson 2006:45).

Four Swiss millstone quarries are described in French or German on the Millstonequarries.eu website (*http://meuliere.ish-lyon.cnrs.fr*). Three of the quarries (Châbles, Châvannes-le-Chêne, and Würenlos) are described above. A fourth quarry, the Murist millstone quarry, was reported by Timothy Anderson. The Murist millstone quarry is located in the Canton de Fribourg region near Murist which exploited a sedimentary rock during the Medieval period (Anderson, Agustoni, Duvauchelle, Serneels, and Castella 2003).

Turkey

A recent study examined Early Neolithic saddle quern production at the Coşkuntepe site located in the Troad region of northwestern Turkey (Takaoğlu 2005:419). Coşkuntepe was a coastal settlement facing the island of Lesbos that appears to have included quern making as part of its economy (Takaoğlu 2005:419, 424). Archaeological evidence for the quern industry found around the site included "tested blocks of stone, discarded and broken preforms, used and broken hammerstones, and percussion flakes of various sizes" (Takaoğlu 2005:426). The raw materials used for the querns consisted of andesite and trachyte-andesite (ca. 80 percent of total debris) and basalt (15 percent of debris) (Takaoğlu 2005:426). Takaoğlu (2005:428–430) defined four major stages for making querns at Coşkuntepe. Currently, no other Neolithic sites in the Troad region or surrounding areas have yielded any of these querns (Takaoğlu 2005:430). However, it is of interest to note that later sites in the Troad region, dating from the Early Bronze I period onward, have produced querns (Takaoğlu 2005:430).

Studies on Millstones and Querns of Ancient Rome

It seems appropriate to include this special section on Ancient Rome, since the influence of Rome extended over much of Europe at one time. The millstone quarries in Italy exploited leucitite outcrops near Orvieto, Umbria (Peacock 1986). The products of these quarries were shipped throughout the areas controlled by the Romans. In Morgantina, Austria millstones were made prior to the Greeks and continued until Roman times (White 1963). They were made from lava. Both the Italian and Austrian quarries yielded millstone forms that predate those used in the United States by several centuries.

Concerning querns used during the Roman period, Bennett and Elton (1898a:157) stated that "Rome, in the age of querns, achieved a triumph in the use of the precise stone which up to recent times remained the finest material for millstones known to Europe. This, derived from the famous quarries at Andernach [Germany], on the Rhine, the Romans used very extensively for both quern and water mills, and popularized its adoption in every country whither their influence extended. Many ancient specimens of querns of the Roman type, including some of the finest examples, found in England and France, are constructed of stone imported from Andernach: probably the articles themselves having been manufactured there."

A geochemical study of millstones recovered from the Red Sea area of Egypt, Jordan, Sudan, and Eritrea was undertaken by Williams and Peacock (2006) to determine the source of the raw material. Initially, the authors set out to determine the source of basalt from ship ballast dumped at Egyptian sea ports (Williams and Peacock 2006:35). Originally, it was thought that some of the dumped basalt was used for millstones. However, geochemical analysis of the samples revealed that the ballast was an alkali basalt whereas the millstones were made from a basaltic andesite; they found no matches for the raw material used for millstones in Near East sources (Williams and Peacock 2006:38). The only match for this material was the islands (Aegina, Melos, Santorini, and Nisyros) in the Aegean volcanic arc of the Mediterranean. A visit to the island of Nisyros by Peacock revealed "unfinished hopper rubber or lever mills ... at quarries on the island, with some evidence for rotary mill production as well" (Williams and Peacock 2006:38–39).

Other studies dealing with various aspects of Roman millstones and millstone quarries include: Antonelli, Lazzarini, and Luni 2005; Williams-Thorpe 1986, 1994; Williams-Thorpe and Thorpe 1987, 1988, 1989, 1991, and 1993. Many of these studies focus on the trade and origin of Roman millstones throughout the Mediterranean (Peacock 1980a; Renzulli, Santi, Nappi, Luni and Vitali 2002; Williams-Thorpe and Thorpe 1988, 1993). For other sources on Roman millstones and querns, see the sections on Great Britain, France, and Germany.

Miscellaneous Millstone Quarries and Studies

There is limited literature about the millstone industries of several countries. Most of these studies were published in languages other than English. Literature is available for Angola (Cruz 1971), Belgium (François 2003), China (Roberts 1989), Cypress (Egoumenidou and Myrianthefs 2003; Elliott, Xenophontos, and Malpas 1986; Williams-Thorpe, Thorpe, and Xenophontos 1991), Denmark (Bloch Jorgensen n.d.), Egypt (Fahmy 2001; Samuel 1993); Georgia (Beduckadze 1956; Bregadze 1989; Relgniez 2003, Tchilakadze 1985), India (Sankalia 1959), Israel (Amiran 1956), Jordan (Philip and Williams-Thorpe 1993), Morocco (El Alaoui 1999, 2003), North Africa (Gast 2003), Norway (Rønneseth 1968), Poland (Dworakowska 1983; Nasz 1950), South Africa (Walton 1953), Saharan Africa (Milburn 1992, 2002), and Yugoslavia (Sprague 1979). Bennett and Elton (1898a:156) mentioned that ancient millstone quarries existed at Magnesia (near Smyrna in western Turkey), in Macedonia (Greece), and at Mona or Anglesea in Britain. Other millstone studies include Croudance and Williams-Thorpe 1988 (x-ray fluorescence analysis); Loubès 1983; Procopiou, Jautee, Vargiolu, and Zahouani 1996 (quern from Syvritos Kephala); and Small and Bruce 1994 (millstones excavated at San Giovanni di Ruoti). Williams-Thorpe and Thorpe (2004) deal with millstones recovered from the 11th century shipwreck *Serce Limani*.

There are several studies that specifically deal with querns. These include: Beranová 1988 (in Czechoslovak Territory); Childe 1943 (in Mediterranean Basin); Crawford 1953 (querns and quern quarries); Crawford and Röder 1955 (quern quarries of Mayen, Germany); Curwen 1937 (querns), Curwen 1941 (querns); Frankel 2003 (querns in Talmudic literature); Hunt and Griffiths 1992 (volcanic stones used for querns); Jecock 1981 (prehistoric querns in Wessex, England); Lundström 1961 (querns from excavations at Helgö I); Lundström and Lindeburg 1964 (querns from excavations at Helgö II); Marshall, James, Bevins, and Horák 2003 (quernstones from Roman Carmarthen); Neri 1993 (saddle querns in antiquity); Philip and Williams-Thorpe 2000, 2001 (groundstone of Southern Levant during 5th to 4th centuries

B.C.); Rogers 1988 (querns); Samuel 1993 (saddle querns in Egypt); Sankalia 1959 (querns from India); Selmeczi Kovács 1990 (querns in Hungary); Smith 1978 (lava querns); and Summer 1976 (ancient middle eastern querns).

Those studies published in English or having an English abstract permit a brief discussion of the millstone or quern research conducted in some of the countries listed above. Roberts (1989) provided information and photographs of many Chinese millstones but unfortunately does not discuss the raw materials used for these millstones. He shared the following information (Roberts 1989:329; his references to figures have been removed):

> We saw — among other things — a remarkably wide variety of querns. Most of these had some common features which distinguished them from those one sees in European mill books. With one exception, all the runner stones had off-centre feed holes in the upper face, leading to a curved tapering opening in the lower (working) face.... All bed stones (except one) were surrounded by a stone trough with a spout.... The dressing in all the querns was based on eight "harps" but the number of furrows per harp varied from six to twelve. All the dresses were for anticlockwise rotation. All the diameters were, except in one case, in the range 38–51 cms.

Other countries include Israel, Mexico, Syria, and Yugoslavia. In the Negev region of southern Israel, excavations at the Camel Site provided evidence for the making and trading of millstones during the early Bronze Age (Rosen 2003). Rosen and Schnider (2001) also reported on the production and exchange of early Bronze Age milling stones in the Negev. Two Mexican studies mention millstones. Durán and Pulido (2007) discussed the use of edge runner millstones in milling agave for mescal manufacture in Mexico. Another Mexican study discussed the use of edge runner millstones to crush maguey pineapples for the juice (Durán-Garcia, Gonzalez-Galvan, and Matadamas 2007). In a study entitled "Charred Plant Remains from a 10th Millennium B. P. Kitchen at Jerf el Ahmar (Syria)," Willcox (2002) discussed saddle querns. Finally, Sprague (1979) wrote a senior honors thesis entitled "The Life Cycle of a Quern: An Analysis of the Grindstones from the Neolithic Site of Selevac, Yugoslavia."

7
Artificial Millstones

There have been several attempts to produce millstones from raw materials other than rocks. Artificial millstones have been constructed from rock emery, glass blocks, wood and emery, porcelain, and cast iron.

Artificial millstones were previously discussed in a book by Apling (1984:17–18). He noted that artificial millstones first appeared as the 19th century concluded. Apling discussed the three basic types of artificial millstones, each of which had a distinctive appearance. These were manufactured from granulated French Burr Stone (a yellow color), granulated rock emery (a brown color), and a basic slag (a black color). The French stone was used for producing food for humans, while the other two types were used for grinding feed for animals. Apling noted that these "materials were bonded together with a magnesium cement, a mixture of magnesite cement and diluted magnesium chloride." Iron bands were placed on the artificial millstones to prevent them from coming apart when operating at high speeds. These stones lasted longer than natural millstones and only had to be redressed as the stone wore away (Apling 1984:17–18).

A brief article appeared in *The Manufacturer and Builder* during 1893 that discussed millstones made from rock emery (Anonymous 1893:231):

> Probably few of our readers have ever seen rock emery and fewer still have heard of millstones made of this hardest of all stones except corundum. Yet rock emery millstones are an accomplished fact, and a long step has been taken towards pulverizing cheaply many hard substances that have heretofore only been reduced at much expense for wear and tear, and by slow and tedious processes. Rock emery is not a common mineral, being found only in a few countries. The best comes from Greece, but the larger importations are from Turkish mines. The consumption is large and its use has become of great importance in many industries, as it grinds all substances with unexampled rapidity. A pure emery face never glazes, but is always sharp and cutting. Rock emery millstones reduce at once the hardest rocks, as well as softer substances, grinding all to any degree of fineness. Heat does emery no harm, and one of the remarkable properties of the emery millstone is their ability to run cool. They form the most rapid grinder known, and are as much more durable than any other millstone as they surpass them in hardness. An emery millstone face is never dressed. They are made to take the place of all other millstones, without any changes in the mill, and whatever other stones are used, rock emery millstones will do better work at less expense, and last much longer. They also grind hard materials that soon destroy all softer millstones. Now that the manufacture of the patent rock emery millstone is understood, they are turned out for all sort of mills and for all purposes, at a moderate price, and are everywhere recognized as wonderful grinders, especially for fine work. These rock emery millstones are ample proof, if any needed, of the progress of American milling.

The following year, *The Manufacturer and Builder* published the following note (Anonymous 1894a:54): "Rock Emery Millstones are said to be rapidly coming into use. It is claimed that they are wonderful grinders, and it seems quite natural blocks of rock emery should cut fast and last longer than anything else."

Parker (1894:671) noted that "the introduction of emery rock millstones will probably cause a still further decline in those made from quartz conglomerate." Additional information was provided by Parker (1895:587) the following year: "During the past few years a new millstone made of emery ore, ground and cemented into solid wheels, has been introduced. It is said to be superior to any of the others, and has certainly been favorably received. The continued decrease in production indicates that the emery-rock millstones have superseded the domestic buhrstones to some extent already."

Schoonhoven (1977:277) discussed artificial millstones called "composition stones." These millstones were comprised of two distinctive layers with different functions. The first layer was for grinding and was usually made from emery or quartz. The second layer was made from sand and gravel and served to strengthen the millstone and give it weight. The percentages of emery and quartz used in the mixture for the grinding layer varied according to the planned purpose of the millstone. For example, a "breakstone" for rye required between 75 and 100 percent of emery while a stone used for wheat was made from 100 percent of quartz. A millstone for grinding feeds, required between 25 and 50 percent of emery and the remainder being composed of quarts. At the time of his article, he noted, "There are, even today, one or two addresses in Holland and Belgium where you can order a composition millstone. They are made to the miller's requirement. The thickness of the grinding layer usually amounts to 150 mm" (Schoonhoven 1977:277):

In 1878, *The Manufacturer and Builder* (Anonymous 1878b:294) reported the use of glass blocks for constructing composite millstones:

> We call the attention of manufacturers who can cast heavy pieces of glass, and also of millers, to a recent German discovery, that the finest flour is produced by those millstones which have the most glassy texture and composition, and the consequent discovery that pieces of glass combined in the same way as the French burr, and similarly grooved on their surfaces, will grind better than the burr millstones. The consequences of this discovery has been the invention of the glass millstones now made by Messrs. Thom, and used in Germany and in Borkendorf with great satisfaction, as it is found that they grind more easily and do not heat the flour as much as is the case with the French burr stone. In grinding grist they run perfectly cold.
>
> In order to make such stone blocks of glass of from 6 to 12 inches sides are cast in a shape similar to the French burrs, but more regular and uniform; they are connected with cement in the same way, and dressed and furrow-cut with picks and pointed hammers; but we believe that diamond-dressing machines might be profitable applied. It is said that these millstones, made of lumps of hard glass, do not wear away faster than the burr stones. Stones of 4½ feet in diameter, driven by 6 horse-power, ground 220 pounds of flour per hour, and did it while remaining cold. The grist is drier, looser, and the hull more thoroughly separated from the kernel than is the case with other stones.
>
> If all this turns out to be correct, it is a valuable discovery, especially when we consider the expensiveness of good blocks of burr.

An artificial millstone was made by a Mr. McKenzie at his barley mill near Malone in Franklin County, New York. His efforts were described by Craik (1877:386–387):

> When this second and last stone went to pieces, McKenzie tried the experiment of a composition stone. For this purpose, he had one made of dry hardwood; it was made of disks or layers of two inch plank, well jointed, and pinned through, the joints being filled and the pins driven with white lead for cement. The outside was then turned off to the size and shape of the other stones and covered over with several coats of glue and emery. This was

> put on in a severe frosty time, and in a hurry, the "stone" being placed near the mill stove to keep it from freezing, and hasten the drying; but parts of the composition were frozen before drying, and other parts overheated and dried too suddenly.
>
> It was hung in place and set to work immediately, and found to hull and scour the barley faster and better than any natural stone, but when examined after running a while, portions of the composition were scaled off. It was afterwards coated anew, the emery this time fastened by a "patent" waterproof glue; this was applied to the stone in its place late in the afternoon, and the next morning it was set to work. Although hot irons were placed under and around the stone to hasten the drying, still it was not so dry as it should have been when set to work, and consequently ... a good deal of the composition again wore off.

Grimshaw (1882:285) made the following comments about artificial millstones: "Discs made up of blocks of porcelain, biscuit, and of glass blocks have been tried, but we have no record of their performance. Cast-iron discs have long been employed for the rougher milling reduction, such as corn grinding, but they are now coming into use for the most delicate reductions of the later and more complicated millings systems. It will be seen that under the head of discs we have a large variety of material, and a still greater number of combinations of methods of mountings, as either the upper-runner, the under-runner, or the disc with horizontal axis, may have either stiff or balanced drive."

Milling machines using iron discs were illustrated by Grimshaw (1882:367–372). These disc mills include machines made by Raymond Brothers' Mill (patented December 30, 1879) and Jonathan Mills' Disc Machines (patented March 28, 1871; December 30, 1875; July 6, 1876; and October 19, 1880) (Grimshaw 1882:368–371). The Raymond Brothers' Mill employed cast iron discs with raised with steel strips or blades. The Jonathan Mills' Disc Machines used iron discs with ridges. For additional discussion on artificial millstone see Hockensmith and Ball (2007).

8

Tools Used in Making and Sharpening Millstones

An important part of millstone making was the specialized tools needed to quarry the stone and then shape it in into millstones. Once the millstones were completed, another set of tools was used by millwrights to cut the grinding furrows into the face of the millstone or re-cut the furrows as the stone dulled.

Tools Used by Millstone Cutters

Various tools were used to quarry and shape millstones. During the past 18 years, the author has had the opportunity to personally examine tools used by millstone makers in Virginia and New York and interview individuals about their manufacture and use. In Montgomery County, Virginia, in 1990, Robert Huston Surface and W. C. Saville were interviewed and the tool kit of Robert Huston Surface was examined (Hockensmith and Coy 1999:45–54; Hockensmith and Price 1999:82–88). In Ulster County, New York, brothers Vincent and Wallace Lawrence and their neighbor Lewis Waruch were interviewed in 1998. Also, a large collection of millstone making tools assembled by Lewis Waruch was documented (Hockensmith and Coy 2008a, 2008b). This section will draw upon the author's previous research and also cite some other sources.

Several tools were used in quarrying bedrock. These tools included sledge hammers, hand drills, wedges and feathers, spoons, and striking hammers. In Virginia, holes were drilled with a steel drill with a flared flat bit (Hockensmith and Coy 1999:51; Hockensmith and Price 1999:84). The drill was hit with eight pound sledge hammers (Hockensmith and Coy 1999:47; McKee 1973:17). Spoons were used to remove the rock dust from the deeper holes (see Brundage 1990:67 and Harvey 1896:40 for illustrations). Once the holes were completed, feathers (L-shaped objects) were placed on opposite sides of a hole and a wedge was put between the feathers. A small hammer was used to tap all the wedges until the stone was split apart. Wedges and feathers were called by other names including plugs and feathers, wedges and slips, as well as wedges and shims (Baker 1889:128; Harvey 1896:45; Hockensmith 1999:91, editor; McKee 1973:16; Rockwell 1993:55).

A specialized group of hammers were used in shaping millstone blanks. These included cutting hammers, a bullset, blocking hammers, chipping hammers, bush hammers, and striking hammers. Blocking hammers were used to remove large pieces of stone during the initial shaping of millstones (Hockensmith 1999:90, editor; Hockensmith and Coy 1999:47; Hockensmith and Price 1999:83). They had rectangular heads with square striking surfaces and came in 6 pound and 16 pound sizes (Hockensmith 1999:90, editor). A bullset was a hammer-like tool with a long handle that was struck with 10 and 12 pound sledge hammers to remove large chunks of stone during the initial shaping of millstones (Hockensmith 1999:90, editor; Hockensmith and Coy 1999:48; Hockensmith and Price 1999:83). Harvey (1896:39) illustrated bullsets designed for granite work. Chipping hammers were stone working hammers designed with wedge-shaped striking faces (Hockensmith 1999:90). Cutting hammers resembled a double bitted axe and were used to level the top of a millstone (Hockensmith 1999:90, editor; Hockensmith and Coy 1999:47; Hockensmith and Price 1999:83). In Virginia, cutting hammers were made in three sizes for use on different sizes of millstones (Hockensmith 1999:90, editor). Bush hammers had small heads with square striking surfaces that contained rows of pointed teeth (Baker 1889:126; Harvey 1896:52; Holmstrom and Holford 1982:136–137; McKee 1971:51; McKee 1973:22). Millstone makers used bush hammers to lower high spots on the grinding faces of millstones. Striking hammers were blunt on both faces and were used to hit chisels, points, and other tools (Hockensmith 1999:91, editor; Hockensmith and Coy 1999:47).

Additional small hand tools included points, pitching tools, chisels, and punches. Points are short iron rods with a cutting edge that is tapered (on four sides) to a point (Baker 1889:127; Hockensmith 1999:91–92, editor; Hockensmith and Price 1999:82; McKee 1971:51; McKee 1973:22). A hammer was used to hit the point in order to remove excess stone (Hockensmith 1999:92, editor). Pitching tools are chisel-like tools with blunt cutting edges that were hit with hammers (Hockensmith 1999:91, editor; Hockensmith and Coy 1999:49; Hockensmith and Price 1999:86; Holmstrom and Holford 1982:133; Rockwell 1993:55). Large pieces of stone could be removed with pitching tools as millstones were shaped (Hockensmith 1999:91, editor). Chisels were cutting tools with a sharp blade on one end that was hit with a hammer (Baker 1889:128; Hockensmith 1999:90, editor; Hockensmith and Coy 1999:51; Hockensmith and Price 1999:82). Punches were hand tools hit with a hammer to cut a groove for the eye of a millstone (Hockensmith 1999:92; Hockensmith and Coy 1999:51).

Other types of tools included crow bars, wooden calipers, and squares. Crow bars were iron bars with flattened ends used to pry up and move stones. Primitive calipers were made from forked tree branches cut to the correct length (Hockensmith 1999:90, editor). The short side of the fork was inserted into a central hole and a piece of coal was held against the other end as the caliper was turned to outline the diameter of the millstone (Hockensmith 1999:90, editor). These primitive wooden calipers were used in New York, Pennsylvania, and Virginia. Metal squares were used to keep the sides of the millstones vertical (Hockensmith and Coy 2008a, 2008b). In New York, the short arms of the squares were cut to different lengths (Hockensmith and Coy 2008a). Harvey (1896:33) advertised "heavy iron squares for quarry work" that came in 2½, 3, 3½, 4, 5, and 6 foot lengths.

An 1843 legislative report mentioned a variety of stone working tools used in Kentucky in 1842 (Apperson and Bullock 1843:385–399). These included hand drills, stone sledges, wedges and slits, pitching tools, bush hammers, stone axes, squares, punches, etc. (Apperson and Bullock 1843:385–399).

Tools Used to Sharpen Millstones

Special skills and specialized tools were required for sharpening millstones. Usually, a pattern of grooves was cut into the grinding surface of each millstone. These grooves were called furrows and the spaces between them were referred to as lands. Dedrick (1924:259) explained, "The furrows are grooves or channels cut out in the face of the stone; the deep side is the 'back edge'; the bottom of furrow is an inclined plane, culminating in what is known as the 'front,' 'fore' or 'feather-edge.' This edge meets the land, and is the cutting edge of the furrow. The lands are that portion of the face left between the furrows and are the true grinding surfaces. The furrows answer a threefold purpose — a means of distribution, ventilation and cutting, because the grain is broken up or cut near the feather-edge in the ascent, thus preparing the material for the grinding action of the lands."

A variety of different styles of dress were used on millstones. Two common patterns were used by American millers, the circular sickle dress and the straight quarter dress (Grassi 2004a:13). Dedrick (1924:263) mentioned "the sickle, scroll, straight-furrow, three or two quarter" styles of millstone dress. Further, Dedrick (1924:260–264) provided great detail on the cutting of the furrows and dressing millstones.

The following discussions of tools used for millstone sharpening is largely based on research by Bob Grassi (2004a, 2004b) and Theodore Hazen (1996a and 1996b). Once cut, the stones had to be "dressed" or re-cut to ensure that the grain was properly ground. Grassi (2004a:12) noted, "With all of the skill required to produce a quality, consistent finished products, the greatest skill required in milling was that of keeping the stones in good order. Millstone dressing (sharpening) was the work of cutting the dress pattern (the grooves) into the grinding faces of the millstones. One pair of 48" diameter stones would typically take between 16 and 20 hours to dress completely and, depending on their use and hardness of stone they were made of, might need dressing within every two weeks."

Millstone dressing tools can be divided into those used to test the stones and those used in cutting the grooves (Grassi 2004b:16). A paint staff was used to rub paint or pigment across the top of millstones to reveal high spots that required lowering (Grassi 2004b:16). Paint staffs were made of maple and were manufactured in 3½, 4, 4½, and 5 foot lengths for use on different diameters of millstones (Hazen 1996b:8). Russell (1949:61) noted that the paint staff is "about 4 ft. long by 3 in. wide by 5 in. deep is sometimes made of red deal or oak and sometimes of mahogany or walnut." Dedrick (1924:262, 267–270) discussed how the paint staff was used to make the millstone face true and discussed other methods for truing the face. The wooden paint staff had to be checked for straightness against a metal proof staff (Grassi 2004b:16). The proof staff is described as being made from cast iron and has a "usual size ... about 4 ft. long by 4 in. wide" (Russell 1949:61).

Howell and Keller (1977:85) described some of the pigments used: "The paint is often made of 'raddle' or 'tiver,' a composition of red oxide powder mixed with water. Other 'paints' were made from red clay, scalded soot, and in later years, domestic washing blue was often utilized."

The furrows were laid out and checked with a wooden furrow stick (Grassi 2004b:16; Hazen 1996b:9). Layouts were stencils made from thin wood or tin that were employed in laying out the furrows that comprised each quarter of a millstone's dress (Hazen 1996b:9). The importance of furrow design and problems that can occur was emphasized by Dedrick (1924:270–272). In order to check the depth of a furrow along its length, a gauge was used (Hazen 1996b:9).

Several basic tools were used to cut the furrows and lower the high spots. Mill picks are perhaps the best known of the sharpening tools. There were companies that sold mill picks (Harvey 1896; Truax 1896) and books that described how to make mill picks (i.e., Wilkie 1874). Grassi (2004b:16) stated that the "tools for cutting were called mill picks. Usually composed of cast steel with harden tips, the mill picks were shaped not with a point but with a flat broad cutting edge like a chisel. Both ends were formed into a cutting edge approximately 1-½" wide. The picks were about 8" to 10" in length and weighed from three to six pounds. Mill picks came in two forms. One type had an eye in the center for an inserted wooden handle similar to a hammer handle. The other type, without an eye, was inserted into a tapered mortised inside the head of a thrift. The thrift was turned out of hardwood and has been described as being similar in form to a wooden sculptor's mallet."

Further information about mill picks was provided by Hazen (1996b:8–9):

> The bills or picks (as they are commonly called in the United States) have two types: the cracking picks and the furrowing picks. Cracking picks are thinner and lighter in weight than furrowing picks. Traditionally, the bill is placed into a wooden handle called a thrift and held in place with strips of leather. The mill pick has a hole in the center for insertion of a wooden handle in the style of a hammer or similar tool.
>
> The wooden handle of the thrift or pick is rested on a bist, which is described as a small cushion of chaff or bran in a cloth sack. The dresser holds the handle against the bist and, with the other hand, raises the tool and allows it to drop on the millstone. All cutting work is done by the dropping action in order that the entire stone will be struck with the same amount of force. The weight of the tool does the work. Thus, the bill and wooden handle thrift work better than the mill pick and wooden hammer handle.

Dedrick (1924:272) discussed the proper use of the cracking and the furrowing pick:

> The cracking pick is used to "crack" and for light facing. It will be noticed that the first pick differs materially in form from the furrowing pick, being lighter and thinner. From the center the taper is gradual until some distance from the ends, when it is drawn out almost straight and nearly equal thickness.
>
> The furrowing pick ... on the other hand, is made heavier and thicker, the taper being carried down almost to the ends or points, thus making the pick rather stubby to withstand the strain of heavier blows in roughing out furrows or for heavy facing. It is provided with an eye for the insertion of a handle.

Other cutting tools included the bushing hammer, the pritchel, and the facing hammer. The bushing hammer had multiple points, in rows, on the striking face that resembled a metal meat tenderizer (Grassi 2004b:16; Hazen 1996b:9). This tool was used to work down high areas and smooth the top of the millstone. The pritchel was pointed on each end and was used to make small holes on the grinding surfaces of those millstones used for hulling buckwheat and oats (Grassi 2004b:16). Howell and Keller (1977:82) indicated that "[i]n later years several types of facing hammers were used to face off the area around the eye of the stones. Some were like large chisels with several points on the cutting edge. Others were a series of several blades set together almost in the form of a suspension spring on a road vehicle."

The tools used for sharpening millstones needed constant maintenance. Hazen (1996a:17) stated, "The old steel mill picks used by the dresser needed constant sharpening. A dresser would sit down to dress a pair of millstones with a pile of picks and go through them quickly. Each of them had to be resharpened several times in order to dress a pair of millstones. A millstone dresser needed the services of a good blacksmith to retemper his picks. When the temper ran out, it would be necessary to occasionally 'draw out' a worn down pick. A pick with a good temper in it produces a good ringing sound when striking the millstone. You know instantly when a pick has lost its temper by the dull sound it makes against the millstone."

As the proceeding sections indicate, specialized tools were required for the manufacture and sharpening of millstones. Some of these tools were purchased from companies that produced stone working tools (i.e., Harvey 1884, 1886, 1896) while other tools were made by blacksmiths. Jäckle (2002, 2003) provides information on tools used for dressing millstones produced by J. C. Kupka of Saxony, Germany. The sharpness and tempering of these tools was essential for working the stone. When the tools were heated by the blacksmith during the sharpening process the tempering had to be just right for the type of stone being cut. If the tools were tempered too hard they would break. On the other hand, if the tools were too soft, they would not cut the stone. Thus, there was an art to making the tools as well as to using them.

9

Working Conditions and Hazards in the Millstone Industry

The manufacture of millstones was a back breaking job with many health hazards. During the author's interview with the last two millstone makers in Virginia (Hockensmith 1999, editor), and in published accounts, some of these problems came to light. The Virginia quarries exploited a conglomerate. Because of the remote locations of the quarries, workers often walked miles to work, worked long hours, and then walked home again (Hockensmith and Coy 1999:33). In the winter months, the workers were subjected to very difficult working conditions. Doctor Serge Jagailloux (2002:151–164) dealt with the accidents and dangers associated with the millstone industry in France in an article entitled "État de Santé des Meuliers, Accidents du travail, Affections Professionnelles." Jagailloux (2002) listed the problems and illustrated his article with x-rays of millstone makers' lungs and pictures of a breathing mask, eye protection, and so forth. In a recent article dealing with the "Occupational Hazards of Milling," Ison (2005:11–12) also mentioned some of the heath problems associated with making millstones.

Tool Related Injuries

Mr. Surface and Mr. Saville indicated that the stone working tools were constantly in need of being sharpened by the quarry blacksmith. In the winter, workers would pick up these hot tools and sometimes their hands would crack open and bleed. Also, tools often broke while being used. Tool fragments could do bodily damage as they flew through the air with great force. Small fragments of steel would break off tools and become embedded in someone's eye or other body area. The following comments deal with these problems associated with tools (Hockensmith and Coy 1999:57–58):

COY: You'd get pieces of steel in your hands, too, quite frequently?
SAVILLE: Uh-hmm.
SURFACE: Yeah. [laughing]
COY: Were your careful about not letting them bruise over too much?
SURFACE: I got four [?] scars in there.
SAVILLE: My father got one right in there. It went right into his heart and every time his heart would beat, it would pump that blood out. He went to the doctor and the doctor, they stopped the blood somewhere and the steel was still in there when he died.
SURFACE: I've laid down a many a time up there, Gaye Smith would take his old overall jacket and lay it down and I'd lay down and he'd take a, one of these big old country

matches about that long and sharpen it. And he, he had a eyesight like a hawk. You think he couldn't take that match and pick that piece of hot steel right off your eyeball. Get that piece of hot steel right off there. First thing you know, sit around a little bit and take, put a little water on your hank or something, hold it over your eye and take the temperature out of it and go back to work. [laughing]

SAVILLE: But on beating on this steel like this it will fold down on you, you know how it will batter up.

HOCKENSMITH: Yes, sir.

SAVILLE: And it was awful bad to fly off.

SURFACE: Ooooh!, yeah, and hit you on the hand.

SAVILLE: It was battered up so bad.

COY: Did you wear glasses when you were doing this or any kind of protection? Did people have any trouble with their eyes?

SAVILLE: Well, they'd get a stone or a little piece of steel or something but usually somebody just took a match and burned it, you know, sulfur and work it out of there.

SURFACE: [laughing] That hot steel would break off of these things sometimes, off of the point or drills, of course this is a chisel here. You had an eighteen inch drill that had a, you put a point on it like this and one's got to hold it, like that. You're drilling the face off or the round. A guy hitting this, standing over here hitting that eighteen inches out here. And that hot steel fly over there and get in your eye, whoooh!

SAVILLE: But that man that was holding that steel knows what he's doing. You can see he'd work his arm around like that every time.

SURFACE: Yeah! He knows exactly what...

SAVILLE: To keep that from hitting you. You get somebody that didn't know how to hold that steel...

SURFACE: He'd throw it all right in your face.

SAVILLE: But he'd always set that steel, he'd work that hand around there before...

HOCKENSMITH: And besides that and silicosis what other types of injuries and problems did you all get?

SAVILLE: There's a piece that has flew off of that one [tool].

SURFACE: Yeah, you see them little old chips [on battered surface].

SAVILLE: Yeah, it will batter up and then those pieces will break off.

SURFACE: And you don't know where they're going. They could go in your eye if you ain't got goggles on or in your hand.

SAVILLE: But now, that's all they preach to you is to wear those safety glasses. Back then didn't nobody wear no glasses.

SURFACE: You didn't know what they were talking about.

McGrain (1991) found a story called "A Lost Art" in the January 1, 1885, edition of the *American Miller* concerning the hazards of picking millstones: "the workman got steel particles embedded in his face, which look like small pox ... pains in the eyes long after retiring...."

Silicosis

The breathing of rock dust caused a major health problem. In the minutes of a public hearing for the Factory Investigating Commission, a doctor listed millstone makers as part of the group of workers being exposed to mineral dust (New York 1912:848). Mr. Saville and Mr. Surface indicated that silicosis took its toll on the Virginia millstone makers after a few years. Mr. Surface shared that his father Robert G. Surface died of silicosis between the age of 50 and 52 years old (Hockensmith and Coy 1999:11). The two men made the following comments about silicosis (Hockensmith and Coy 1999:57):

HOCKENSMITH: Was that a problem on the conglomerate quarries?
SURFACE: Oh, yeah, the worst, that dust, wasn't it, Dub?
HOCKENSMITH: I know I've read articles about the French millstone makers and that was a major problem there.
SAVILLE: This here, this here had a lot of glass in it, too, this millstone.
HOCKENSMITH: Did you all have any health problems as a result of working at the quarries?
SURFACE: Well, in the old days, this old man here [in historic photograph], I remember, poor Uncle Tom Cromer, I'd go to him when I was just a kid and he'd be sitting on the porch and you could hear him getting his breath out in the yard.
SAVILLE: Yeah, a lot of the older ones had problems.
HOCKENSMITH: Did you all develop any or did you not work at the quarries long enough?
SAVILLE: No, I didn't work there long enough, which I worked at it all my life but this here...
SURFACE: This here was before our time, wasn't it, Dub?
SAVILLE: But this here's got so much of that silicon in it, just like sand rock.
SURFACE: Glass.
SAVILLE: Now, the limestone rock that you work around here don't have all that in it where we worked in most of the time. Of course, we got them on our lungs, you know, got the rock dust on them.

Flory's (1951a:78) interview with Pennsylvania millstone maker William D. Nagle mentioned this problem. He stated, "Mr. Nagle informed me that this dust was very harmful to the lungs, and was one of the reasons he did not continue at the millstone trade, as it was common knowledge that millstone cutters died comparatively young" (Flory 1951a:78). Ward (1984b:31) stated that "silicosis takes fifteen years to destroy a strong, vigorous man. But, long before this his body wastes away, he is racked with spasms of coughing, he loses all appetite, his strength ebbs away until he is no longer able to raise his tools." Silicosis was also a major problem among the English and French millstone makers (Tucker 1977:12; Tomlinson 1981:19; Ward 1984b:31).

The April 1872 issue of the *Manufacturer and Builder* carried a story entitled the "The Ventilation of Unwholesome Manufactories" which contained the following quote on French millstone making (Manufacturer and Builder 1872:81): "Another unwholesome and laborious operation is that of dressing millstones by hand—a process now much facilitated by the use of modern diamond dressing-machines. The working population of La Ferté-sous-Jouarre, the well-known quarry of French burr millstones, is said to be decimated by the considerable disengagement of stone and steel dust which takes place in dressing the stones. Ten years or so of work of this kind is said to often result in mortal disorders."

Other Injuries

Other injuries received by millstone makers as related by Mr. Surface and Mr. Saville (Hockensmith and Coy 1999:58–59):

HOCKENSMITH: Were there other types of ways you could get hurt at the quarry, a rock falling on your or anything when you were blasting or roll over the hill on you?

SAVILLE: No. The only thing if you weren't careful you'd mash your foot or something like that. It was easy to, every body kept a toe mashed.

SURFACE: Yeah, you'd mash your foot, you know. Or turn one over on it ... [laughing]

SURFACE: Yeah, he hit his fist [?]. [laughing] Yeah, he didn't want to try that anymore.

Saville: I tell you, you beat them up when you start learning to..

SURFACE: Huh?

SAVILLE: You beat your fists up when you first start learning.

SURFACE: Oooh! Yeah! That's, when I'd go to bed, I'd have to lay there. I've had my hand, I'm left-handed. I've had this hand all bruised and millered up until I had to lay it out here to the side when I'd go to sleep. [laughter] Yeah, you'd touch it, it would be just like a nerve in a tooth, you know.

SAVILLE: Another thing is, old people are a whole lot tougher than they are now.

SURFACE: Oh, me.

SAVILLE: But to handle that hot steel, you'd get it from a blacksmith, you know, and it would be warm in the winter time and their hands would crack open. You've seen people's hands crack open.

SURFACE: Yeah, handling that hot steel.

SAVILLE: One old fellow would get a needle and threads to sew those cracks back up in his. Didn't have no gloves or nothing. Blood would just run out of them.

SURFACE: Yeah, just crack open.

SAVILLE: It run out of mine many a time. You know, just from handling that hot steel in cold weather would dry your hands out so much it would just crack open. That one old man would always sew his up. [laughter] But the skin was dead, I reckon there wasn't any feeling in there.

SAVILLE: But you used to blacksmith them tools, they'd really crack up, wouldn't they?...

SURFACE: Uh-hmm, mercy.

SAVILLE: They wear gloves anymore. They didn't have them back then. They didn't know what gloves was.

In the summer months, poisonous snakes such as copperheads or rattlesnakes were probably a hazard in remote quarries in the eastern United States. Snakes were a problem on Brush Mountain in Virginia. Even today, these snakes are present in the area containing the Powell County, Kentucky, quarries. It is assumed that snakes would have been a hazard in most millstone quarries since they were usually located in remote, forested, rocky areas that would have been excellent habitats for snakes. However, no published literature was encountered on this potential health hazard. It is assumed that most poisonous snakes were killed by quarry workers when encountered.

The European millstone quarry literature mentions several other health hazards. Ward (1984b:31) noted that many French millstone makers died of pneumonia after getting chilled from cold winter winds blowing across their sweaty bodies. Another problem was that the "stone and steel [from the mill pick] riddle the stomach, the thighs, the knees and the hands of the millstone-maker, thus tattooed with the stigmata of his profession" (Ward 1984b:31). Also, falling stones and equipment caused many broken arms and legs (Ward 1984b:30). Because of the various health hazards, especially silicosis, the men who continuously made millstones rarely lived beyond the age of 45 (Tucker 1977:12).

Injuries were common in the sharpening as well as the manufacture of millstones. Hazen (1996a:16) stated, "Many millstone dressers had beards. This may not have been just a fashion statement or to protect them from the cold as the miller did, but beards did double duty to protect their faces from flying stone chips. The miller was always glad to turn this laborious job of dressing his millstones over to a millstone dresser. They wore the mark of their trade in their hands and forearms which were speckled blue from embedded steel chips lodged under the skin. A miller or mill owner would ask a millstone dresser to demonstrate his skill by pulling up his shirt sleeves, 'To Show His Metal.'"

Injuries were also possible when millstones were turned over for dressing (American Miller 1885; Hazen 1996a). Hazen (1996a:17) explained, "The first task in dressing millstones is to uncover the millstones and lift the runner stone. For many centuries, mills did not have millstone cranes (some small rural or custom mills still do not have cranes). Without a crane, many a miller or millstone dresser has gotten hurt or killed when the millstone got away from them. Also, it was not unusual for the millstone to fall through the floor."

Conclusion

This study has demonstrated tremendous diversity and longevity for the millstone industry in Europe and the United States. Not surprisingly, different cultures discovered the suitability of various rock types for manufacturing querns and millstones. Conglomerates were used in Canada, England, France, Luxembourg, Scotland, Slovenia, Spain, the United States, and Wales. Sandstones were exploited in Austria, England, France, Germany, Greece, Hungary, Scotland, Spain, Sweden, and the United States. Granite deposits were selected in Austria, Canada, England, Scotland, Spain Sweden, and the United States. Various volcanic deposits were quarried in Albania, Austria, the Czech Republic, Germany, Greece, Hungary, Italy, Spain, and Turkey. Some limestone formations were found suitable for grinding in England, France, Spain, Switzerland, and the United States. Other stones such as flint (France and the United States) and various combinations of garnet and schist (Norway and Sweden) were less commonly used.

Querns and millstones were used during different time periods. During the Neolithic period, querns and grinding stones were used in France, Germany, Greece, Italy, Turkey, and Yugoslavia. Bronze Age usage was reported in Greece, Israel, and Italy. Greek period grinding stones were used in Albania, Austria, and Greece. La Tène period usage was recorded for the Czech Republic and Germany. In England and Wales, querns were produced during the Iron Age. Wide distribution of querns and millstones occurred during the Roman period: Albania, Austria, England, France, Germany, Greece, Italy, Scotland, Spain, and Switzerland. Norway and Sweden produced querns and millstones during the Viking era. Querns and millstones were manufactured during the Middle Ages in the Czech Republic, England, France, Germany, Greece, Italy, Norway, Spain, Sweden, and Switzerland. Finally, millstone and quern production continued into the modern era in Austria, Canada, England, France, Germany, Greece, Hungary, Ireland, Italy, Mexico, Norway, Scotland, Slovenia, Spain, Sweden, Wales, and the United States.

The quarrying of querns and millstones date to antiquity. Ironically, modern technology has permitted us to develop a more comprehensive understanding of these ancient quarries. Websites provide researchers opportunities to read about and view images of specific millstone quarries. Likewise, we can benefit from internet sites that provide the on-line text of old and obscure books as well as journals. These sites allow us to find difficult to locate information on querns and millstones through key word searches. In the recent past, it was difficult to locate millstone and quern literature in other countries that was not published in major journals. However, websites such as Millstonequarries.eu (http://meuliere.ish-lyon.cnrs.fr) have permitted scholars, for the first time, the ability to access text and photographs for numerous millstone and quern quarries in Europe. The author anticipates that the internet will greatly facilitate the dissememination of information on such quarries in coming years.

Abandoned quarry sites have tremendous potential for addressing research questions about the manufacture of millstones and querns. For the earliest quarries, there are no written records. Even in more modern times, few records are available about the quarrying and manufacture of millstones and querns. Typically, millstone makers were poorly educated and did not write about their profession. As a consequence, much of our understanding of millstone and quern quarrying must be largely derived from archaeological studies. The abandoned quarries and rejected millstones and querns have an important story to tell.

Unfortunately, in many countries millstone and quern quarries have failed to receive attention from archaeologists. In other countries, archaeologists are just beginning to study millstone and quern quarries. Consequently, I would like to encourage archaeologists and other interested researchers to seek out and study millstone quarries in their respective countries. As the population increases, modern development is beginning to threaten once remote areas of the landscape. Thus, it is important that studies be conducted while the quarries are still extant.

Only when the archaeological community realizes the research potential of millstone and quern quarries will we begin to understand and adequately document this industry that was once so vital to mankind's production of bread and other products. As we work together, we can compile a global history on how mankind has created the technology to grind grain for his sustenance.

Glossary

Bedstone: The lower millstone that remains stationary during the grinding process.

Blocking Hammer: Stoneworking hammer used for removing large pieces of stone during the initial shaping of millstones. These hammers had rectangular heads with square striking surfaces and came in 6 pound and 16 pound sizes.

Boulder: Large pieces of conglomerate that are usually scattered across the landscape and not part of an outcrop. In Powell County, Kentucky, the term has been used to designate large pieces of conglomerate found at the millstone quarries.

Brush Mountain Millstones: Formerly well known white conglomerate millstones made on Brush Mountain in Montgomery County, Virginia, near Blacksburg.

Buhrstone *see* **Burrstone**

Bullrigging: A segment of a small hickory sapling that has been notched in the center. A drill or other tool was held where the sapling was folded over at the notches. This wooden tool holder served as an extension of the arm to allow the worker clearance from sledge hammers hitting the drill.

Bullset: A hammer-like tool with a long handle that was struck with a sledge hammer to remove large chunks of stone during the initial shaping of millstones.

Burrs: Millstones or grinding stones.

Burrstone: A stone with suitable characteristic for being used as a burr millstone. This often refers to a porous flint type stone. "Burr" is also spelled "buhr."

Bush Hammer: A small stoneworking hammer with square striking surfaces containing rows of pointed teeth. These hammers were used to pulverize the high spots on the grinding faces of millstones.

Caliper: A tool used to measure the diameter of a millstone. Simple calipers could be made from a forked branch cut to the desired length. The short side of the fork was placed in a shallow central hole and a piece of coal was held to the opposite end to outline a circle as the caliper was rotated.

Chasers: Vertical-running millstones attached to a shaft that turned in a circle to crush ores and other materials. One stone chases or follows the other stone around the circle.

Chipping Hammer: A stoneworking hammer that had wedge-shaped striking surfaces on both ends.

Chisel: An iron bar with a sharp flat cutting blade on one end. It is hit with a hammer on the other end.

Cocalico Millstones: Conglomerate millstones once made in Lancaster County, Pennsylvania, near Cocalico Township.

Composite Millstones: Millstones made from several carefully shaped blocks of stone that were cemented together. Iron bands were placed around these millstones to help hold them together.

Conglomerate: A type of sandstone containing rounded quartz pebbles. The color of the sandstone and size of pebbles can vary greatly. These geological deposits are usually associated with ancient streambeds where the sand and pebbles become cemented together to form rock.

Cross Grain: Grain refers to the bedding plane in conglomerate. When the stone was broken across this grain, rather than with the grain, the resulting break would be irregular.

Cullin Millstones: Monolithic German millstones made from dark blue basalt.

Cutting Hammer: A special stone hammer resembling a double-bitted axe that was used in leveling the face of a millstone. Cutting hammers were made in different sizes to accommodate different diameters of millstones.

Derrick: A device for lifting and moving stones at a quarry. The derrick was usually composed of a large vertical timber secured with guy wire connected to the top. A hinged timber boom was attached to its base and could be raised and lowered with a crank. The derrick also rotated to facilitate the movement of stone.

Drill: A steel rod with a flared cutting bit for drilling holes in conglomerate. The drill was hit with sledge hammers and rotated between blows. Longer drills with smaller bits were used as the holes became deeper.

Drill Holes: This term has been used in Powell County, Kentucky, to refer to either complete holes or the profiles of drill holes remaining after the conglomerate was split.

Edge Runners: Millstones designed to operate vertically (on their edges or sides). They range from single stones working alone to two stones, attached to the same axle, that follow one another.

Esopus Millstones: Famous white conglomerate millstones made at several communities in Ulster County, New York.

Eye: The central hole in a millstone where the grain is fed in and where the shaft is attached to turn the runner stone. Eyes in runner stones are usually round while eyes in bedstones are usually square. Notches were often added on either side of the eyes on runner stones to facilitate their attachment to the power source.

Face: The leveled grinding surface of a millstone on which the furrows are cut.

Face Grinders: Millstones designed to work in pairs and grind in a horizontal position.

Flint Ridge Millstones: Early monolithic and later composite millstones made from flint in Licking and Muskingum counties, Ohio.

French Burr Millstones: Millstones quarried at La Ferté-sous-Jouarre east of Paris and other nearby areas of France. The early French millstones were monolithic while the later millstones were composite.

Furrows: Shallow grooves cut into the grinding surfaces of millstones to facilitate the grinding of grains as the upper stone rotated. Several patterns of furrows were used by millers.

Georgia Burr Millstones: Composite flint millstones once quarried in Burke County, Georgia.

Goose Creek burrs: Millstones possibly produced in western Kentucky in the early 19th century. The name Goose Creek may refer to the former name of a stream near the quarry.

Laurel Hill Millstones: Conglomerate millstones once quarried on Chestnut Ridge in the Allegheny Mountains of Fayette County, Pennsylvania.

Leveling Crosses: Two trough-like depressions at right angles on the top (face) of an unfinished millstone that intersect in the center of the stone. These crosses provided level surfaces across both axes of a millstone. They also provided sighting lines for removing the pie-shaped high areas in between.

Meuliers: The French word for millstones.

Millstone Cutters: Individuals who cut or shaped millstones at a quarry.

Millstone Grit: An early geological term that refers to coarse sand-and-pebble conglomerates that have long been connected with millstones.

Millstone Pick: A tool for cutting furrows in millstones. This hammer-like tool had a horizontal head resembling a chisel that was sharp on both ends.

Millstones: Disk-shaped stones made for the grinding and/or crushing of grains, ores and other materials. They were commonly made from conglomerate, granite, and flint in the United States. Millstones were usually used in pairs.

Mineral Resources of the United States: An early series of annual reports on minerals in the United States published by the U.S. Bureau of Mines.

Minerals Yearbook: Annual series of reports by the U.S. Bureau of Mines on U.S. mineral industries. It was a successor to the earlier *Mineral Resources of the United States.*

Monolithic Millstones: Millstones made from a single piece of stone.

Moore County Grit Millstones: Blue granite millstones with white flint inclusions once quarried in Moore County, North Carolina, near the community of Parkewood.

Mühlstein: The German word for millstone.

Outcrops: Exposed areas of in situ bedrock that can be exploited for quarrying.

Paint Staff: A straight-edged board that was coated with pigment to rub across the grinding face of a millstone. This action left a residue of pigment on the high spots of a millstone, showing areas that had to be worked down to level the grinding surface.

Peak Millstones: English monolithic millstones quarried from millstone grit in the Peak District of Derbyshire.

Pitching Tool: A chisel-like tool with a very blunt cutting edge. It is hit with a hammer to remove large pieces of stone while shaping a millstone.

Pits: In Powell County, Kentucky, a term referring to shallow oval depressions that were associated with excavations to expose the bases of large boulders.

Plugs and Feathers: Plugs are metal wedges that fit between two L-shaped "feathers" placed on either side of drill holes. As the wedges are tapped, the pressure is increased on the sides of the holes, causing the stone to split in a line along the drill holes.

Point: A short iron rod that is tapered on one end (on four sides) to form a pointed cutting edge. This tool was hit with a hammer to remove excess stone during the shaping of a millstone.

Proof Staff: A metal staff used to check wooden paint staffs for straightness.

Quarry: An area where bedrock or boulders were removed by splitting with hand tools or by blasting. If substantial amounts of stone were removed, a depression was produced in the landscape.

Querns: Small millstones designed to be turned by hand power. They originated in antiquity and were used in many countries.

Raccoon Creek Buhr Millstones: Composite flint millstones made in Athens, Licking, Muskingum, and Vinton counties, Ohio. Their name is derived from the creek that flowed through the area.

Red River Millstones: Light brown conglomerate millstones quarried in present day Powell County, Kentucky, near Pilot Knob or Rotten Point. They were named for the Red River, which drains this area.

Roller Mills: Mills that use grooved steel rollers to grind grain. This new technology led to the demise of the millstone industry.

Runner Stone: An upper millstone that rotates during the grinding process. A shaft attached to the millstone was connected to the power source to turn it.

Sandstone: A sedimentary rock primarily formed by grains of sand that are cemented together.

Shaping Debris: Pieces of conglomerate or other stone chipped or broken from millstone blanks and boulders during the shaping process. These discarded rocks were either left where they fell or piled out of the way.

Sharpening: Recutting furrows in millstones that have become worn or dulled by frequent grinding.

Sledge Hammer: A large heavy hammer (usually eight pounds or more) used to strike a rock or another tool. These hammers can be blunt on both ends or blunt on one end and wedge-shaped on the other end.

Striking Hammers: Hammers with blunt faces that were used to hit chisels, points, and other tools.

Stone Cutter: An individual who cuts and works stone at a quarry. The millstone makers were usually known as stone cutters.

Subterranean Millstone Quarries: Underground cave-like quarries created by the removal of suitable stone for millstones.

Tool Marks: Visible marks in stone resulting from tool use.

Turkey Hill Millstones: Monolithic conglomerate millstones once made on Turkey Hill in Lancaster County, Pennsylvania.

U.S. Bureau of Mines and Minerals: The federal agency formerly charged with the collection and reporting of information on U.S. mines and mineral resources. The U.S. Geological Survey in Reston, Virginia, currently maintains information on mineral related industries.

Welsh Millstones: Quartz conglomerate millstones quarried in Gloucestershire, Wales.

Bibliography

During the preparation of this book, the author spent many years searching for sources on millstones, querns, and their quarries, both in the United States and elsewhere. Many of the sources uncovered in that search are cited in this book and listed below. The author has personally examined many of these foreign millstone studies, but many additional studies cited here were found in the bibliographies of other publications and were not examined. The author's goal was to make this listing as extensive as possible in order to guide the serious scholar to most of the available millstone literature.

Undoubtedly, many other millstone studies exist, contained within the pages of obscure publications, or published in languages not accessible to this author. Hopefully, many of these studies will come to light in the future as communication increases between millstone researchers in different parts of the world.

The millstone and quern studies listed here were published in several languages (English, French, German, Norwegian, Spanish, Swedish, and others). To make the bibliography more useful, the author has briefly annotated many of these sources, at least to the extent of identifying (in brackets, at the end of the entry) the country being discussed. In some cases he has provided more detailed information, including translations of titles (a bracketed phrase in quotation marks is a translation; other bracketed material simply annotates the entry). In some cases further information may be found in other publications: For the French millstone industry, see Belmont and Hockensmith's (2006) bibliography that provides English annotations for numerous articles and books published in French. A detailed bibliography of the German millstone literature with English annotations is currently in preparation (Kling and Hockensmith 2008). Sources published in English that have clear descriptive titles are not annotated. The reader should note that a few foreign citations contained abbreviations in the journal titles, but no other information was available for these sources.

Addison, Jon 1995. Querns, Millstones, and Trade in Roman and Anglo-Saxon Britain. Honors thesis, University of Adelaide, Department of Classics, Adelaide, Australia. [England]

Agapain 2002. *Les Meuliers. Meules et Pierres Meulières.* Presses du Village, Étrépilly, France. [Millstone industry in the Paris Basin]

Agard, Patrick 2003. Les Meulières de Charente. *Aquitaine Historique* 65:12–15. [France]

Agard, Patrick 2005. Les Meulières de Charente. *Le Monde des Moulins* 11:7–8. [France]

Aldsworth, F. G. 1984. Roman-British Quern Fragment from Alfoldean, Slinfold. *Sussex Archaeological Collections* cxxii:221.

Allabert, Blandine, R. Marty, and J. Zanca 2000. Extraire les Meules. In Marie-Christine Bailly-Maitre and Marie-Elise Gardel, *La Pierre, le Métal et le Feu: Économie Castrale en Territoire audois (XIe-XIVe s.)*, pp. 13–26. CNRS, Projet Collectif de Recherche 25, Rapport 2000. [France]

Allen, John W. 1949. *Pope County Notes.* Museum of Natural and Social Sciences, Southern Illinois University, Carbondale, Illinois.

Alsvik, E., K. Sognnes, and A. Stalsberg 1978. Kulturhistoriske Undersøkelser ved Store Kvernfjellvatn, Selbu, Sør-Trøndelag. *Rapport Arkeologisk Serie* 1981:1, 146 pages. [Selbu millstone quarry in Norway]

Ambrose, Paul M. 1964. Abrasive Materials. In *Minerals Yearbook, 1963*, pp. 187–206. Department of the Interior,

U.S. Bureau of Mines, Government Printing Office, Washington, D.C.
American Miller 1885. Hazards of Millstone Pickings. *American Miller* 13:25.
Amiran, R. 1956. The Millstone and the Potter's Wheel. *Eretz-Israel* 4:46–49. [In Hebrew with English abstract]
Amouric, Henry 1990. Carrières de Meules et Approvisionnement de la Provence au Moyen Age et à l'Epoque Moderne. *Carrière et Constructions,* pp. 443–464, 115e Congrès National des Sociétés Savantes. [Provence, France]
Amouric, Henry 1991. Carrières de Meules et Approvisionnement de la Provence au Moyen Âge et à l'Epoque Moderne. *Actes du 115e Congrès National des Sociétés Savantes, Section d'Histoire des Sciences et Techniques, Avignon, 1990,* pp. 443–467. Paris, Éditions du CTHS.
Anderson, Timothy 2005. The Three Roman Rotary Handmill Quarries in Switzerland. Paper presented at the Colloque International "Les Meulières, Recherche, Protection et Valorisation d'un Patrimoine Industriel Européen (Antiquité–XXIe s.)," Grenoble, France, September 22–25, 2005.
Anderson, Timothy, Clara Agustoni, Anika Duvauchelle, Vincent Serneels, and D. Castella 2003. Des Artisans à la Campagne: Carrière de Meules, Forge et voie Gallo-Romaines à Châbles (FR.) *Archéologie Fribourgeoise* 19, Fribourg. [Roman era millstone quarry in eastern Switzerland near Bern.]
Anderson, Timothy, D. Castella, C. Doswald, and Damien Villet 2004. Meules à Bras et Meules "Hydrauliques" en Suisse Romaine: Répartition et Pétrographie. *Minaria Helvetica* 24a: 3–16. [Roman era millstone quarry in eastern Switzerland near Bern.]
Anderson, Timothy, Anika Duvauchelle, and Clara Agustoni 2001. Carrière et Forgeron Gallo-Romains à Châbles. *Cahiers d'Archéologie Fribourgeoise/Freiburger Hefte für Archäologie* 3:2–13. [Roman era millstone quarry in eastern Switzerland near Bern.]
Anderson, Timothy, Anika Duvauchelle, Vincent Serneels, and Clara Agustoni 2000. Stone and Metal Working on the Site of Châbles-Les Saux. In *Iron, Blacksmiths and Tools: Ancient European Crafts, Acts of the Instrumentum Conference at Podsreda (Slovenia) 1999,* edited by M. Feugère and M. Gustin. *Monographie Instrumentum* 12, pp. 103–108, Montagnac.
Anderson, Timothy, Damien Villet, and C. Doswald 2002. Production and Distribution of Iron Age and Roman Handmills in Switzerland. In *ASMOSIA VI, Interdisciplinary Studies on Ancient Stone, Proceedings of the 6th International Conference of the Association for the Study of Marble and Other Stones in Antiquity, Venice, June 15–18, 2000,* edited by L. Lazzarini, pp. 79–84, Padova. [Roman era millstone quarry in eastern Switzerland near Bern.]
Anderson, Timothy, Damien Villet, and Vincent Serneels 1999. La Fabrication des Meules en Grès Coquillier sur le Site Gallo-Romain de Châbles-Les Saux (FR). *Archéologie Suisse* 22:182–189. [Roman era millstone quarry in eastern Switzerland near Bern.]
Anderson, Timothy J. 2006. Three Roman Quern Quarries in Switzerland. In *Mühlsteinbrüche. Erforschung, Schutz und Inwertsetzung eines Kulturerbes europäischer Industrie (Antike–21. Jahrhundert),* edited by Alain Belmont and Fritz Mangartz, pp. 41–46. Römisch-Germanisches Zentralmuseum, Tagungen Band 2, Mainz, Germany.
Andrews, E. B. 1870. *Report of Progress in the Second District.* Geological Survey of Ohio, Columbus Printing Company, Columbus.
Andrieu, Gaby 1990. Les Meulières de Langeac (43). *Le Gonfanon,* 9e année, No. 36:7–13. [France]
Anonymous 1878a. French Millstones — A Warm Dispute. *The Miller* 4:646.
Anonymous 1878b. Millstones Made of Glass. *The Manufacturer and Builder* 10 (7):294.
Anonymous 1879. *The History of McLean County, Illinois....* W. LeBaron, Jr. & Company, Chicago.
Anonymous 1882. *History of DeWitt County, Illinois: With Illustrations Descriptive of the Scenery, and Biographical Sketches of Some of the Prominent Men and Pioneers.* W. R. Brink & Company, Philadelphia.
Anonymous 1888a. *History of Franklin, Jefferson, Washington, Crawford & Gasconade Counties, Missouri.* Goodspeed Publishing Company, Chicago.
Anonymous 1888b. *Worcester, Its Past and Present: A Brief Historical Review of Two Hundred Years....* O. B. Wood, Worcester, Massachusetts.
Anonymous 1890. *Western North Carolina, Historical and Biographical.* A. D. Smith, Charlotte, North Carolina.
Anonymous 1893. Rock Emery Millstones. *The Manufacturer and Builder* 25 (10):231.
Anonymous 1894a. Rock Emery Millstones. *The Manufacturer and Builder* 26 (3):54.
Anonymous 1894b. History of Texas. Unknown date and publisher. On HeritageQuest.
Anonymous 1896. *Commemorative Biographical Encyclopedia of Dauphin County, Pennsylvania: Containing Sketches of Prominent and Representative Citizens, and Many of the Early Scotch-Irish and German Settlers.* J. M. Runk, Chambersburg, Pennsylvania.
Anonymous 1898. *The History of Will County, Illinois....* W. LeBaron, Jr. & Company, Chicago.
Anonymous 1905. *Historical Encyclopedia of Illinois.* Munsell Publishing Company, Chicago.
Anonymous 1910. *History of De Witt County, Illinois: With Biographical Sketches of Prominent Representative Citizens of the County.* Pioneer Publishing Company, Chicago.
Anonymous 1922. *A Modern History of New London County, Connecticut.* Lewis Historical Publishing Company, New York, New York.
Anonymous 1962. Stone-Age Skill Cut Millstones from Mountain: Boulders Scattered Over 50 Acres in Northeast Lancaster County Show Stonecutters' Work. *The Sunday News,* June 3, 1962, page 11. Newspaper clipping in vertical file at Lancaster County Historical Society, Lancaster, Pennsylvania.
Anonymous 1967. *History of Gallatin, Saline, Hamilton, Franklin, and Williamson Counties, Illinois....* Unigraphic, Evansville, Indiana. Reprint of 1887 version by Goodspeed Publishing Company of Chicago.
Anonymous 1971. A Rotary Quern Still in Use. *Old Kilkenny Review* 23:35. [Ireland]
Anonymous 2002a. Les Pierres Monolithes. In *Les Meuliers. Meules et Pierres Meulières* by Agapain, pp. 47–52. Presses du Village, Étrépilly, France. [France]
Anonymous 2002b. Les Carreaux [burrstones]. In *Les Meuliers. Meules et Pierres Meulières* by Agapain, pp. 53–62. Presses du Village, Étrépilly, France. [France]
Anonymous 2002c. La Manufacture de Meules à Carreaux. In *Les Meuliers. Meules et Pierres Meulières* by Agapain, pp. 63–84. Presses du Village, Étrépilly, France. [France]
Anonymous 2002d. L'Evolution de l'Industrie Meulière. In *Les Meuliers. Meules et Pierres Meulières* by Agapain, pp. 85–108. Presses du Village, Étrépilly, France. [France]
Anonymous 2002e. Grandeur et Ruine d'une Industrie. In

Les Meuliers. Meules et Pierres Meulières by Agapain, pp. 165–170. Presses du Village, Étrépilly, France. [France]

Anonymous 2005. *Millstone Country—Stone Cutting in Hyllestad.* Booklet for a walking trail at Hyllestad, Norway.

Anonymous n.d. *Arrowheads to Atoms: The Story of Millstone Point.* Booklet published by Northeastern Utilities, Connecticut.

Antonelli, Fabrizio, F. Bernardini, S. Capedri, Lorenzo Lazzarini, and E. Montagnari Kokelj 2004. Archaeometric Study of Protohistoric Grinding Tools of Volcanic Rocks Found in the Karst (Italy-Slovenia) and Istria (Croatia). *Archaeometry* 46 (4):537–552.

Antonelli, Fabrizio, Lorenzo Lazzarini, and Mario Luni 2005. Preliminary Study on the Import of Lavic Millstones in Tripolitania and Cyrenaica (Libya). *Journal of Cultural Heritage* 6 (2):137–145.

Antonelli, Fabrizio, G. Nappi, and Lorenzo Lazzarini 2000. Sulla 'Pietra da Mole' della Regione di Orvieto. Caratterizzazione Petrografica e Studio Archeometrico di Macine Storiche e Protostoriche dall'Italia Centrale. In *Proceedings of 'I Congresso Nazionale di Archeometria,' Verona, 1999,* pp. 195–207. Patron Editore.

Antonelli, Fabrizio, G. Nappi, and Lorenzo Lazzarini 2001. Roman Millstones from Orvieto (Italy): Petrographic and Geochemical Data for a New Archaeometric Contribution. *Archaeometry* 43 (2):167–189.

Apling, Harry 1984. *Norfolk Corn Windmills.* The Norfolk Windmill Trust, Norfolk, England.

Apperson, Richard, and James M. Bullock 1843. Report of the Commissioners of the Sinking Fund, in Relation to the Penitentiary. In *Reports Communicated to Both Branches of the Legislature of Kentucky, at the December Session, 1842,* pp. 385–399. A. G. Dodge, Frankfort.

Aris, R. 1963. L'Industrie du Basalte dans l'Antiquité à Agde. Fédération Historique du Languedoc Méditerranéen et du Rousillon 36th Congrès (Lodève, 1963). Fédération Historique du Languedoc Méditeranéen et du Rousillon, Faculté des Lettres et Sciences Humaines, Montpellier. [France]

Aris, R. 1974. Le Site Préromain d'Embonne: Une Antique Fabrique de Meules sous la Nouvelle Ville du Caps d'Agde. *Études sur Pézenas et sa Région* 5:3–18. [France]

Arndt, Karl J. R. 1975. *A Documentary History of the Indiana Decade of the Harmony Society 1813–1824.* Volume 1, 1814–1819, Indiana Historical Society, Indianapolis.

Arnow, Harriette S. 1983. *Seedtime on the Cumberland.* The University Press of Kentucky, Lexington.

Arnow, Harriette S. 1984. *Flowering of the Cumberland.* The University Press of Kentucky, Lexington.

Atwater, Caleb 1838. *A History of the State of Ohio, Natural and Civil.* Glezen and Shepard, Cincinnati, Ohio.

Azéma, Jean-Pierre Henri 2006. Prospection et Inventaire des Meulières Souterraines du Tarn et de L'Aveyron. In *Mühlsteinbrüche. Erforschung, Schutz und Inwertsetzung eines Kulturerbes europäischer Industrie (Antike–21. Jahrhundert),* edited by Alain Belmont and Fritz Mangartz, pp. 155–161. Römisch-Germanisches Zentralmuseum, Tagungen Band 2, Mainz, Germany. [Millstone mines in Tarn and Aveyron in southwest France]

Azéma, Jean-Pierre Henri, Roger Meucci, and Georges Naud 2003. Carrières et Diffusion des Meules de Moulins dans le Département de l'Archdè (début du XIXe s.). In *Meules à Grains: Actes du Colloque International de La Ferté-sous-Jouarre 16–19 Mai 2002,* edited by Mouette Barboff, François Sigaut, Cozette Griffin-Kremer, and Robert Kremer, pp. 239–257. Éditions Ibis Press and Éditions de la Maison des Sciences de l'Homme, Paris, France. [Millstone quarries at Ardèche, Coux, and Aubenas, France, which includes sandstone millstones.]

Bailly-Maître, Marie-Christine 2006. Techniques Meulières—Techniques Minières? Exemple des Carrières Souterraines du Sud de la France ["Millstone Quarrying Techniques—Mining Techniques? The Case of Millstone Mines in Southern France"]. *Mühlsteinbrüche. Erforschung, Schutz und Inwertsetzung eines Kulturerbes europäischer Industrie (Antike–21. Jahrhundert),* edited by Alain Belmont and Fritz Mangartz, pp. 163–170. Römisch-Germanisches Zentralmuseum, Tagungen Band 2, Mainz, Germany.

Bajulaz, Lucien 1987. Les Anciennes Carrières de Pierre des Voirons et du Vouan. *Bulletin de l'ESPI (Étude et Sauvegarde du Patrimoine Industriel),* No. 3, pp. 19–34. [France]

Baker, Ira O. 1889. *A Treatise on Masonary Construction.* John Wiley & Sons, New York.

Ball, Donald B., and Charles D. Hockensmith 2005. Early Nineteenth Century Millstone Production in Tennessee. *Ohio Valley Historical Archaeology* 20:1–15.

Ball, Donald B., and Charles D. Hockensmith 2007a. *Millstone Studies: Papers on Their Manufacture, Evolution, and Maintenance.* Special Publication No. 1. Jointly published by the Symposium on Ohio Valley Historic Archaeology, Murray, Kentucky, and the Society for the Preservation of Old Mills, East Meredith, New York.

Ball, Donald B., and Charles D. Hockensmith 2007b. Preliminary Directory of Millstone Makers in the Eastern United States. In *Millstone Studies: Papers on Their Manufacture, Evolution, and Maintenance* by Donald B. Ball and Charles D. Hockensmith, 1–98. Special Publication No. 1. Jointly published by the Symposium on Ohio Valley Historic Archaeology, Murray, Kentucky, and the Society for the Preservation of Old Mills, East Meredith, New York.

Ball, Donald B., and Charles D. Hockensmith 2007c. Occupational Disease and Work Place Hazards Associated with Millstone Making. In *Millstone Studies: Papers on Their Manufacture, Evolution, and Maintenance,* by Donald B. Ball and Charles D. Hockensmith, pp. 99–105. Special Publication No. 1. Jointly published by the Symposium on Ohio Valley Historic Archaeology, Murray, Kentucky, and the Society for the Preservation of Old Mills, East Meredith, New York.

Ball, Donald B., and Charles D. Hockensmith 2007d. Nineteenth Century Millstone Production in Tennessee. In *Millstone Studies: Papers on Their Manufacture, Evolution, and Maintenance,* by Donald B. Ball and Charles D. Hockensmith, pp. 117–133. Special Publication No. 1. Jointly published by the Symposium on Ohio Valley Historic Archaeology, Murray, Kentucky, and the Society for the Preservation of Old Mills, East Meredith, New York.

Ball, Donald B., and Charles D. Hockensmith 2007e. Bibliography of American Millstone Studies. In *Millstone Studies: Papers on Their Manufacture, Evolution, and Maintenance,* by Donald B. Ball and Charles D. Hockensmith, pp. 208–222. Special Publication No. 1. Jointly published by the Symposium on Ohio Valley Historic Archaeology, Murray, Kentucky, and the Society for the Preservation of Old Mills, East Meredith, New York.

Baltimore American 1822. News item concerning millstones quarried in Montgomery County, Virginia. *Baltimore American,* July 9, 1822.

Barbera Miralles, Benjami 2003. La Provincia de Castello: Molinos Harineros y Muelas. In *Meules à Grains: Actes du Colloque International de La Ferté-sous-Jouarre 16–19 Mai 2002*, edited by Mouette Barboff, François Sigaut, Cozette Griffin-Kremer, and Robert Kremer, pp. 188–196. Éditions Ibis Press and Éditions de la Maison des Sciences de l'Homme, Paris, France. [Catalan millstones in Spain]

Barboff, Mouette 2002. Carriers et Meuliers: Mémoires Enfouies. In *Les Meuliers. Meules et Pierres Meulières* by Agapain, pp. 185–212. Presses du Village, Étrépilly, France. [France]

Barboff, Mouette, François Sigaut, Cozette Griffin-Kremer, and Robert Kremer (editors) 2003. *Meules à Grains* ["Grain Millstones"]*: Actes du Colloque International de La Ferté-sous-Jouarre 16–19 Mai 2002.* Éditions Ibis Press and Éditions de la Maison des Sciences de l'Homme, Paris, France.

Barford, P. M. 1984. Some Possible Quern Quarries in the Bristol Area — A Preliminary Survey. *Bristol and Avon Archaeology* 3:13–17. [England]

Barnes & Pearsol 1869. *Directory of Lancaster County: Embracing a Full List of all the Adult Males and Heads of Families, ... and a Classified Business Directory ... 1869–70.* Pearsol & Geist, Printers, Lancaster, Pennsylvania.

Bauer, Herbert 2003. *Im Reich der Mühlsteinbrüche.* Olberdof. [Millstone quarries at Jonsdorf, Germany]

Baug, Irene 2002. Kvernsteinbrota I Hyllestad. Arkeologiske Punktundersøkingar i Steinbrotsområdet i Hyllestad, Sogn og Fjordane. Bergverksmuseet. Skrift nr. 22, Kongsberg. [Millstone quarries at Hyllestad and Sogn, Norway.]

Baug, Irene 2004. *Kvernsteinbrota i Hyllestad (Millstone Quarries in Hyllestad)*, Hyllestad kommune, 20 p. [Bilingual booklet on the millstone quarries at Hyllestad, Norway.]

Baug, Irene 2005. Who Owned the Products? Production and Exchange of Quernstones, Hyllestad I Sogn, Western Norway. In *Utmark. The Outfield as Industry and Ideology in the Iron Age and the Middle Age,* edited by H. Ingunn and S. Inselset, Bergen.

Baug, Irene 2006. The Quarries at Hyllestad — Production of Quern Stones and Millstones in Western Norway. In *Mühlsteinbrüche. Erforschung, Schutz und Inwertsetzung eines Kulturerbes europäischer Industrie (Antike–21. Jahrhundert)*, edited by Alain Belmont and Fritz Mangartz, pp. 55–59. Römisch-Germanisches Zentralmuseum, Tagungen Band 2, Mainz, Germany.

Baug, Irene [Hyllestad, Norway] 2008. E-mail to Charles D. Hockensmith, February 10. [Millstone quarries in Norway; includes comments on Hockensmith's section on Norwegian quarries.]

Baug, Irene, T. Heldal, A. Englert, A. Marøy Hansen, S. Gullbekk, and C. Kilger 2006. Stein som Handelsvare. *Hyllestadseminaret* No. 3, Hyllestad, 62 pages. [Millstone quarry near Hyllestad, Norway.]

Beach, L. M., and A. T. Coons 1922. Abrasive Materials. In *Mineral Resources of the United States, 1919*, pp. 381–386. Part II — Nonmetals. Department of the Interior, U.S. Bureau of Mines, Government Printing Office, Washington, D.C.

Beach, L. M., and A. T. Coons 1923. Abrasive Materials. In *Mineral Resources of the United States, 1920*, pp. 155–159. Part II — Nonmetals. Department of the Interior, U.S. Bureau of Mines, Government Printing Office, Washington, D.C.

Beach, L. M., and A. T. Coons 1924. Abrasive Materials. In *Mineral Resources of the United States, 1921,* pp. 15–18. Part II — Nonmetals. Department of the Interior, U.S. Bureau of Mines, Government Printing Office, Washington, D.C.

Beach, L. M., and A. T. Coons 1925. Abrasive Materials. In *Mineral Resources of the United States, 1922*, pp. 221–225. Part II — Nonmetals. Department of the Interior, U.S. Bureau of Mines, Government Printing Office, Washington, D.C.

Beadle, H. L. n.d. Carr Crags Quarry: Newbiggin-in-Teasdale (NY 918 316). *North England Industrial Archaeological Society Bulletin*, pp. 24–28, 32. [England]

Beaman, Lela 1985. Parkewood, NC, Millstone Quarry. *Old Mill News* 13 (3):18–19.

Beauvois, Jacques 1978. Les Meules. *Bulletin Fédération Française des Amis des Moulins* 1er semestre 1978, 10 pages. [France]

Beauvois, Jacques 1980. Quand Epernon Produisait des Meules à Moulin. *Les Moulins* 4:5–13. [France]

Beauvois, Jacques 1982. Millstone Making in France: When Epernon Produced Millstones. *Wind and Water Mills* 3:32–35. [France]

Beauvois, Jacques 2002a. Les Grèves Chez les Meuliers. In *Les Meuliers. Meules et pierres Meulières* by Agapain, pp. 213–230. Presses du Village, Étrépilly, France. [France]

Beauvois, Jacques 2002b. Une Vie au Service de la Meulière. In *Les Meuliers. Meules et Pierres Meulières* by Agapain, pp. 231–234. Presses du Village, Étrépilly, France. [France]

Beauvois, Jacques 2002c. Moulins et Meuniers. Autour des Meules, le Moulin. In *Les Meuliers. Meules et Pierres Meulières* by Agapain, pp. 235–242. Presses du Village, Étrépilly, France. [France]

Beck, Lewis C. 1823. *A Gazetteer of the States of Illinois and Missouri: Containing a General View of Each State, a General View of Their Counties, and a Particular Description of Their Towns, Villages, Rivers, &c.: With a Map, and Other Engravings.* C. R. and G. Webster, Albany.

Beckwith, H. W., and P. S. Kennedy 1881. *History of Montgomery County: Together with Historic Notes on the Wabash Valley, Gleaned from Early Authors, Old Maps and Manuscripts, Private and Official Correspondence, and Other Authentic, Though, For the Most Part, Out-of-the-way Sources.* H. H. Hill and Iddings, Chicago.

Bedukadze, Sarah 1956. Le Développement des Meules à la Main. In *Revue du Musée de l'Etat de Géorgie*, 19 (V):153–169.

Beiron, Ingemar 2006. The Millstone Quarry "Minnesfjället" in Lugnås, Sweden. In *Mühlsteinbrüche. Erforschung, Schutz und Inwertsetzung eines Kulturerbes europäischer Industrie (Antike–21. Jahrhundert)*, edited by Alain Belmont and Fritz Mangartz, pp. 203–206. Römisch-Germanisches Zentralmuseum, Tagungen Band 2, Mainz, Germany.

Beitzell, Edwin W. 1968. *Life on the Potomac River.* Privately printed, Abell, Maryland.

Bell, Edward L. [Massachusetts Historical Commission, Boston] 1991. Letter to Charles D. Hockensmith. [Millstone quarry at Medfield, Massachusetts]

Belmont, Alain 2001. La Pierre et le Pain: Les Carrières de Meules de Moulin de Quaix-en-Chartreuse (XVI e–XVIII e siècle). *Histoire et Sociétés Rurales* 16 (2):45–79. [Quarries at Quaix, France, in the Alps that produced millstones during the 17th and 18th centuries]

Belmont, Alain 2002a. Le Salaire de la Pierre. *L'Alpe*, No. 17, Autumn, pp. 48–53. [General overview of millstones quarries in the Alps of France.]

Belmont, Alain 2002b. Une Mine Pour la Farine: Les Meulières de Berland (38) sous l'Ancien Régime. *La Pierre et l'Écrit* 13:37–68. [A study of subterranean millstone quarries in the Savoy region of France that produced stone from the Middle Ages to the early 19th century.]

Belmont, Alain 2002c. Les Carrières de Meules de Moulins en France à l'Époque Moderne. In *Moulins et Meuniers dans les Campagnes Européennes (IXᵉ–XVIIIe siècle). Actes des XXIᵉ Journées Internationales d'historie de Flanan, Toulouse*, edited by Mireille Mousnier, pp. 147–165. [A general overview of the French millstone quarries during the 16th, 17th, and 18th centuries]

Belmont, Alain 2002d. Un Patrimines à Faire Valoir: Les Carrières de Meules de Moulin da le Parc Naturel Régional de Chartreuse (partie isèroise). Rapport de Recherches, Parc Naturel Régional de Chartreuse, Décembre 2002, 97 p. [France]

Belmont, Alain 2003a. Commerce et Diffusion des Meules de La Ferté-sous-Jouarre aux XVᵉ et XVIᵉ siècles["Trade and Diffusion of La Ferté-sous-Jouarre Millstones during the 15th and 16th Centuries"]. In *Meules à Grains: Actes du Colloque International de La Ferté-sous-Jouarre 16–19 Mai 2002*, edited by Mouette Barboff, François Sigaut, Cozette Griffin-Kremer, and Robert Kremer, pp. 282–288. Éditions Ibis Press and Éditions de la Maison des Sciences de l'Homme, Paris, France. [France]

Belmont, Alain 2003b. La Pierre à Pain. Artisanat et Industrie des Meules de Moulin en France sous l'Ancien Régime. [France]

Belmont, Alain 2003c. L'Épopée des Meules Françaises de La Ferté-sous-Jouarre, 15ᵉ-19ᵉ siècles. *Pour la Science*, No. 308, June 2003, pp. 68–73. [General history of La Ferté-sous-Jouarre millstone quarries.]

Belmont, Alain [Université de Grenoble II, Grenoble, France] 2004. E-mail to Charles D. Hockensmith, November 8. [Millstone quarries near Grenoble, France.]

Belmont, Alain 2005. Die Mühlsteinbrüche. Les carrières de meules de moulins dans le massif des Vosges. In *Mélanges offerts à Jean-Michel Boehler*. [Millstone quarries in the mountain mass "Vosges," near the German border.]

Belmont, Alain 2006a. *La Pierre à Pain: Les Carrières de Meules de Moulins en France du Moyen Age à la Revolution Industrielle*. Presses Universitaries de Grenoble, Grenoble. [Two-volume study on the millstone quarries of France dealing with quarries dating from the Middle Ages to the Industrial Revolution.]

Belmont, Alain 2006b. Les Meulières Médiévales. Résultats d'Une Moisson Dauphinoise ["The Medieval Millstone Quarries. Results of a Survey in the Dauphinois"]. In *Mühlsteinbrüche. Erforschung, Schutz und Inwertsetzung eines Kulturerbes europäischer Industrie (Antike–21. Jahrhundert)*, edited by Alain Belmont and Fritz Mangartz, pp. 81–90. Römisch-Germanisches Zentralmuseum, Tagungen Band 2, Mainz, Germany. [French Alps]

Belmont, Alain 2006c. Introduction: Antique and Medieval Millstone Quarries. In *Mühlsteinbrüche. Erforschung, Schutz und Inwertsetzung eines Kulturerbes europäischer Industrie (Antike–21. Jahrhundert)*, edited by Alain Belmont and Fritz Mangartz, p. 4. Römisch-Germanisches Zentralmuseum, Tagungen Band 2, Mainz, Germany.

Belmont, Alain [Université de Grenoble II, Grenoble, France] 2007. E-mail to Charles D. Hockensmith, August 3. [Millstone quarries on the Millstonequarries.eu website.]

Belmont, Alain, Blandine Allabert, Robert Marty, and Jean Zanca 2003. Extraire les Meules. In Marie-Christine Bailly-Maitre, Marie-Elise Gardel (editors), *La Pierre, le Métal et le Feu: Économie Castrale en Territoire Audois (XIe-XIVe s.)*, CNRS, Projet Collectif de Recherche 25, pp. 20–37 [Official report of a survey made on millstone quarries in the Carcassonne area of France near the Spanish border.]

Belmont, Alain, and Michèle Bois 1998. *Les Carrières de Meules de Moulin de Quaix-en-Chartreuse (Isère): Document Final de Synthèse*. Service Régional de l'Archéologie (Rhône-Alpes), Juillet 1998, 44 p. [Report on the excavations at the Quaix-en-Chartreuse millstone quarries, France.]

Belmont, Alain, and Charles D. Hockensmith 2006. Millstones, Querns, and Millstone Quarry Studies in France: A Bibliography. *International Molinology* 72:2–11, Watford, England.

Belmont, Alain, and Fritz Mangartz (editors) 2006. *Mühlsteinbrüche. Erforschung, Schutz und Inwertsetzung eines Kulturerbes europäischer Industrie (Antike–21. Jahrhundert)*, Römisch-Germanisches Zentralmuseum, Tagungen Band 2, Mainz, Germany.

Belmont, Alain, and Fritz Mangartz 2006. Introduction. In *Mühlsteinbrüche. Erforschung, Schutz und Inwertsetzung eines Kulturerbes europäischer Industrie (Antike–21. Jahrhundert)*, edited by Alain Belmont and Fritz Mangartz, pp. xxv–xxviii. Römisch-Germanisches Zentralmuseum, Tagungen Band 2, Mainz, Germany.

Bennett, Richard, and John Elton 1898a. Chapter V. The Quern. In the *History of Corn Milling. Volume I: Handstones, Slave & Cattle Mills* by Richard Bennett and John Elton, pp. 128–176. Burt Franklin, New York.

Bennett, Richard, and John Elton 1898b. *History of Corn Milling. Volume II: Watermills and Windmills*. Burt Franklin, New York.

Bennett, Richard, and John Elton 1900. *History of Corn Milling, Volume III: Feudal Laws and Customs*. Burt Franklin, New York.

Beranová, Magdalena 1988. Manual Rotation Grain Mills on Czechoslavak Territory Up to the Incipient 2nd Millennium AD. *Ethnolgia Slavica* 19:15–43, Bratislava. [Czech Republic]

Berg, Thomas N. 1986. A Sesqicententennial Story: Early Millstone Quarry in Tioga County. *Pennsylvania Geology* 17 (1):3–6.

Biddle, M., and D. Smith 1990. The Querns. In *Object and Economy in Medieval Winchester*, edited by M. Biddle. Winchester Studies 7, Oxford.

Bishop, John Leander 1866. *A History of American Manufacturers from 1608 to 1860 ... Comprising Annals of the Industry of the United States in Machinery, Manuafactures and Useful Arts*. E. Young & Co, Philadelphia.

Blackburn, Glen A. (compiler) 1942. *The John Tipton Papers. Volume I: 1809–1827.* The Indiana Historical Bureau, Indianapolis.

Bloch Jorgensen, A. n.d. Investigations of Danish Rotary Querns from the Iron Age: Archaeological Evidence and Practical Experiments. The British Library Document Supply Centre, Paper CN060122221.

Boisvert, Richard 1992. Personal communication to Charles D. Hockensmith concerning a Shaker granite millstone quarry on Moose Mountain, Enfield, New Hampshire. New Hampshire Division of Historical Resources, Concord.

Boisvert, Richard 2006. Personal communication to Charles

D. Hockensmith concerning a Shaker granite millstone quarry on Moose Mountain, Enfield, New Hampshire. New Hampshire Division of Historical Resources, Concord. March 27, 2006.

Bonneff, Léon, and Bonneff, Maurice 2002. Vie et Mort des Meuliers. In *Les Meuliers. Meules et Pierres Meulières* by Agapain, pp. 141–150. Presses du Village, Étrépilly, France.

Bonson, Tony 1999. *Millstones to Megawatts: A Bibliography for Industrial Historians; The Publications of Dr. D. G. Tucker.* Midland Wind & Water Mills Group, Congleton, England.

Bonson, Tony 2001. Mow Cop Millstones. *Proceedings of the 19th Mills Research Conference*, pp. 17–24. Manningtree, England.

Bosset, L. 1943. Châvannes-le-Chêne: Une Nécropole Burgonde dans une Ancienne Carrière Romane. *La Suisse Primitive* 7:34–41. [Châvannes-le-Chêne is a Roman era quern quarry in western Switzerland.]

Bost, Gerald 2002. The Millstone Quarries at Jonsdorf. *International Molinology* 64:30–32, Holland. [Germany]

Bottin, C. 1905. Rapport sur la Découverte de Deux Meules Gallo-Romaines par M. le Colonel Noir. Étude Historique de l'Antelier de ces Mueles Situé sur le Plateau du Rocher de l'Aigle et à la Guérade. *Bulletin de l'Académie du Var* 73:193–213, Toulon. [France]

Bowles, Oliver 1930. Abrasive Materials. In *Mineral Resources of the United States, 1928*, pp. 237–252. Part II — Nonmetals. Department of the Interior, U.S. Bureau of Mines, Government Printing Office, Washington, D.C.

Bowles, Oliver 1932. Abrasive Materials. In *Mineral Resources of the United States, 1929*, pp. 65–81. Part II — Nonmetals. Department of the Interior, U.S. Bureau of Mines, Government Printing Office, Washington, D.C.

Bowles, Oliver 1939. *The Stone Industries*. McGraw-Hill Book Company, New York. Second edition.

Bowles, Oliver, and A. E. Davis 1934. Abrasive Materials. In *Minerals Yearbook, 1934*, pp. 889–906. Department of the Interior, U.S. Bureau of Mines, Government Printing Office, Washington, D.C.

Boyd, Andrew 1872. *Boyd's New York State Directory, 1872–1873...* Truair, Smith & Company, Syracuse, New York.

Boyer, François 2003. Données Géologiques et Critères de Choix des Pierres Meulières ["Geological Factors and Criteria for the Choice of Stones for Millstones"] [Abstract]. In *Meules à Grains: Actes du Colloque International de La Ferté-sous-Jouarre 16–19 Mai 2002*, edited by Mouette Barboff, François Sigaut, Cozette Griffin-Kremer, and Robert Kremer, p. 441. Éditions Ibis Press and Éditions de la Maison des Sciences de l'Homme, Paris, France.

Boyer, François, and Olivier Buchsenschutz 1998. Les Conditions d'une Interprétation Fonctionnelle des Moulins "Celtiques" Rotatifs à Main sont-elles Réunies? *Revue Archéologique du Centre de la France* 37: 197–206. [Study of Celtic querns in France.]

Boyer, François, and Olivier Buchsenschutz 2000. Les Meules Rotatives Manuelles. In G. Berthaud, *Mazières-en-Mauges Gallo-Romain: Un Quartier à Vocation Artisanale et Domestique*, Angers, ARDA-AFAN, pp. 171–186 [Study of a collection of Roman querns from the Loire area of France.]

Boyer, François, Olivier Buchsenschutz, Caroline Hamon, Luc Jaccottey, Jean-Paul Lagadec, Annabelle Milleville, Émilie Thomas, and Bertrand Triboulot 2006. Production et Diffusion des Meules du Néolithique à L'Antiquité: Quelques Exemples Français. In *Mühlsteinbrüche. Erforschung, Schutz und Inwertsetzung eines Kulturerbes europäischer Industrie (Antike–21. Jahrhundert)*, edited by Alain Belmont and Fritz Mangartz, pp. 5–13. Römisch-Germanisches Zentralmuseum, Tagungen Band 2, Mainz, Germany. ["Production and Diffusion of Querns from the Neolithic to Antiquity: Some French Examples."]

Bradley, P., F. Roe, and G. Wait 2004. Stone. In *Gravelly Guy, Stanton Harcourt, Oxfordshire: The Development of a Prehistoric and Romano-British Community*, by George Lambrick and T. Allen, pp. 368–376, 398–400. Oxford Archaeology, Thames Valley Landscapes Monograph No. 21. [This discussion includes a large number of querns dating from the Iron Age and Roman periods.]

Bragg, M. M. 1892. Memorial History of Utica, N. Y.: From Its Settlement to the Present Time. D. Mason, Syracuse, New York.

Brainard, Jehu 1854. *Origin of the Quartz Pebbles of the Sandstone Conglomerate, and the Formation of the Stratified Sand Rocks.* Jehu Brainard, Cleveland, Ohio.

Branson, George 1926. Early Flour Mills in Indiana. *Indiana Magazine of History* 22:20–27.

Bregadze, Nelli 1989. Production des Meules à Main en Géorgie avant la Révolution, Tbilissi, Académie des Sciences, Série d'Economie et de Droit, pp. 76–85.

Briggs, S. 1988. The Lithology, Origin and Disposal of Some Quernstones from Yorkshire. In *Recent Research in Roman Yorkshire*, edited by J. Price, pp. 289–313. BAR British Series 193.

Broadhead, G. C., F. B. Meek, and B. F. Shumard 1873. *Reports on the Geological Survey of the State of Missouri, 1855–1871.* Missouri Bureau of Mines, Metallurgy and Geology. Regan & Carter, Jefferson City.

Broadhead, Garland C. 1874. *Report on the Geological Survey of the State of Missouri, Including Field Work of 1873–1874.* Missouri Bureau of Mines, Metallurgy and Geology. Regan & Carter, Jefferson City.

Brown, Bruce B. 1990. *John Deeter: Allegheny Mountain Pioneer.* Closson Press, Apollo, Pennsylvania.

Brown, Bruce B. [Greencastle, Pennsylvania] 1991. Letter to Charles D. Hockensmith, May 25. On file at the Kentucky Heritage Council, Frankfort.

Brown, James B. 1851. *Views of Canada and the Colonists.* 2nd edition. Adam and Charles Black, Edinburgh.

Brown, Lisa 1992. Quernstones from Suddern Farm, Nether Wallop, Hants. *Quern Study Group Newsletter* 2:3–4. England. [England]

Brown, Lisa, and Kathy Laws 1991. A Cache of Querns from Bury Hill, Hants. *Quern Study Group Newsletter* 1:2–3. England. [England]

Brown, Robin 2000. Get a Glimpse of Delaware's Milling History. *The News Journal*, September 23, 2000, Wilmington, Delaware (on NewsBank website). [Article mentions a deserted millstone quarry at White Clay Creek State Park.]

Browne, Tony 2005. Millstones from Mow Cop. *Manchester Geological Association Newsletter* 5–6.

Bruggeman, Jean 2003. Les Meules en Flandre s'après les Comptes et Prisées du Moyen-Age à la Révolution. ["Millstones in Flanders from the Middle Ages to the Renaissance."] In *Meules à Grains: Actes du Colloque International de La Ferté-sous-Jouarre 16–19 Mai 2002*, edited by Mouette Barboff, François Sigaut, Cozette Griffin-Kremer, and Robert Kremer, pp. 231–238. Éditions Ibis Press and Éditions de la Maison des Sciences de l'Homme, Paris, France. [Flanders is on the border of northwest France and Belgium.]

Brundage, Larry 1990. Sawing Through the Grit, Part II. *The Chronicle of the Early American Industries Association* 43 (3):65–67.

Buchanan, Mungo 1912. Notice of a Pair of Quernstones Found at Highland Dykes Near Falkirk. *Proceeding of the Society of Antiquities of Scotland 1911/12*, XLVI, pp. 367–369. [Scotland]

Buchsenschutz, Olivier, and Claudine Pommepuy 2002. Les Enjeux d'une Recherche sur les Meules Rotatives dans le Monde Celtique. In Hara Procopiou and René Treuil, *Moudre et Broyer. L'interprétation Fonctionnelle de l'Outillage de Mouture et de Broyage dans la Préhistoire et l'Antiquité*, Paris, CTHS, t. II, pp. 177–182.

Buckley, D. G. 2001. Querns and Millstones. In *The Romano-British Small Town at Wanborough* by A.S. Anderson, John S. Wacher, and A. P. Fitzpatrick, pp. 156–160. Britannia Monograph Series 19. Society for the Promotion of Roman Studies, London.

Buckman, J. 1866. On the Materials of Roman Querns. *Wiltshire Archaeological Magazine* IX:291–294.

Buentke, Holger 2003. Millstones-literature Scandinavia. Manuscript bibliography produced by the author in Lugnås, Sweden.

Buentke, Holger, and Hans Gustafsson 2006. Ein Experiment zu Abbau und Bearbeitung von Mühlsteinen in einem historischen Untertagebau in Lugnås (Schweden). ["An Experiment on the Quarrying and Shaping of Millstones in a Historical Mine in Lugnås (Sweden)."] In *Mühlsteinbrüche. Erforschung, Schutz und Inwertsetzung eines Kulturerbes europäischer Industrie (Antike–21. Jahrhundert)*, edited by Alain Belmont and Fritz Mangartz, pp. 171–176. Römisch-Germanisches Zentralmuseum, Tagungen Band 2, Mainz, Germany.

Buffone, Luigi, Serjio Lorenzoni, Mauro Pallara, and Eleconora Zanettin 2003. The Millstones of Ancient Pompei: A Petro-Archaeological Study. *European Journal of Mineralogy* 15 (1):207–215.

Bulleid, Arthur 1917. Millstones and Querns. In *The Glastonbury Lake Village Volume 2*, edited by A. Bulleid and H. St. George Gray, pp. 608–621. Glastonbury Antiquarian Society, Glastonbury. [England]

Bulliot, J. G. 1888. Les Carriers Gallo-Romain du Plateau de St. Emilan. *Mémoires de la Sociéte Éduenne, Nouvelle Série* 16:214–227. [Sandstone millstones from St. Emilon near Autun, Saône-et-Loire, in east central France dating from the Gallo-Roman period to the Middle Ages]

Burdett, Charles 1860. *Margaret Moncrieffe; The First Love of Aaron Burr. A Romance of the Revolution. With an Appendix Containing the Letters of Colonel Burr to "Kate" and "Eliza," and from "Leonora," etc.* Derby & Jackson, New York.

Burdett, Charles 1865. *The Beautiful Spy. An Exciting Story of Army and High Life in New York in 1776.* J. E. Potter, Philadelphia.

Burnham, Barry, and Helen Burnham 2004. Early Millstones Recovered from the Dolaucothi Area. In *Dolaucothi-Pumsaint, Survey and Excavations at a Roman Gold-Mining Complex 1987–1999,* by Barry Burnham and Helen Burnham, pp. 286–290. Oxbow Books, Oxford. [Wales]

Butcher, L. H. 1970. Wharncliffe: Quern Workings. *Archaeological Field Guide*, Prehistoric Society, pp. 36–37, Sheffield. [England]

Butler, Brian M. 1971. *Hoover–Beeson Rockshelter, 40Cn4, Cannon County, Tennessee.* Miscellaneous Paper No. 9. Tennessee Archaeological Society, Knoxville.

Butt, John 1967. *The Industrial Archaeology of Scotland.* Augustus M. Kelley, New York.

Cabezuelo, Ulysse 2006. Le Site de la Zac des "Meules" à Vic-le-Comte (Puy-de-Dôme). ["The Site in the 'Les Meules' Commercial District at Vic-le-Comte (Puy-de-Dôme)."] In *Mühlsteinbrüche. Erforschung, Schutz und Inwertsetzung eines Kulturerbes europäischer Industrie (Antike–21. Jahrhundert)*, edited by Alain Belmont and Fritz Mangartz, pp. 109–114. Römisch-Germanisches Zentralmuseum, Tagungen Band 2, Mainz, Germany.

Cabezuelo, Ulysse, Yves Connier, and Fabrice Gauthier 2001. Les Carrières de Meules de Moulins dans la Région de Vic-le-Comte: Premier État de la Recherche. *Revue d'Auvergne*, No. 1–2 ("Nouvelles Archéologiques"), pp. 184–200. [Millstone quarries in central France]

Campbell, E. 1987. A Cross-Marked Quern from Dunadd and Other Evidence for Relations Between Dunadd and Iona. *Proceedings of the Society of Antiquaries of Scotland* 117:105–117.

Campbell, Marius R., et al. 1925. *The Valley Coal Fields of Virginia.* Virginia Geological Survey, Bulletin Number 25, Charlottesville, Virginia.

Cantet, Jean-Pierre 1995. La Grotte de l'Esquérou à Faget-Abbatial (Gers). Étude Préliminaire. *Actes des 15ᵉ et 16ᵉ Journées des Archéologues Gersois. Société Archéologique, Historique et Scientifique du Gers*, Auch, pp. 7–12. [France]

Capedri, S., G. Venturelli, and G. Grandi 2000. Euganean Trachytes: Discrimination on Quarried Sites by Petrographic and Chemical Parameters and by Magnetic Susceptibility and Its Bearing on the Provenance of Stones of Ancient Artefacts. *Journal of Cultural Heritage* 1:34–64.

Carcauzon, C. 1986. Les Tailleries de Meules de Saint-Crépin-de-Richemont. *Revue Archéologique* 36–37:25–40. [Millstone quarries near Saint-Crépin-de-Richemont in the Aquitaine region of France]

Carelli, P., and P. Kresten 1997. Give Us This Day Our Daily Bread: A Study of Late Viking Age and Medieval Quernstones in South Scandinavia." *Acta Archaeologica* 68:109–137.

Caron, C. K. 1871. *Caron's Annual Directory of the City of Louisville, for 1871.* Bradley and Gilbert, Louisville.

Caron, C. K. 1872. *Caron's Annual Directory of the City of Louisville, for 1872.* Bradley and Gilbert, Louisville.

Caron, C. K. 1873. *Caron's Annual Directory of the City of Louisville, for 1873.* Bradley and Gilbert, Louisville.

Caron, C. K. 1874. *Caron's Annual Directory of the City of Louisville, for 1874.* Bradley and Gilbert, Louisville.

Caron, C. K. 1875. *Caron's Annual Directory of the City of Louisville, for 1875.* Bradley and Gilbert, Louisville.

Caron, C. K. 1876. *Caron's Annual Directory of the City of Louisville, for 1876.* Bradley and Gilbert, Louisville.

Caron, C. K. 1877. *Caron's Annual Directory of the City of Louisville, for 1877.* Bradley and Gilbert, Louisville.

Caron, C. K. 1878. *Caron's Annual Directory of the City of Louisville, for 1878.* Bradley and Gilbert, Louisville.

Caron, C. K. 1879. *Caron's Annual Directory of the City of Louisville, for 1879.* Bradley and Gilbert, Louisville.

Caron, C. K. 1880. *Caron's Annual Directory of the City of Louisville, for 1880.* Bradley and Gilbert, Louisville.

Casseday, Ben 1852. *The History of Louisville, from Its Earliest Settlement Till the Year 1852.* Hull and Brother, Louisville.

Castella, D., and Timothy Anderson 2004. Les Meules du Musée Romain d'Avenches. *Bulletin Pro Aventico* 46:115–169.

Cattani, M., L. Lazzarini, and R. Falcone 1997. Macine Protostoriche dall'Emilia e dal Veneto: Note Archeologiche, Caratterizzazione Chimico–Petrografica e Determinazioni Della Provenienza. *Padusa* 31:105–137.

Caufield, John J. 1966. The Rotary Quern in Ireland. M. A. thesis, University College Dublin.

Caulfield, Séamas 1977. The Beehive Quern in Ireland. *Journal of the Royal Society of Antiquaries of Ireland* 107:104–138.

Caulkins, Frances Manwaring 1852. *History of New London, Connecticut: From the First Survey of the Coast in 1612 to 1852.* F. M. Caulkins, New London, Connecticut.

Chalkousaki, Elena 2003. Les Pierres Meulières de la Grèce, Le Cas de l'Île de Milos. In *Meules à Grains: Actes du Colloque International de La Ferté-sous-Jouarre 16–19 Mai 2002,* edited by Mouette Barboff, François Sigaut, Cozette Griffin-Kremer, and Robert Kremer, pp. 184–187. Éditions Ibis Press and Éditions de la Maison des Sciences de l'Homme, Paris, France. [Millstone quarrying and work in Milos, Greece.]

Chambers 1872. *Chamber's Encyclopaedia: A Dictionary of Universal Knowledge for the People...* J. B. Lippincott & Company, Edinburgh.

Chambon, R. 1954. La Trouée de l'Oise avant et pendant la Domination Romaine. *Documents et Rapports de la Société Archéologique de Charleroi* 49:1–64. [Sandstone millstones made by the Romans in the L'Oise Valley on the France/Belgium border.]

Chandler, Henry P., and Annie L. Marks 1954. Abrasive Materials. In *Minerals Yearbook, 1951,* pp. 111–127. Department of the Interior, U.S. Bureau of Mines, Government Printing Office, Washington, D.C.

Chandler, Henry P., and Annie L. Marks 1955. Abrasive Materials. In *Minerals Yearbook, 1952,* pp. 99–114. Department of the Interior, U.S. Bureau of Mines, Government Printing Office, Washington, D.C.

Chandler, Henry P., and Annie L. Marks 1956. Abrasive Materials. In *Minerals Yearbook, 1953,* pp. 127–142. Department of the Interior, U.S. Bureau of Mines, Government Printing Office, Washington, D.C.

Chandler, Henry P., and Gertrude E. Tucker 1953. Abrasive Materials. In *Minerals Yearbook, 1950,* pp. 90–107. Department of the Interior, U.S. Bureau of Mines, Government Printing Office, Washington, D.C.

Chandler, Henry P., and Gertrude E. Tucker 1958a. Abrasive Materials. In *Minerals Yearbook, 1954,* pp. 117–132. Department of the Interior, U.S. Bureau of Mines, Government Printing Office, Washington, D.C.

Chandler, Henry P., and Gertrude E. Tucker 1958b. Abrasive Materials. In *Minerals Yearbook, 1955,* Volume 1 (Metals and Minerals), pp. 121–140. Department of the Interior, U.S. Bureau of Mines, Government Printing Office, Washington, D.C.

Chandler, Henry P., and Gertrude E. Tucker 1958c. Abrasive Materials. In *Minerals Yearbook, 1956,* pp. 139–157. Department of the Interior, U.S. Bureau of Mines, Government Printing Office, Washington, D.C.

Chandler, Henry P., and Gertrude E. Tucker 1958d. Abrasive Materials. In *Minerals Yearbook, 1957,* pp. 145–164. Department of the Interior, U.S. Bureau of Mines, Government Printing Office, Washington, D.C.

Chandler, Henry P., and Gertrude E. Tucker 1959. Abrasive Materials. In *Minerals Yearbook, 1958,* pp. 129–145. Department of the Interior, U.S. Bureau of Mines, Government Printing Office, Washington, D.C.

Chandler, Henry P. and Gertrude E. Tucker 1960. Abrasive Materials. In *Minerals Yearbook, 1959,* pp. 137–153. Department of the Interior, U.S. Bureau of Mines, Government Printing Office, Washington, D.C.

Chandler, Henry P., and Gertrude E. Tucker 1961. Abrasive Materials. In *Minerals Yearbook, 1960,* pp. 145–163. Department of the Interior, U.S. Bureau of Mines, Government Printing Office, Washington, D.C.

Chapman, Hilary S. 1990. Beehive Querns: Technological Marvels of the Iron Age. In the National Library of Wales. [Wales]

Chazin, Daniel 2003. Hike of the Week. *The Record,* March 13, 2003, page 2. [Bergen County, New Jersey]

Chazin, Daniel 2005. Hiking: Ramapo Mountain State Forest and Camp Glen Gray. *The Record,* July 1, 2005, page G 31 [Bergen County, New Jersey]

Chevillot, C., et al. 2005. Prospection-Inventaire (Vallée de la Dronne): VII. Le Triangle Lisle/Saint-Pardoux-la-Rivière/Thiviers (2005), Vallée de l'Isle et de la Double. *Documents d'Archéologie et d'Histoire Périgourdines* 20:206–210. [Includes information on millstone quarries near Saint-Crépin-de-Richemont in the Aquitaine region of France.]

Child, Hamilton 1871. *Gazetteer and Business Directory of Ulster County, N.Y. for 1871–2.* H. Child, Syracuse, New York.

Childe, V. G. 1943. Rotary Querns on the Continent and in the Mediterranean Basin. *Antiquity* 17 (65):19–26.

Cinadr, Thomas J., and David S. Brose 1978. *Archaeological Excavations in Caesar's Creek Lake, Ohio: Section II, The Carr Mill Race Site (33Wa75).* Report submitted to the National Park Service, Atlanta, by the Cleveland Museum of Natural History, Cleveland.

Cincinnati Gazette Company 1852. *The Ohio Railroad Guide; Illustrated and Descriptive.* Cincinnati Gazette Company, Cincinnati.

Cisneros Cunchillos, M., M.P. Lapuente Mercadal, M.A. Magallón Botaya, and M. Ortiga Castillo 1985. Estudio Arqueológico–Geológico de Cerro Redondo (Pardos, Zaragoza) ["Archaeological-Geological Study of Round Hill (Pardos, Zaragoza)"]. *Turiaso* 6:137–164. Centro de Estudios Turiasonenses, Tarazona. [Spain]

Cist, Charles 1841. *Sketches and Statistics of Cincinnati in 1841.* Charles Cist, Cincinnati.

Cist, Charles 1851. *Sketches and Statistics of Cincinnati in 1851.* W. H. Moore & Co., Cincinnati.

Cist, Charles 1859. *Sketches and Statistics of Cincinnati in 1859.* W. H. Moore & Co., Cincinnati.

Claracq, Paul 1994. Recherche sur les Meules de Moulin des Hautes-Pyrénées. *Bulletin de la Société Ramond,* 129e année, pp. 61–121. [Early millstones (sandstone, granite, and flint) in France near the Spanish border.]

Clark, Victor S. 1929. *History of Manufactures in the United States, Volume I:1607–1860.* Carnegie Institution of Washington. McGraw Hill Book Company, New York.

Clark County, Kentucky 1797. Law Suit filed in Clark County, Kentucky Circuit Court. Valentine Huff versus Peter DeWitt Concerning a Plea of Debt. Kentucky State Archives, Frankfort. [This document of April 1, 1797, mentions that Peter DeWitt was a millstone cutter by trade.]

Clark County, Kentucky 1803. Law Suit filed in Clark County, Kentucky Circuit Court. James Daniel versus Martin DeWitt Concerning a Plea of Covenant Broken over 25 Grindstones on November 16, 1801. Kentucky State Archives, Frankfort.

Clark County, Kentucky 1810. Law Suit filed in Clark County, Kentucky Circuit Court. Joseph Wilkerson versus Spencer Adams Concerning a Plea of Covenant Broken over a Pair of Millstones at the Red River Quarry in 1807. Kentucky State Archives, Frankfort.

Clark County, Kentucky 1822. Law Suit filed in Clark County, Kentucky Circuit Court. Martin Johnson Administration versus Spencer Adams Regarding Work Done by Martin Johnson at the [Red River] Millstone Quarry in 1819. Kentucky State Archives, Frankfort.

Clark County, Kentucky 1826. Law Suit filed in Clark County, Kentucky Circuit Court. Cornelius Summers versus Spencer Adams Concerning a Plea of Covenant broken over a Pair of Millstones at the Red River Quarry in 1825. Kentucky State Archives, Frankfort.

Clarke, Robin 2003. Millstone Quarries at Whittle Hills, Lancashire. *Wind & Watermills* 22: 15–23. Midland Wind and Watermills Group.

Clay, R. Berle 1994. Brutus' Industrial Complex, Mill and Saw Mill. On file at Kentucky Heritage Council, Frankfort. [Manuscript containing dated journal entries and other information about grist and saw mills owned by Brutus Clay on his farm near Paris in Bourbon County, Kentucky.]

Cleare, John 1988. Stanage and the Hathersage Heights: Peak District. In *Fifty Best Hill Walks of Britain*. Webb & Bower, Devon.

Clifford, J. D. 1926. Geological and Mineralogical Observations. *Tennessee Historical Magazine* 9 (4):275–278.

Clochard, Delphine 2004. *Les Meulières de Claix/Chaumes du Vignac: Mémoire d'un Site*. Rapport à Monsieur le Président du Conservatoire d'Espaces Naturels et des Sites de Poitou-Charentes. Association Régionale pour la Promotion de l'Ethnologie en Poitou-Charentes (ARPE), Gençay , 88 p. [Millstone quarry in France]

Close-Brooks, J. 1983. Some Early Querns. *Proceedings of the Society of Antiquaries of Scotland* 113:282–289.

Clouse, Jerry A. [Pennsylvania Historical and Museum Commission, Harrisburg] 1991. Letter to Charles D. Hockensmith, February 2. On file at the Kentucky Heritage Council, Frankfort.

Cohen, Ronald, and Curtis N. Runnels 1981. The Source of the Kitsos Millstones. In *La Grotte Préhistorique de Kitsos (Attique)* edited by Nicole Lambert, pp. 233–239. Recherche sur les Grandes Civilisations, Synthèse #7, Éditions A.D.P.F.-École Française d'Athènes, Paris. [Millstones from Kitsos Cave in Greece.]

Collins, Gabriel 1848. *Gabriel Collins' Louisville and New Albany Directory and Annual Advertiser for 1848*. G. H. Monsarrat, Louisville.

Collins, Lewis 1847. *Historical Sketches of Kentucky*. Lewis Collins, Maysville and J. A. & U. P. James, Cincinnati.

Collins, Lewis, and Richard H. Collins 1874. *History of Kentucky*. Collins and Company, Covington.

Colyer, Charles 1824. Contract with Sidney Payne Clay, February 28, 1824. Sidney Payne Clay Papers (A\C621a, folder 5) Special Collections, The Filson Historical Society, Louisville. [Handwritten contract between Colyer, of Rockcastle County, and Clay, of Bourbon County, for the delivery of two millstones to Clay]

Conley, James F. 1962. *Geology and Mineral Resources of Monroe County, North Carolina*. Bulletin 76, North Carolina Department of Conservation and Development, Division of Mineral Resources, Raleigh.

Cookson, Mildred M. 2003. Practical Experience in Using French Millstones. In *Meules à Grains: Actes du Colloque International de La Ferté-sous-Jouarre 16–19 Mai 2002*, edited by Mouette Barboff, François Sigaut, Cozette Griffin-Kremer, and Robert Kremer, pp. 344–353. Éditions Ibis Press and Éditions de la Maison des Sciences de l'Homme, Paris, France.

Cookson, Mildred M. 2005. International Millstone Congress Grenoble 22–25 September 2005. *International Molinology* 71:30–31, Watford, England.

Coons, A. T., and B. H. Stoddard 1929. Abrasive Materials. In *Mineral Resources of the United States, 1926*, pp. 245–253. Part II — Nonmetals. Department of the Interior, U.S. Bureau of Mines Government Printing Office, Washington, D.C.

Coons, A. T., and B. H. Stoddard 1930. Abrasive Materials. In *Mineral Resources of the United States, 1927*, pp. 91–98. Part II — Nonmetals. Department of the Interior, U.S. Bureau of Mines, Government Printing Office, Washington, D.C.

Cooper, James D., and Gertrude E. Tucker 1962. Abrasive Materials. In *Minerals Yearbook, 1961*, pp. 215–231. Department of the Interior, U.S. Bureau of Mines, Government Printing Office, Washington, D.C.

Cooper, James D., and Gertrude E. Tucker 1963. Abrasive Materials. In *Minerals Yearbook, 1962*, pp. 197–212. Department of the Interior, U.S. Bureau of Mines, Government Printing Office, Washington, D.C.

Coquebert-Montbret 1796. Description des Carrières de Pierre à Meules qui Existent dans la Commune des Molièrers, Seine-et-Oise. *Journal des Mines*, pp. 25–36. [France]

Cordier, Louis 1807. Des Carrières à Meules de Moulin. *Journal des Mines*, pp. 57–58. [France]

Cossons, Neil 1975. *The BP Book of Industrial Archaeology*. David and Charles, London.

Cothren, William 1872. History of Ancient Woodbury, Connecticut, from the First Indian Deed in 1659 to 1872, Including the Present Towns of Washington, Southbury, Bethlem, Roxbury, and Part of Oxford and Middlebury. Volume II. Published by the author, Woodbury, Connecticut.

Coustet, Robert, Bernard Valette, and Didier Alos 2001. *Inventaire des Souterrains du Tarn*. Volume 3. Lavaur, Centre Régional d'Étude et de Documentation des Souterrains, 285 p. [France]

Couston, Jean-Claude 1982. Les Anciennes Carrières de Meules pour Moulins prède Rocamadour. *Quercy-Recherches* 45/46:16 19. [France]

Coutard, André 1998. Les Meules de Villaines-la-Gonais [Sarthe]. *Moulins de la Sarthe. Bulletin de l'Association des Amis des Moulins de la Sarthe*, January, pp. 8–12. [France]

Coutard, André 1998. Millstones from Villaines-la-Gonais. *International Molinology* 57:6–9. [Translated from French to English by Owen Ward. Flint millstone quarries at the village of Villaines-la-Gonais in France dating ca. 1806–1931]

Coxe, Tench 1814. *A Statement of the Arts and Manufacturers of the United States of America, for the Year 1810*. Printed by A. Corman, Philadelphia.

Craik, David 1870. *The Practical American Millwright and Miller*. H. C. Baird, Philadelphia.

Craik, David 1877. *The Practical American Millwright and Miller*. Henry C. Baird, Philadelphia.

Craik, David 1882. *The Practical American Millwright and Miller*. Henry Carey Baird and Company, Philadelphia.

Crawford, H. S. 1909. Some Types of Quern, or Hand-Mill. *Journal of the Royal Society of Antiquaries of Ireland* 39:393–396.

Crawford, O. G. S. 1953. Querns and Quern-Quarries.

Chapter 9 in *Archaeology in the Field* by Osbert Guy Stanhope Crawford, pp. 98–106. Phoenix House, London.

Crawford, O. G. S., and J. Röder 1955. The Quern-Quarries of Mayen in the Eifel. *Antiquity* 29:68–76. [Germany]

Črnilec, M. 2003. *Izdelovalci Mlinskih Kamnov, Umetniki Izdelkov iz Konglomerata.* Občina Naklo, Naklo, 12 p. ["The Manufacturers of Millstones, the Artists of Products Made of Conglomerate"].

Cross, Alice 1996. Meet You at the Station(s). *The Accordian* 10 (1):6–10, Rochester, New York.

Croudance, I. W., and O. Williams-Thorpe 1988. A Low-Dilution, Wavelength-Dispersive X-ray Fluorescence Procedure for the Analysis of Archaeological Rock Artefacts. *Archaeometry* 30:227–236.

Cruse, John 2006. Current & Future Activities of the Yorkshire Quern Survey. *Quern Study Group Newsletter* 7: 2–3. Oxford, England.

Cruz, Alberto Machado 1971. Etoko (pia de moer). *Relatorios e Comunicacoes* 15. Instituto de Investigacao Cientifica de Angola, Luanda. [Millstones from Angola in southwest Africa]

Cuming, F. 1810. Sketches of a Tour to the Western Country Through the States of Ohio and Kentucky; a Voyage Down the Ohio and Mississippi Rivers, and a Trip Through the Mississippi Territory, and Part of West Florida Commenced at Philadelphia in the Winter of 1807, and Concluded in 1809. Reprinted in *Early Western Travels 1748–1846*, edited by Reuben Gold Thwaites (1904), Volume 4, pp. 17–377. Arthur H. Clark Company, Cleveland.

Cumming, David A. 1984. John MacCulloch's Millstone Survey and Its Consequences. *Annals of Science* 41 (6):567–591. [Millstone survey conducted in western Great Britain by geologist John MacCulloch]

Curran, J. 1908. An Ancient Quern, or Millstone, County Kerry. *Journal of the Royal Society of Antiquaries of Ireland* 38:74–75.

Curwen, E. Cecil 1937. Querns. *Antiquity* 11:133–151.

Curwen, E. Cecil 1941. More About Querns. *Antiquity* 15 (42):15–32.

Curwen, E. Cecil 1950. The Querns. In *A Prehistoric and Romano-British Site at West Blatchington, Hove* by N. E. S. Norris and G. P. Burstow. *Sussex Archaeological Collections* lxxxix:50–52.

Dana, James D. 1857. *Manual of Mineralogy, Including Observations on Mines, Rocks, Reduction of Ores, and the Applications of the Science to the Arts, with 260 Illustrations. Designed for the Use of Schools and Colleges.* Durrie & Peck, Philadelphia.

Darton, N. H. 1894. Shawangunk Mountain. *National Geographic Magazine*, March 17, 1894.

Davies, David Llewelyn 1997. *Watermill: Life Story of a Welsh Cornmill, Being the History of Felin Lyn, Dyffryn Ceiriog, Denbighshire, North Wales.* Ceiriog Press, Cardiff, Wales.

Davis, A. E. 1935. Abrasive Materials. In *Minerals Yearbook, 1935*, pp. 995–1010. Department of the Interior, U.S. Bureau of Mines, Government Printing Office, Washington, D.C.

Davis, Robert S., Jr. 1990a. The Georgia Buhrs: A Forgotten Millstone. *Old Mill News* 18 (4):6–7.

Davis, Robert S., Jr. 1990b. "As Good as the French": A History and Comparison of Georgia's Buhr Stone Industry with the French Buhr Stone, 1810–1854. Papers: 19th Annual Conference, Society for Industrial Archeology, Philadelphia, Pennsylvania, May 31–June 3, 1990.

Day, David T. 1886. Abrasive Materials. In *Mineral Resources of the United States, Calendar Year 1885*, pp. 428–430. Part II — Nonmetals. U.S. Bureau of Mines, U.S. Geological Survey, Government Printing Office, Washington, D.C.

Day, David T. 1888. Abrasive Materials. In *Mineral Resources of the United States, Calendar Year 1887*, pp. 552–553. Part II — Nonmetals. U.S. Bureau of Mines, U.S. Geological Survey, Government Printing Office, Washington, D.C.

Day, David T. 1890. Abrasive Materials. In *Mineral Resources of the United States, Calendar Year 1888*, pp. 576–577. Part II — Nonmetals. U.S. Bureau of Mines, U.S. Geological Survey, Government Printing Office, Washington, D.C.

Day, David T. 1892. Abrasive Materials. In *Mineral Resources of the United States, Calendar Years 1889–1890*, pp. 556–557. Part II — Nonmetals. U.S. Bureau of Mines, U.S. Geological Survey, Government Printing Office, Washington, D.C.

Dean, Lewis S. [Geological Survey of Alabama, Tuscaloosa] 1996. Letter to Charles D. Hockensmith, November 21. [The millstone industry in Alabama.]

Dean, Lewis S. (editor) 1995. *Michael Tuomey's Reports and Letters on the Geology of Alabama, 1847–1856.* Information Series 77. Geological Survey of Alabama, Tuscaloosa.

Deaton, J., and Suzanne G. Cherau 2003. Intensive (Locational) Archaeological Survey, Worcester Vocational High School, Worcester, Massachusetts. Technical Report No. 1400.01. Report submitted to Lamoureux Pagano Associates Architects, Worcester, Massachusetts, by PAL, Pawtucket, Rhode Island.

Dedrick, B. W. 1924. *Practical Milling.* National Miller, Chicago. Reprinted in 1989 by the Society for the Preservation of Old Mills, Manchester, Tennessee.

Deering, Richard 1859. *Louisville: Her Commercial, Manufacturing and Social Advantages.* Hanna & Co., Louisville.

Deffontaines, Benoît 2002. La Production et la Commercialisation de Meules de Moulins à Cinq-Mars-la-Pile (Indre-et-Loire) ["The Production and Marketing of Millstones at Cinq-Mars-la-Pile"]. *Les Cahiers de l'Association des Amis des Moulins de Touraine,* No. 1 (May 2002), 32 p. [France]

Deffontaines, Benoît 2003. La Production et la Commercialisation de Meules de Moulins à Cinq-Mars-la-Pile (Indre & Loire) ["The Production and Marketing of Millstones at Cinq-Mars-la-Pile"]. In *Meules à Grains: Actes du Colloque International de La Ferté-sous-Jouarre 16–19 Mai 2002*, edited by Mouette Barboff, François Sigaut, Cozette Griffin-Kremer, and Robert Kremer, pp. 294–311. Éditions Ibis Press and Éditions de la Maison des Sciences de l'Homme, Paris, France. [France]

Delmas, Jean 1975. Carte des Meulières de l'Aveyron, Procès Verbaux. *Lettres de l'Aveyron* 42:134–150. [Millstones from the l'Aveyron area of southern France]

Dendaletche, Claude, and Nanou Saint-Lebe 1997. L'Artisanat Lapidaire sur le Massif d'Artzamendi de 1741 à nos jours. 1— Extraction et Évacuation des Meules. *Bulletin du Musée Basque*, 147:1–20. [France]

Deschamps, Marise 2003. La Façonnage des Meules en Savoie ["Millstone Making at Savoie, France"] [Abstract]. In *Meules à Grains: Actes du Colloque International de La Ferté-sous-Jouarre 16–19 Mai 2002*, edited by Mouette Barboff, François Sigaut, Cozette Griffin-Kremer, and Robert Kremer, p. 449. Éditions Ibis Press and Éditions de la Maison des Sciences de l'Homme, Paris, France.

Desirat, Guy 1985. Tailleries de Meules pour Moulins à Huile et à blé à Bagnols-en-ForéVar. *Moulins de Provence Bulletin* 1:13–15. [Millstones from Bagnols-en-ForéVar in southern France.]

Dohm, B. 1950. Die Eis- und Mühlsteinhöhlen von Roth bei Gerolstein. *Die Eifel* 45., S. 113. Düren. [Germany]

Doswald, Cornel 1994. Herkunft und Verbreitung der Römerzeitlichen Mühlsteine im Kanton Aargau. *Minaria Helvetica* 14a:22–23. [Würenlos is a Roman era quern quarry in northern Switzerland.]

Dubech, Pascal, and Thibaut Gaborit 2006. Les Espaces Naturels Protégés: Des Outils de Protection des Pierres Meulières? In *Mühlsteinbrüche. Erforschung, Schutz und Inwertsetzung eines Kulturerbes europäischer Industrie (Antike–21. Jahrhundert)*, edited by Alain Belmont and Fritz Mangartz, pp. 183–189. Römisch-Germanisches Zentralmuseum, Tagungen Band 2, Mainz, Germany. [Combined papers including Dubech's "Nature Reserves: Possibilities for the Conservation of Millstone Quarries?" "The Nature Reserve of Pinail at Vouneuil-sur-Vienne" and "Nature Reserves: Possibilities for the Conservation of Millstone Quarries? The Millstone Quarries of Claix, Dept. Charente."]

Duc, Jean-Paul 2003. Situation Géographique du Gisement d'Epernon et Son Histoire. In *Meules à Grains: Actes du Colloque International de La Ferté-sous-Jouarre 16–19 Mai 2002*, edited by Mouette Barboff, François Sigaut, Cozette Griffin-Kremer, and Robert Kremer, pp. 289–293. Éditions Ibis Press and Éditions de la Maison des Sciences de l'Homme, Paris, France. [Quarrying and production of millstones at Epernon, France.]

Duc, Jean-Paul 2005. *Carriers et Meuliers de Région d'Epernon*. L'Association Epernon Patrimoine et Alentours en collaboration avec la Mairic d'Epernon, Epernon, France. [Excellent book on the millstone industry at the community of Epernon in France.]

Ducassé, Élie 1995. Meules et Moulins en Lomagne. *Bulletin de la Société Archéologique, Historique, Littéraire et Scientifique e du Gers*, 3e trimestre 1995, pp. 280–292. [Millstones from Lomagne in southern France.]

Dufrénoy, Ours Pierre Armand 1834. Rapport sur le Gisement des Pierres Meulières des Environs de Paris, leur Exploitation et les Différentes Circonstances de leur Commerce. *Bulletin de la Société d'Encouragement pour l'Industrie Nationale*, October 1834, pp. 397–405; November 1834, pp. 411–423. [France]

Dufrénoy, Ours Pierre Armand 2002. Rapport sur le Gisement des Pierres Meulière des Environs de Paris. In *Les Meuliers. Meules et Pierres Meulières* by Agapain, pp. 121–140. Presses du Village, Étrépilly, France. [France]

Duhard, Jean-Pierre 1996. Les Meules de Grès d'Artzamendia (Commune d'Itxassou, Pyrénées-Atlantiques). *Bulletin Trimestriel. Société d'Anthropologie du Sud-Ouest* 31 (1–2):59–82. [France]

Durán, H. M., and J. L. Pulido 2007. Análisis de la Molienda en la Proceso de Elaboración de Mezcal. *Información Tecnológica* 18 (1): 47–52. [Use of millstones in milling agave for mescal manufacture in Mexico.]

Durán-Garcia, H. M., E. J. Gonzalez-Galvan, and P. O. Matadamas 2007. Use of the Rolling Mill in the Extraction of Maguey's Juice. *Journal of Food, Agriculture and Environment* 5 (2):13–16. [The use of millstones to crush maguey pineapples for juice in Mexico.]

Durand-Vaugaron, L. 1969. Technologie et Terminologie du Moulin à eau en Bretagne. *Annales de Bretagne* 76 (2): 285–353. [Mills in Brittany during 17th and 18th centuries, with information on millstones from Moulière forest, La Ferté-sous-Jouarre, and Cinq-Mars.]

Durham, Emma 2006. Stone Finds from Excavations at Le Yaudet, Ploulec'h, Brittany. *Quern Study Group Newsletter* 7: 3–4. Oxford, England. [Discusses a small group of granite querns dating to the Iron Age, early Roman, and Medieval periods from excavations at Le Yaudet at Côtes-d'Armor in France.]

Dworakowska, A. 1983. *Quarries in Roman Provinces*. Polish Academy of Sciences, Institute of History of Material Culture, Wrodaw.

Ebert, Milford 1993. On the Right Track with the Accord Train Station. *The Accordian* 7 (3):1–5, Rochester, New York.

Eck, T. 1901. Note Sur Moulins à Grain de Vermand (Aisne). *Bulletin Archéologique du Comité des Travaux Historiques et Scientifiques* 1901, 225. [Gallo-Roman millstones between Paris and Rheims, Seine-et-Marne, France]

Edwards, Richard 1855. *Statistical Gazetteer of the State of Virginia, Embracing Important Topographical and Historical Information from Recent and Original Sources, Together with the Results of the Last Census Population, in Most Cases, to 1854*. Richard Edwards, Richmond.

Edwards, Richard (editor) 1864. *Edwards' Annual Directory ... City of Louisville for 1864–65*. Richard Edwards, Louisville.

Edwards, Richard, and M. Hopewell 1860. *Edwards's Great West and Her Commercial Metropolis: Embracing a General View of the West and a Complete History of St. Louis, from the Landing of Ligueste, in 1764, to the Present Time: With Portraits and Biographies of Some of the Old Settlers, and Many of the Most Prominent Business Men*. Edwards's Monthly, St. Louis.

Edwards, Greenough & Deved 1866. *Edwards's Annual Directory of the Inhabitants, Institutions, Incorporated Companies, Manufacturing Establishments, Business Firms, Etc., Etc. in the City of Louisville for 1866*. Edwards, Greenough & Deved, Louisville.

Edwards, Greenough & Deved 1870. *Edwards Sixth Annual Directory of the City of Louisville for 1870*. Southern Publishing Company, Louisville.

Egle, William H. 1876. *An Illustrated History of the Commonwealth of Pennsylvania*. W. C. Goodrich & Co., Harrisburg.

Egleston, Thomas 1872. *Lectures on Mineralogy*. D. Van Nostrand, New York.

Egoumenidou, Euphrosyne, and Diomedes Myrianthefs 2003. Trade and Use of Millstones in Cyprus during the Recent Past (18th–20th Century). In *Meules à Grains: Actes du Colloque International de La Ferté-sous-Jouarre 16–19 Mai 2002*, edited by Mouette Barboff, François Sigaut, Cozette Griffin-Kremer, and Robert Kremer, pp. 175–183. Éditions Ibis Press and Éditions de la Maison des Sciences de l'Homme, Paris, France.

Eibensteiner, F. K. 1933. Geschichte der Mühlsteinindustrie in Perg. *Bilder-Woche der Tagespost, Linz a. d' Donau, Februar 1933*. [Austria or vicinity.]

Ekroll, Ø. 1997 Med Kleber og Kalk. Norsk Steinbygging I Mellomalderen, Oslo. [Deals with millstone quarries in Norway.]

El Alaoui, Narjys 1999. Le Soleil dans la Main, une Meules à Grains de l'Anti-Atlas. Maroc, in *Trésors Méconnus du Musée de l'Homme*, Muséum National d'Histoire Naturelle-le Cherche midi éds, Paris.

El Alaoui, Narjys 2003. Meules à Bras le Sud Marocain. Différences Suivant les Produits Broyés. In *Meules à*

Grains: Actes du Colloque International de La Ferté-sous-Jouarre 16–19 Mai 2002, edited by Mouette Barboff, François Sigaut, Cozette Griffin-Kremer, and Robert Kremer, pp. 51–60. Éditions Ibis Press and Éditions de la Maison des Sciences de l'Homme, Paris, France. [Hand querns in south Morocco.]

Elfwendahl, Magnus, and Peter Kresten 1993. *Geoarkeologi Inom Kvarteret Bryggaren. Arkeologiska Artefakter av sten från det Medeltida Uppsala.* Riksan-tikvarieämbetet och Statens Historiska Museer. Rapport UV 1993:5, Uppsala. [Geoarchaeology of Medevial querns from the Uppsala area of Sweden.]

Elliott, C., C. Xenophontos, and J. G. Malpas 1986. Petrographic and Mineral Analysis Used in Tracing the Provenance of Late Bronze Age and Roman Artefacts from Cyprus. *Report of the Department of Antiquities of Cypress* 1986:80–96.

Elliott, Daniel T. 1987. Traveling from Door to Door: Two Years of Historic Site Survey in South Carolina's National Forest. Paper presented at the joint meeting of the Southeastern Archaeological Conference and Eastern States Archaeological Federation, Charleston, South Carolina.

Elton, John 1904. *History of Corn Milling. Volume IV: Some Feudal Mills.* Burt Franklin, New York.

Emerson, Edgar C. 1898. *Our County and Its People: A Descriptive Work on Jefferson County, New York.* Boston History Company, Boston.

Emmons, Ebenezer 1852. *Report of Professor Emmons, on His Geological Survey of North Carolina.* S. Gales, Raleigh.

Emmons, Ebenezer 1856. *Geological Report of the Midland Counties of North Carolina.* George P. Putnam & Company, New York.

Emmons, Ebenezer 1858. *Agriculture of the Eastern Counties; Together with Descriptions of the Fossils of the Marl Beds.* Report of the North Carolina Geological Survey. Henry D. Turner, Raleigh.

Erpelding, Émile 1982. Les Pierres à Moulins et l'industrie Meulière de la Ferté sous Jouarre. *Fédération Française des Amis des Moulins* 7:83–111. [France]

Evans, J. 1994. Worked Stone and Quernstones. In *Roman Alcester: Southern Extramural Area. 1964–1966 Excavations. Part 2 Finds and Discussion*, edited by S. Cracknell and C. Mahany, pp. 231–248. CBA Research Report 97.

Fabre, Denis, Elisabeth Carrio, Yves Orengo, and Claudine Malacour 2006. Analyses Pétrographiques et Mécaniques d'un Ensemble de Roches Meulières Françaises ["Petrographic and Mechanical Analyses of a Group of French millstone Rocks."]. In *Mühlsteinbrüche. Erforschung, Schutz und Inwertsetzung eines Kulturerbes europäischer Industrie (Antike–21. Jahrhundert)*, edited by Alain Belmont and Fritz Mangartz, pp. 91–97. Römisch-Germanisches Zentralmuseum, Tagungen Band 2, Mainz, Germany.

Fahmy, A.G. 2001. Palaeoethnobotanical Studies of the Neolithic Settlement in Hidden Valley, Farafra Oasis, Egypt. *Vegetation History and Archaeobotany* 10 (4):235–246. [This Neolithic site contained querns.]

Falkenstein, Franz 1983. The Millstone Mine at Waldshut, Tiengen. Edited by Arthur Dunn. *Appendix of the International Molinological Society Newsletter* 24:1–5. [Germany]

Falkenstein, Franz 1986. Der Ehemalige Mühlsteinbergbau bei Waldshut. *Badische Heimat* 66 (2):291–294. [Germany]

Falkenstein, Franz 1987. Der ehemalige Mühlsteinbergbau im Schmitzinger Tal bei Waldshut. *Heimat am Hochrhein* (Jb. Landkrs. Waldshut) 12:168–193. [Germany]

Falkenstein, Franz 1988a. The Millstone Mines of Waldshut, Germany. Edited by Arthur Dunn. *Old Mill News* 16 (3):21. [Germany]

Falkenstein, Franz 1988b. Millstone Mine at Waldshut-Tiengen, Baden-Wurttemberg, West Germany. Edited by Arthur Dunn. *Bulletin Subteranea Britannica* 24:19–22. [Germany]

Falkenstein, Franz 1989. Die Mühlsteingräberei in der Umgebung von Dogern. *Heimat am Hochrhein* (Jb. Landkrs. Waldshut) 14:121–141. [Germany]

Falkenstein, Franz 1997. Eine fast vergessene Mühlsteinindustrie im Südöstlichen Schwarzwald. *Der Mühlstein* 14 (2):20–23, DGM Minden/Westfalen. [Germany]

Falkenstein, Franz 2001. Ganz schön bunt, der Buntsandstein in der Umgebung von Waldshut am Südöstlichen Schwarzwaldrand. *Aufschluss* 52:227–241. [Germany]

Falkenstein, Franz, and U. Koerner 1989. Hangrutschrisiken, Buntsandsteinstratigraphie und alter Mühlsteinbergbau bie Waldshut. *Iber. Mitt. Oberrhein geol.* Ver. N.F. 71:19–26. [Germany]

Farmer, David L. 1992. Millstones for Medieval Manors. *Agricultural History Review* 11:97–111.

Fassbind-Bacq, Andrée 2002. Une Lignée de Meuliers: les Gueuvin et les Dupety. In *Les Meuliers. Meules et Pierres Meulières* by Agapain, pp. 171–178. Presses du Village, Étrépilly, France. [France]

Faujas de Saint-Fond, Barthélemy 1802. Description des Carrières Souterraines et Volcaniques de Niedermennich, à Trois Lieues d'Andernach, d'où l'on Tire des Laves Poreuses, Propres à Faire d'Excellentes Meules de Moulins. *Annales du Muséum National d'Histoire Naturelle* 3: 181–193. [Germany]

Fayette County, Kentucky 1804. Law Suit filed in Fayette County, Kentucky Circuit Court. John Higbee versus Absolom Hanks Concerning a Plea of Covenant Broken over a Pair of Millstones to Be Made at the Red River Quarry According to an Agreement on December 10, 1799. Kentucky State Archives, Frankfort.

Feachem, R. W. 1958. A Quern from Mullochard, Duthil, Inverness-shire. *Proceedings of the Society of Antiquaries of Scotland* 91:189–190.

Ferla, P., R. Alaimo, and F. Spatafora 1984. Studio Petrografico delle Macine di Etá Arcaica e Classica da Monte Castellazzo di Poggioreale (Sicila Occidentale). *Sicilia Archeologica* 56:1–30. [Study of saddle querns from five sites in Sicily.]

Fess, Simeon D. 1937. *Ohio: A Four-Volume Reference Library on the History of a Great State.* Volume 3: Historical Gazetteer of Ohio. Lewis Publishing Company, Chicago.

Flick, Alexander C. 1927. *The Papers of Sir William Johnson.* Volume V. The University of the State of New York, Albany.

Flick, Alexander C. 1931. *The Papers of Sir William Johnson.* Volume VII. The University of the State of New York, Albany.

Flick, Alexander C. 1933. *The Papers of Sir William Johnson.* Volume VIII. The University of the State of New York, Albany.

Flint, James 1822. Letters from America, Containing Observations on the Climate and Agriculture of the Western States, the Manners of the People, the Prospects of Emigrants, &c., &c. (1822) [for the years 1818–1820]. Reprinted in *Early Western Travels 1748–1846*, edited by Reuben Gold Thwaites, Volume 9, pp. 15–333. Arthur H. Clark Company, Cleveland.

Flory, Paul B. 1951a. Old Millstones. *Papers Read Before the Lancaster County Historical Society*, 55 (3):73–86, Lancaster, Pennsylvania.

Flory, Paul B. 1951b. Millstones and Their Varied Usage. *Papers Read Before the Lancaster County Historical Society*, 55 (5):125–136. Lancaster, Pennsylvania.

Fontaine, William M. 1869. The Building Stone and Slate of Virginia. *The Manufacturer and Builder* 1 (2):46–47. Western and Company, New York.

Foster, John W. 1869. *The Mississippi Valley: Its Physical Geography, Including Sketches of Topography, Botany, Climate, Geology, and Mineral Resources: and of the Progress of Development in Population and Material Wealth*. S. C. Griggs, Chicago.

Fox, H. S. A. 1994. The Millstone Makers of Medieval Dartmoor. *Devon and Cornwall Notes and Queries* 37 (5):153–157. [England]

François, Léon 2003. Les Moulins Belges et Leurs Meulles au Fil des Breverts durant le XIXe Siècle. In *Meules à Grains: Actes du Colloque International de La Ferté-sous-Jouarre 16–19 Mai 2002*, edited by Mouette Barboff, François Sigaut, Cozette Griffin-Kremer, and Robert Kremer, pp. 380–386. Éditions Ibis Press and Éditions de la Maison des Sciences de l'Homme, Paris, France. [Belgian mills and millstones as seen through 19th century patents.]

Frankel, R. 2003. Mills and Querns in the Talmudic Literature — A Reappraisal in Light of Archaeological Evidence. *Cathedra* 110: 43–60, 188. [Published in Hebrew with an English abstract.]

Freshwater, Tom 1996. A Lava Quern Workshop in Late Saxon London. *London Archaeologist* 8 (2):39–45.

Fries, Robert 1995a. Handmade, from the Last of the Stonecutters: Recollections of the Brush Mountain Quarries. *New River Current* 7 (300):16–18; *Roanoke Times & World-News*, February 5, 1995. [Story about millstone makers Robert Houston Surface and W. C. Saville.]

Fries, Robert 1995b. Millstone Memories: All that Remains of the Brush Mountain Quarries Are the Recollections of Two Montgomery County Men, the Last of the Stonecutters Who Made Millstones by Hand. *Roanoke Times & World-News* (New River Valley Edition), February 5, 1995, page 16. On NewsBank and ProQuest websites. [Story about millstone makers Robert Houston Surface and W. C. Saville.]

Fries, Robert 1998. Montgomery County Stonecutter Dies at 84; Several Tech Buildings Bear His Mark. *Roanoke Times & World-News* (New River Valley Edition), May 15, 1998, page 3. On NewsBank and ProQuest websites. [Death of millstone maker Robert Houston Surface.]

Galetti, Paola 2006. Moulins, Meules, Meulières et Carrières de Meules dans L'Italie Médiévale ["Mills, Millstones, Millstone Rocks and Millstone Quarries in Medieval Italy."]. In *Mühlsteinbrüche. Erforschung, Schutz und Inwertsetzung eines Kulturerbes europäischer Industrie (Antike–21. Jahrhundert)*, edited by Alain Belmont and Fritz Mangartz, pp. 71–80. Römisch-Germanisches Zentralmuseum, Tagungen Band 2, Mainz, Germany.

Galer, Greg [Easton, Massachusetts] 2006. E-mails to Charles D. Hockensmith, April 27 and May 2. [Millstone quarry at Stonehill College campus in Easton, Massachusetts.]

Gandilhon, René 1986. Les Meules à Moulin de Saint-Martin d'Ablois (XVIIe-XVIIIe Siècle). *Mémoires de la Société d'Agriculture, Commerce, Sciences et Arts de la Marne*, 101:183–201. [Study of a quarry located near La Ferté-sous-Jouarre during 17th and 18th centuries.]

Gannett, Henry 1883. Abrasive Materials. In *Mineral Resources of the United States, Calendar Year 1882*, pp. 476–481. Part II — Nonmetals. Department of the Interior, U.S. Bureau of Mines, Government Printing Office, Washington, D.C.

Garber, D. W. 1970. *Waterwheels and Millstones: A History of Ohio Gristmills and Milling*. The Ohio Historical Society, Columbus.

Gary, Margaret, Robert McAlfee, Jr., and Carol L. Wolf (editors) 1974. *Glossary of Geology*. American Geological Institute, Washington, D.C.

Gast, Marceau 2003. Meules et Molettes Sahariennes ["Millstones and Mullers of the Sahara."] In *Meules à Grains: Actes du Colloque International de La Ferté-sous-Jouarre 16–19 Mai 2002*, edited by Mouette Barboff, François Sigaut, Cozette Griffin-Kremer, and Robert Kremer, pp. 61–66. Éditions Ibis Press and Éditions de la Maison des Sciences de l'Homme, Paris, France. [North Africa]

Gaucheron, André 1981. Statistiques — Cerclage des Meules. *The International Molinological Society Transaction of the 5th Symposium on Molinology*, pages 83–91. [France]

Gaucheron, André 1982. La Ferté-sous-Jouarre. Notes and statistics prepared for TIMS, April 1982, 3 pages. [France]

Gaucheron, André 1985. A Milestone in Milling. *The International Molinological Society Transaction of the 6th Symposium on Molinology*, pp. 51–61, Gent, Belgium. [France]

Gaucheron, André 1991a. La Révolution de la Meunerie au XIXe siècle. Les Moulins 10 (1985/86):15–21. [France]

Gaucheron, André 1991b. *Société Générale Meulière*, ARAM Provence. [Catalogue of 1930s with introductory leaflet; France.]

Gaucheron, André 2001. Des Assiettes et des Meules: l'Extraction des Meules de la Ferté-sous-Jouarre, Vue par un Artiste Peintre de la Manufacture Royale de Sèvres. *Revue de la Fédération Française des Amis des Moulins* 45:4–12. [France]

Gaucheron, André 2002. Meuniers en Beauce et Alentours. In *Les Meuliers. Meules et Pierres Meulières* by Agapain, pp. 243–271. Presses du Village, Étrépilly, France. [France]

Geist, Henri 1991. L'Exploitation des Meulières en Montagne de Reims. *Actes du 115e Congrès National des Société Savantes, Section d'Histoire des Sciences et Techniques, Avignon, 1990*, pp. 107–116. Paris, Éditions du CTHS. [France]

Geist, Henri 1995. Une Carrière de Meules sur le Rivage du Cap d'Ail (Alpes Maritimes). *Archéam (Cahiers du Groupe Archéologique, Cercle Historique des Alpes Maritimes)*, 1995–96, No. 3, pp. 41–46. [France]

Geist, Henri 2003. Une Carrière de Meules sur le Rivage de Cap d'Ail ["A Millstone Quarry on the Shore of Cap d'Ail"] (A.M.). In *Meules à Grains: Actes du Colloque International de La Ferté-sous-Jouarre 16–19 Mai 2002*, edited by Mouette Barboff, François Sigaut, Cozette Griffin-Kremer, and Robert Kremer, pp. 218–230. Éditions Ibis Press and Éditions de la Maison des Sciences de l'Homme, Paris, France. [France]

Gerke, Tammie, Sharon R. Stocker, Jack L. Davis, J. Barry Maynard, and Craig Dietsch 2006. Sourcing Volcanic Millstones from Greco–Roman Sites in Albania. *Journal of Field Archaeology* 31 (2):137–146.

Gibbings, Chris 2006. The Trade in Millstones in the Côtes d'Armor c. 1400 to 1800. *International Molinology* 73:32–33, Watford, England. [France]

Gibbons, Harry 1994. Millstone Dressing. *Tiverton Museum Newsletter* 5:12–15.

Gibert, Louis-François 1987. L'Extraction des Pierres Meulières de la Plaine de Bord au XVIIIe Siècle, in Sarlat et le Périgord, 38e Congrès de la Fédération Historique du Sud-Quest, Sarlat, 1986. *Bulletin de la Société Historique et Archéologique du Périgord*, supplément au t. 114, pp. 511–523. [France]

Gifford, Alan 1998. A Mysterious Millstone Found at Bishop's Castle, Shropshire. *Wind & Watermills* 17:2–14, Midland Wind and Watermills Group.

Gilpin, Joshua 1926. Journal of a Tour from Philadelphia Thro' the Western Counties of Pennsylvania in the Months of September and October, 1809. *The Pennsylvania Magazine of History and Biography* (1926) 50:64–78, 163–178, 380–382; (1927) 51:172–190, 351–375; and (1928) 52:29–58.

Gleisberg, H. 1977. Millstone Quarries. *Transactions of the 4th Symposium, the International Molinological Society*, pages 177–180. [England]

Goffin, R. 2003. The Quernstones. In *Middle Saxon London. Excavations at the Royal Opera House 1989–99*, edited by G. Malcolm and D. Bowsher, pp. 204–209. MoLAS Monograph 15.

Gradie, Robert R. III 1982. State of Connecticut Historic Resources Inventory Form for Pachaug State Forest Quarry. Connecticut Historical Commission, Hartford.

Graham, Alan H., and D. O. Farmer 1990. Medieval and Post-Medieval Millstones from the Old Malthouse, Abbotsbury, Dorset. *Proceedings of the Dorset Natural History and Archaeological Society* 112:141–142. [England]

Grassi, Robert 2004a. The Miller and Millstones — Part I. *Old Mill News* 32 (1):12–13.

Grassi, Robert 2004b. The Miller and Millstones — Part II. *Old Mill News* 32 (2):16–17.

Grillo, Paolo 1993. Il Commercio delle Mole nel Piemonte Medievale. In *Mulini da Grano nel Piemonte Medievale: Secoli XII–XV*, edited by Rinaldo Comba. Società per gli Storici, Archeologicied Artistici delle Provincia di Cuneo, Cuneo.

Grimshaw, Robert 1882. *The Miller, Millwright, and Millfurnisher: A Practical Treatise*. H. Lockwood, New York.

Grypari, Maria, and Louisa Karapidakis 2003. Les Meules de Moulins Grecs. In *Meules à Grains: Actes du Colloque International de La Ferté-sous-Jouarre 16–19 Mai 2002*, edited by Mouette Barboff, François Sigaut, Cozette Griffin-Kremer, and Robert Kremer, pp. 86–88. Éditions Ibis Press and Éditions de la Maison des Sciences de l'Homme, Paris, France. [Greek millstones.]

Guérin, Hubert 1985. L'Exploitation des Meulières en Montagne de Reims de 1792 à 1918. *Mémoires de la Société d'Agriculture, Commerce, Sciences et Arts du Département de la Marne* 100:159–177. [France]

Guérin, Hubert 1990. L'Exploitation des Meulières en Montagne de Reims. *Actes du 115e Congrès National des Société Savantes, Amiens 1990*. Carrière et Constructions, pp. 107–116. [France]

Guérin, Hubert 1991. L'Exploitation des Meulières en Montagne de Reims. *Actes du 115e Congrès National des Société Savantes, Section d'Histoire des Sciences et Techniques, Avignon, 1990*, pp. 107 à 116. Paris, Éditions du CTHS. [France]

Guettard, Jean-Etienne 1758. Mémoire sur la Pierre Meulière. *Histoire de l'Académie Royale des Sciences*, pp. 203–236. [French millstones.]

Guettard, Jean-Etienne 2002. Mémoire sur la Pierre Meulière. In *Les Meuliers. Meules et Pierres Meulières* by Agapain, pp. 109–120. Presses du Village, Étrépilly, France. [French millstones.]

Gwilt, Adam, and David Heslop 1995. Iron Age and Roman Querns from the Tees Valley. In *Moorland Monuments: Studies in the Archaeology of North-east York in Honor of Raymond Hayes and Don Spratt*, edited by Blaise E. Vyner. Council for British Archaeology Research Report 101.

Haberbosch, P. 1938. Römischer Steinbruch bei Würenlos. *Badener Neujahrsblätter* 1938:57–61. [Würenlos is a Roman era quern quarry in northern Switzerland.]

Haddock, John A. 1895. *The Growth of a Century: As Illustrated in the History of Jefferson County, New York, from 1793 to 1894*. Weed-Parsons Printing Company, Albany, New York.

Halcomb, Clarence [Hamilton, Ohio] 1992. Letter to Charles D. Hockensmith, June 7. On file at the Kentucky Heritage Council, Frankfort.

Haliburton, Thomas C. 1829. *An Historical and Statistical Account of Nova-Scotia: In Two Volumes*. J. Howe, Halifax.

Hall, James 1852. *A Chart Giving an Ideal Section of the Successive Geological Formations with an Actual Geological Section, from the Atlantic, to the Pacific Ocean. The Whole Illustrated by the Characteristic Fossils of Each Formation*. R. H. Pease, Albany, New York.

Hands, Edmund C. 1995. *Easton's Neighborhoods*. Easton Historical Society, Easton, Massachusetts.

Hange-Persson, Berit 1979. Om Kvarnstensbrytningen i Lugnås.-Ordlista med gamla Lugnåsord.-Uppgifter om Kvarnstenshuggarekassan i Lugnås.-Medlemsblad f Lugnås Hembygdsförening 7, s 28–41. [Deals with millstone quarrying at Lugnås, Sweden.]

Hange-Persson, Berit 1982a. *Mühlsteinbrüche in Lugnås*. Förlag Verlag Skaraborgs Länsmuseum. [Millstone quarries at Lugnås, Sweden.]

Hange-Persson, Berit 1982b. Kvarnstensbrotten i Lugnås. Förlag Skaraborgs länsmuseum. [Millstone quarries at Lugnås, Sweden.]

Hansen, P. B. 1989. Møller og kværne I Næstveds middelalder. *Liv og Levn* 3. [Study of quernstones in southern Scandinavia.]

Hansen, Steffen Stummann, and Anne-Christine Larsen 2000. Miniature Quern- and Millstones from Shetland's Scandinavian Past. *Acta Archaeologica* 71 (2):105–121.

Hanson, J. W. 1848. History of the Town of Danvers: From Its Early Settlement to the Year 1848. J. W. Hanson, Danvers, Massachusetts.

Harms, Eduard, and Fritz Mangartz 2002. *Vom Magma zum Mühlstein: Eine Zeitreise durch die Lavaströme des Bellerberg-Vulkans*. Römisch-Germanisches Zentralmuseum, Vulkanpark-Forschungen, Band 5, Mainz, Germany. [Book that deals with the basalt millstone industry in the vicinity of Mayen in northwestern Germany.]

Harper, L. 1857. *Preliminary Report on the Geology and Agriculture of the State of Mississippi*. By order of the Legislature of Mississippi. E. Barksdale, Jackson.

Harrison, Emma 2002. Quernstones from a Survey Around Nájera, Spain. *Quern Study Group Newsletter* 6:7–8, Reading, England.

Harrison, Glenn 1996a. Erratics to Millstones. *Linn County Historical Society Newsletter*, March issue, pages 2–3, Albany, Oregon.

Harrison, Glenn 1996b. Erratics to Millstones. *Old Mill News* 24 (4):6.

Hartnagel, C. A. 1927. *Mining and Quarry Industries of New York 1919–1924*. New York State Museum Bulletin Number 273, Albany.

Hartnagel, C. A., and John G. Broughton 1951. *Mining and Quarry Industries of New York 1937–1948*. New York State Museum Bulletin Number 343, Albany.

Harverson, Michael 2006. Review of *La Pierre à Pain: Les Carrières de Meules de Moulins en France du Moyen Age à la Revolution Industrielle* by Alain Belmont. *International Molinology* 72:43, Watford, England.

Harvey, H. H. 1884. *H. H. Harvey's Illustrated Catalogue and Price List for 1884 & 1885 of Stone Cutters,' Quarrymen's, Miners,' Railroad, Grist Mill, and Blacksmiths' Hammers, Sledges, and Tools, Sleds, &c., &c.* Kennebec Journal Printers, Augusta, Maine.

Harvey, H. H. 1886. *H. H. Harvey's Illustrated Catalogue and Price List for 1886 of Stone Cutters,' Quarrymen's, Miners,' Railroad, Grist Mill, and Blacksmiths' Hammers, Sledges, and Tools, Sleds, &c., &c.* Kennebec Journal Printers, Augusta, Maine.

Harvey, H. H. 1896. *H. H. Harvey's Illustrated Catalogue and Price List for 1896–1897 of Stone Cutters,' Quarrymen's, Miners,' Railroad, Grist Mill, Coopers,' Blacksmiths,' and Slaters' Hammers, Sledges, Tools, and Outfits; Also, Contractors' Supplies, Handles, Iron, Steel, etc., etc.* H. H. Harvey, Augusta, Maine. Reprinted in 1973 by Early American Industries Association.

Hatmaker, Paul, and A. E. Davis 1932. Abrasive Materials. In *Mineral Resources of the United States, 1930*, pp. 151–169. Part II — Nonmetals. Department of the Interior, U.S. Bureau of Mines, Government Printing Office, Washington, D.C.

Hatmaker, Paul, and A. E. Davis 1933a. Abrasive Materials. In *Mineral Resources of the United States, 1931*, pp. 111–130. Part II — Nonmetals. Department of the Interior, U.S. Bureau of Mines, Government Printing Office, Washington, D.C.

Hatmaker, Paul, and A. E. Davis 1933b. Abrasive Materials. In *Minerals Yearbook, 1932–33*, pp. 647–667. Part II — Nonmetals. Department of the Interior, U.S. Bureau of Mines, Government Printing Office, Washington, D.C.

Hawes, George W. 1859. *George W. Hawes' Kentucky State Gazetteer and Business Directory for 1859 and 1860.* George W. Hawes, Louisville.

Hawes, George W. 1860. *Ohio State Gazetteer and Business Directory for 1860–61.* G. W. Hawes, Indianapolis.

Hawksley, Jeff 2006. TIMS Mid-Term Excursion to South-West France, June 2006. *International Molinology* 73:2–11, Watford, England. [France]

Hayes, R. H., J. E. Hemingway, and D. A. Spratt 1980. The Distribution and Lithology of Beehive Querns in North-east Yorkshire. *Journal of Archaeological Science* 7:297–324. [England]

Haywood, John 1823. *The Natural and Aboriginal History of Tennessee: Up to the First Settlements Therein by the White People, in the Year 1768.* G. Wilson, Nashville.

Hazard, Samuel 1850. *Annals of Pennsylvania from the Discovery of the Delaware, 1609–1682.* Kite and Walton, Philadelphia.

Hazen, Theodore R. 1996a. Millstone Dressing — Part 1. *Old Mill News* 25 (1):16–17.

Hazen, Theodore R. 1996b. Millstone Dressing — Part 2. *Old Mill News* 25 (3):8–9, 11.

Hedges, Henry P., William S. Pelletreau, and Edward H. Foster 1874. *The First Book of Records of the Town of Southampton with Other Ancient Documents of Historical Value, Including All the Writings in the Town Clerk's Office from 1639 to 1660.* J. H. Hunt, Sag Harbor, New York.

Hehnly & Wike 1880. Broadside advertisement for Turkey Hill millstones. Hehnly & Wike, Durlach, Pennsylvania. Hagley Museum and Library, Wilmington, Delaware.

Heller, William Jacob 1920. *History of Northampton County (Pennsylvania) and the Grand Valley of the Lehigh.* American Historical Society, Boston.

Hemphill, Robert 1999. Transcription of advertisement for John Murray's Millstone Quarry in Burke County, Georgia, from the August 15, 1792, edition of the *The Chronicle and State Messenger*, Augusta, Georgia. Available on the web site RootsWeb.com (http://archiver.rootsweb.com/th/read/ORANGEBURGH_SC/1998-11/0910299005).

Henderson, A. H. 1959. Millstone and Grindstone Making in the Sheffield District. *Sorby Record* II:34–36. [England]

Herrscher, Estelle, Renée Colardelle, and Frédérique Valentin 2006. Meulières et Pathologies Humaines: Un Rapport Effectif? Analyse d'une Documentation Bucco-Dentaire entre le XIIIe et le XVIIIe Siècle à Grenoble ["Millstone Rocks and Human Pathologies: A True Relationship? Analysis of Oral-dental Evidence between the 13th and the 18th Century in Grenoble"]. In *Mühlsteinbrüche. Erforschung, Schutz und Inwertsetzung eines Kulturerbes europäischer Industrie (Antike–21. Jahrhundert)*, edited by Alain Belmont and Fritz Mangartz, pp. 99–108. Römisch-Germanisches Zentralmuseum, Tagungen Band 2, Mainz, Germany.

Heslop, D. H. 1988. The Study of Beehive Querns. *Scottish Archaeological Review* 5 (1 & 2):59–65.

Heverly, Clement F. 1915. *Pioneer and Patriot Families of Bradford County, Pennsylvania.* Volume 2. Bradford Star Printer, Towanda, Pennsylvania.

Higham, Arthur P. 1907. *The Millstones of Muscovy.* Sisley's London. [England]

Hildreth, Samuel P. 1828. Miscellaneous Observations on the Coal, Diluvial and Other Strata of Certain Portions of the State of Ohio. *American Journal of Science* 13 (1):38–40.

Hirsch, Steve [Kingston, New York] 2005. E-mails to Charles D. Hockensmith, January 2 and 10. [Ulster County, New York, millstone quarries.]

Hitchcock, Edward 1841. *Final Report on the Geology of Massachusetts: Vol. 1.* J. H. Butler. Northhampton.

Hockensmith, Charles D. 1988. The Powell County Millstone Quarries. *Kentucky Archaeology Newsletter* 6 (1):4.

Hockensmith, Charles D. 1990a. The Ware Millstone Quarry. *Kentucky Archaeology Newsletter* 7 (1):5.

Hockensmith, Charles D. 1990b. The Lower McGuire and Ewen Millstone Quarries. *Kentucky Archaeology Newsletter* 7 (2):5–6.

Hockensmith, Charles D. 1990c. Interview with Virginia's Last Millstone Makers. *Kentucky Archaeology Newsletter* 7 (2):6.

Hockensmith, Charles D. 1993a. *The Pilot Knob Millstone Quarry: A Self-Guided Trail.* Pilot Knob State Nature Preserve, Powell County, Kentucky. Booklet (limited edition) published by the Kentucky Nature Preserves Commission and the Kentucky Heritage Council, Frankfort.

Hockensmith, Charles D. 1993b. Millstone Quarrying in the Eastern United States: A Preliminary Overview. *Ohio Valley Historical Archaeology* 7 & 8:83–89.

Hockensmith, Charles D. 1993c. Study of American Millstone Quarries. *Old Mill News* 21 (1):5–7; 21 (2):4–8.

Hockensmith, Charles D. 1994a. *The Pilot Knob Millstone Quarry: A Self-Guided Trail.* Pilot Knob State Nature Preserve, Powell County, Kentucky. Booklet (second edition with run of 5,000 copies) published by the Kentucky Nature Preserves Commission and the Kentucky Heritage Council, Frankfort.

Hockensmith, Charles D. 1994b. European Millstone Quarries: A Bibliography. *Old Mill News* 22 (1):16–17.

Hockensmith, Charles D. 1994c. European Millstone Quarries: A Bibliography. *International Molinology* 49:30–32, Holland.

Hockensmith, Charles D. 1998. Old World Millstone and Quern Studies: A Supplemental Bibliography. *Old Mill News* 26 (2):22–24.

Hockensmith, Charles D. 1999a. The Millstone Industry in Southwest Virginia. In *Millstone Manufacture in Virginia: Interviews with the Last Two Brush Mountain Millstone Makers*, edited by Charles D. Hockensmith, pp. 1–3. Society for the Preservation of Old Mills, Newton, North Carolina.

Hockensmith, Charles D. 1999b. The Manufacture of Conglomerate Millstones at the Powell County, Kentucky, Quarries. Paper presented at the Annual Meeting of the Society for the Preservation of Old Mills, Newton, North Carolina.

Hockensmith, Charles D. 2000. *An Archaeological Reconnaissance of a 6.4 Mile Long Segment of the Proposed Smith to Stanton Power Line in Powell County, Kentucky.* Draft Report for the Kentucky Heritage Council Occasional Reports in Archaeology, Number 3, Frankfort.

Hockensmith, Charles D. 2002. The Conglomerate Millstone Industry in the Eastern United States. Paper presented at the Colloque International "Extraction, Façonnage, Commerce et Utilisation des Meules de Moulin — Une Industrie dans la Longue Durée," La Ferté-sous-Jouarre, France. May 16–19, 2002.

Hockensmith, Charles D. 2003a. The International Millstone Conference at La Ferté-sous-Jouarre, France. *Old Mill News* 31 (1):14–18.

Hockensmith, Charles D. 2003b. The Conglomerate Millstone Industry in the Eastern United States. In *Meules à Grains: Actes du Colloque International de La Ferté-sous-Jouarre 16–19 Mai 2002*, edited by Mouette Barboff, François Sigaut, Cozette Griffin-Kremer, and Robert Kremer, pp. 197–216. Éditions Ibis Press and Éditions de la Maison des Sciences de l'Homme, Paris, France.

Hockensmith, Charles D. 2003c. The Millstone Industry in Kentucky: Brief Glimpses from Archival Sources. *The Millstone* 2 (1):6–17. Kentucky Old Mill Association, Clay City, Kentucky.

Hockensmith, Charles D. 2003d. The Ohio Buhr Millstones: The Flint Ridge and Raccoon Creek Quarries. *Ohio Valley Historical Archaeology* 18:135–142.

Hockensmith, Charles D. 2004a. *Early American Documents and References Pertaining to Millstones: 1628–1829.* Kentucky Old Mill Association, Clay City, Kentucky.

Hockensmith, Charles D. 2004b. Additional Information on the Georgia Burr Millstone: The Lafayette Burr Mill Stone Years. *Old Mill News* 32 (3):22–25.

Hockensmith, Charles D. 2004c. The Historic Millstone Industry in Eastern Kentucky: The Red River Quarries. Paper presented at the 22nd Annual Symposium on Ohio Valley Urban and Historic Archaeology, New Harmony, Indiana.

Hockensmith, Charles D. 2004d. The Millstone Industry of Tennessee. *The Millstone* 3 (1):9–15. Kentucky Old Mill Association, Clay City, Kentucky.

Hockensmith, Charles D. 2004e. The Millstone Industry of North Carolina. *Old Mill News* 32 (4):17–24.

Hockensmith, Charles D. 2004f. The Millstone Industry of Missouri. *The Millstone* 3 (2):29–34. Kentucky Old Mill Association, Clay City, Kentucky.

Hockensmith, Charles D. 2005a. The Millstone Industry of Alabama. *The Mill Monitor*, Special Edition, No. 2, pp. 1–9 [Winter 2005]. The International Molinology Society of America, Sterling, Virginia.

Hockensmith, Charles D. 2005b. The Preservation, Ownership, and Interpretation of American Millstone Quarries. Paper presented at the Colloque International "Les Meulières, Recherche. Protection et Valorisation d'un Patrimoine Industriel Européen (Antiquité–XXIe s.)," Grenoble, France, September 22–25, 2005.

Hockensmith, Charles D. 2006a. The Millstone Quarries of Missouri. *Old Mill News* 34 (3):20–23.

Hockensmith, Charles D. 2006b. The Second International Millstone Conference at Grenoble, France. *Old Mill News* 34 (4):21–25.

Hockensmith, Charles D. 2006c. The Preservation, Ownership, and Interpretation of American Millstone Quarries. In *Mühlsteinbrüche. Erforschung, Schutz und Inwertsetzung eines Kulturerbes europäischer Industrie (Antike–21. Jahrhundert)*, edited by Alain Belmont and Fritz Mangartz, pp. 193–204. Römisch-Germanisches Zentralmuseum, Tagungen Band 2, Mainz, Germany.

Hockensmith, Charles D. 2006d. The American Conglomerate Millstone Industry: An Archival Overview of Its Quarries and Products. Draft manuscript.

Hockensmith, Charles D. 2006e. American Millstones Similar to the French Burr: 19th Century Attempts to Find Substitutes. Draft manuscript.

Hockensmith, Charles D. 2006f. Sandstone Millstones in the United States. Draft manuscript.

Hockensmith, Charles D. 2006g. Limestone Millstones in the United States. In preparation.

Hockensmith, Charles D. 2006h. The Millstone Trade in Colonial Virginia: The Use and Sale of American and Foreign Millstones from 1739 to 1779. Draft manuscript.

Hockensmith, Charles D. 2007a. The Ohio Buhr Millstones: The Flint Ridge and Raccoon Creek Quarries. In *Millstone Studies: Papers on Their Manufacture, Evolution, and Maintenance,* edited by Donald B. Ball and Charles D. Hockensmith, pp. 134–143. Special Publication No. 1. Jointly published by the Symposium on Ohio Valley Historic Archaeology, Murray, Kentucky, and the Society for the Preservation of Old Mills, East Meredith, New York.

Hockensmith, Charles D. 2007b. The Granite and Gneiss Millstone Quarries of the United States. In *Millstone Studies: Papers on Their Manufacture, Evolution, and Maintenance,* edited by Donald B. Ball and Charles D. Hockensmith, pp. 144–159. Special Publication No. 1. Jointly published by the Symposium on Ohio Valley Historic Archaeology, Murray, Kentucky, and the Society for the Preservation of Old Mills, East Meredith, New York.

Hockensmith, Charles D. 2008a. The French Burr Millstone in Kentucky: Insights from Early Ads, 1792–1890. In *Foreign and Domestic Millstones Used In Kentucky: Papers Examining Archival Records*, compiled by Charles D. Hockensmith, pp. 5–38. Kentucky Old Mill Association, Clay City, Kentucky.

Hockensmith, Charles D. 2008b. Millstones from Ohio and Pennsylvania Imported into Kentucky: Raccoon Buhrs and Laurel Hill Stones. In *Foreign and Domestic Millstones Used In Kentucky: Papers Examining Archival Records*, compiled by Charles D. Hockensmith, pp. 39–54. Kentucky Old Mill Association, Clay City, Kentucky.

Hockensmith, Charles D. 2008c. Early References to Red River Millstones in Kentucky: Newspaper Ads, 1803–1839. In *Foreign and Domestic Millstones Used In Kentucky: Papers Examining Archival Records*, compiled by Charles D. Hockensmith, pp. 55–62. Kentucky Old Mill Association, Clay City, Kentucky.

Hockensmith, Charles D. 2009. *The Millstone Quarries of Powell County, Kentucky*. McFarland, Jefferson, North Carolina.

Hockensmith, Charles D. 2008e. The Millstone Industry in New York. In *The Historic Millstone Industry in New York State with an Emphasis on Ulster County*, edited by Charles D. Hockensmith. Society for the Preservation of Old Mills. Draft manuscript.

Hockensmith, Charles D. 2008f. A Bibliography of Quern Studies, Quern Quarries, and Hand Mills. *Quern Study Group Newsletter* 8. Oxford, England. In press.

Hockensmith, Charles D. (editor) 1999. *Millstone Manufacture in Virginia: Interviews with the Last Two Brush Mountain Millstone Makers*. Society for the Preservation of Old Mills. Marblehead Publishing, Raleigh, North Carolina.

Hockensmith, Charles D. (editor) 2008a. *Foreign and Domestic Millstones Used In Kentucky: Papers Examining Archival Records*. Kentucky Old Mill Association, Clay City, Kentucky.

Hockensmith, Charles D. (editor) 2008b. *The Historic Millstone Industry in New York State with an Emphasis on Ulster County*. Society for the Preservation of Old Mills, East Meredith, New York. Draft manuscript.

Hockensmith, Charles D., and Donald B. Ball 2007. Ahead of Their Time: Artificial Millstones. In *Millstone Studies: Papers on Their Manufacture, Evolution, and Maintenance*, by Donald B. Ball and Charles D. Hockensmith, pp. 193–198. Special Publication No. 1. Jointly published by the Symposium on Ohio Valley Historic Archaeology, Murray, Kentucky, and the Society for the Preservation of Old Mills, East Meredith, New York.

Hockensmith, Charles D., and Fred E. Coy, Jr. 1999. Early Twentieth Century Millstone Manufacture in Southwest Virginia: An Interview with Millstone Makers Robert Huston Surface and W. C. Saville. In *Millstone Manufacture in Virginia: Interviews with the Last Two Brush Mountain Millstone Makers*, edited by Charles D. Hockensmith, pp. 5–62. Society for the Preservation of Old Mills. Marblehead Publishing, Raleigh, North Carolina.

Hockensmith, Charles D., and Fred E. Coy, Jr. 2008a. Twentieth Century Conglomerate Millstone Manufacture Near Accord in Ulster County, New York: An Interview with Vincent and Wallace Lawrence. In *The Historic Millstone Industry in New York State with an Emphasis on Ulster County*, edited by Charles D. Hockensmith. Society for the Preservation of Old Mills. Draft manuscript.

Hockensmith, Charles D., and Fred E. Coy, Jr. 2008b. The Esopus Millstone Industry at Accord, Ulster County, New York: An Interview with Lewis Waruch. In *The Historic Millstone Industry in New York State with an Emphasis on Ulster County*, edited by Charles D. Hockensmith. Society for the Preservation of Old Mills. Draft manuscript.

Hockensmith, Charles D., and Larry G. Meadows 1996. Historic Millstone Quarrying in Powell County, Kentucky. *Ohio Valley Historical Archaeology* 11:95–104.

Hockensmith, Charles D., and Larry G. Meadows 1997a. Conglomerate Millstone Quarrying in the Knobs Region of Powell County, Kentucky. *Old Mill News* 25 (2):17–20; 25 (3):24–26.

Hockensmith, Charles D., and Larry G. Meadows 1997b. Millstone Quarrying in the Knobs Region of Powell County, Kentucky. Paper presented at the 14th Annual Kentucky Heritage Council Archaeological Conference, Natural Bridge State Park, Slade, Kentucky.

Hockensmith, Charles D., and Larry G. Meadows 2006. Red River Millstone Quarries in Lawsuits. Part I. *The Millstone* 5 (2):9–17. Kentucky Old Mill Association, Clay City, Kentucky.

Hockensmith, Charles D., and Larry G. Meadows 2007. Red River Millstone Quarries in Lawsuits. Part 2. *The Millstone* 6 (1):31–38. Kentucky Old Mill Association, Clay City, Kentucky.

Hockensmith, Charles D., Larry G. Meadows, and Lief Meadows 2008. William S. Webb's Historic Millstone Collection: An Overview and Analysis of Specimens Collected by Kentucky's Pioneer Archaeologist. Kentucky Old Mill Association, Clay City, Kentucky. Draft manuscript.

Hockensmith, Charles D., and Jimmie L. Price 1999. Conglomerate Millstone Making in Southwestern Virginia: An In-depth Interview with Millstone Maker Robert Huston Surface. In *Millstone Manufacture in Virginia: Interviews with the Last Two Brush Mountain Millstone Makers*, edited by Charles D. Hockensmith, pp. 63–89. Society for the Preservation of Old Mills. Marblehead Publishing, Raleigh, North Carolina.

Hockensmith, Charles D., and Owen Ward 2007. Millstones, Querns and Millstone Studies in the United Kingdom: A Bibliography. *International Molinology* 74.24–31, Watford, England.

Hodgman & Co. 1865. *Kentucky State Gazetteer and Business Directory*. George H. Hodgman, Louisville.

Holmstrong, John G., and Henry Holford 1982. *American Blacksmithing and Twentieth Century Toolsmith and Steelworker*. Greenwich House, New York.

Holloway, William R. 1870. *Indianapolis: A Historical and Statistical Sketch of the Railroad City, a Chronicle of its Social, Municipal, Commercial and Manufacturing Progress, with Full Statistical Tables*. Indianapolis Journal Printer, Indianapolis.

Holtmeyer-Wild, Vera 2000. *Vorgeschichtliche Reibsteine aus der Umgebung von Mayen: Reibsteine aus Basaltlava*. Römisch-Germanisches Zentralmuseum,Vulkanpark-Forschungen, Mainz, Germany. [Publication that deals with archaeological excavations at Mayen, Germany which includes information on basalt millstones.]

Homans, I. Smith 1859. *A Cyclopedia of Commerce and Commercial Navigation*. 2nd ed. Harper & Brothers, New York.

Hörmann, P. K., and A. Richter 1983. Vergleichende mineralogisch-petrographische Untersuchungen an Mühlsteininresten aus Haithabu und Bruchsteinproben aus der Eifel. *Beriche über Ausgrabungen in Haithabu* 18:93–107. [Germany]

Hörter, F., Jr. 1950. Die Arbeitsverhältnisse in der Basaltlavaindustrie des Kreises Mayen. Examensarbeit Berufspädagogisches Institut. Solingen-Ohligs. [Germany]

Hörter, F., Sr. 1942. Neue Ergebnisse über die vor- und

Frühgeschichtliche Mühlsteinindustrie. Unpublished manuscript. Mayen Museum. [Germany]
Hörter, F., Sr., F. X. Michels, and J. Röder 1950–. Die Geschichte der Basaltlava-Industrie von Mayen und Niedermendig, 1951. I:Vor- und Frühgeschichte. *Jahrbuch für Geschichte und Kultur des Mittelrheims und seiner Nachbargebiete* 2–3, S. 1–32, Koblenz. [Germany]
Hörter, F., Sr., F. X. Michels, and J. Röder 1954. Die Geschichte der Basaltlava-Industrie von Mayen und Niedermendig, 1955. II: Mittelalter und Neuzeit. *Jahrbuch für Geschichte und Kultur des Mittelrheims und seiner Nachbargebiete* 5–6, S. 7–32, Koblenz. [Germany]
Hörter, Fridolin 1979. Ein Vermauerter Mühlstein in Mendig als Zeuge Mittelalterlicher Handwerksordnung. *Eifeljahrbuch*, pp. 77–82, Düren. [Germany]
Hörter, Fridolin 1990. Drei Mahlsteine aus dem Grafschafter-Museum in Moers. *Rheinische Heimatpflege* N.F. 1, S. 59–61. [Germany]
Hörter, Fridolin 1994. *Getreidereiben und Mühlsteine aus der Eifel* ["Querns and Millstones from the Eifel"]. Geschichte und Altertumsverein fur Mayen un Umgebung, Mayen. [Germany]
Hörter, Fridolin 2003. Gewinnung und Handel Rheinisher Mühlsteine in Schriftbelegen vom 9. bis 16. Jahrhundert. In *Meules à Grains: Actes du Colloque International de La Ferté-sous-Jouarre 16–19 Mai 2002*, edited by Mouette Barboff, François Sigaut, Cozette Griffin-Kremer, and Robert Kremer, pp. 169–174. Éditions Ibis Press and Éditions de la Maison des Sciences de l'Homme, Paris, France. [Written sources on the quarrying of Rhineland millstones in Germany between the 9th and 16th centuries.]
Hörter, P. 1914. Die Basaltlava-Industrie bei Mayen (Rheinland) in vorrömischer und römischer Zeit. *Mannus* 6, S. 283–294, Würzburg. [Germany]
Hörter, P. 1917. Vorgeschichtliche Werkzeuge der Basaltlava-Industrie bei Mayen (Rheinland) in vorrömischer und römischer Zeit. *Mannus* 9, S. 83–86, Leipzig und Würzburg. [Germany]
Hörter, P. 1922. Geschichte der Basaltlavaindustrie von Mayen und Umgebung. Zeitsch. Rh. Ver. Denkmalpfl. *U. Heimatschutz* 15 Jg.: 68–75. [Germany]
Horton, John T., Edward T. Williams, and Harry S. Douglass 1947. *History of Northwestern New York: Erie, Niagara, Wyoming, Genesee and Orleans Counties*. Lewis Historical Publishing Company, New York.
Horvat, A. and M. Župančič 1987. Prazgodo-Vinske in Rimske Žrmlje v Zahodni Sloveniji: Prvi Rezultati Petrografske Analize ["Prehistoric and Roman Querns in Western Slovenia: First results of the Petrographic Analysis"]. *Geološki Zbornik* 8:105–110. Ljubljana.
Hough, Franklin B. 1860. *A History of Lewis County, in the State of New York*. Munsell & Rowland, Albany.
Howard, George W. 1873. *The Monumental City, Its Past History and Present Resources*. J. D. Ehlers, Baltimore.
Howe, Henry 1851. *Historical Collections of Ohio*... Published by H. Howe at E. Morgan & Co., Cincinnati.
Howe, Henry 1875. *Historical Collections of Ohio*... R. Clarke & Co., Cincinnati.
Howe, Henry 1888. *Historical Collections of Ohio: An Encyclopedia of the State*... H. Howe & Son, Columbus.
Howell, Charles 1985. Colonial Watermills in the Wooden Age. In *America's Wooden Age: Aspects of Its Early Technology*, edited by Brooke Hindle, pp. 120–159. Sleepy Hollow Press, Tarrytown, New York.
Howell, Charles 1993. International Mini-Mill Symposium. *Old Mill News* 21 (1):15–17.
Howell, Charles 1997. Millstones: An Introduction. *Old Mill News* 25 (4):18–22.
Howell, Charles, and Allan Keller 1977. *The Mill at Philipsburg Manor and a Brief History of Milling*. Sleepy Hollow Restorations, Tarrytown, New York.
Hudson, Phil 1989. Old Mills and Millstone Quarries in the Forest of Lancaster. *Contrebis* 15:35–64.
Hudson, Phil 1996. Mills, Millstones and Millstone Quarries in the Lancaster Area. *North-West Mills Group Newsletter* 40 (3):9–10. [England]
Hudson, Phil 1998. Millstone Making in Quernmore. *Lancashire History Quarterly* 2 (4):165.
Hudson, Philip John 1995. Landscape and Economic Development of Quernmore Forest, Lancaster: An Upland Marginal Area in North West Lancashire to c 1850. Unpublished master's thesis, University of Lancaster, Lancaster, England.
Hughes, William C. 1869. *The American Miller, and Millwright's Assistant*. Henry C. Baird, Philadelphia.
Hume, Abraham 1851. Remarks on Querns, Ancient and Modern, from Notes of a Lecture Addressed to the "Cambrian Archaeological Association" during Its Congress at Dolgelly, on Thursday, August 29, 1850. W. Pickering, London. [England]
Hunold, Angelika 2006. Introduction: Protection and Valorization of the Monuments. In *Mühlsteinbrüche. Erforschung, Schutz und Inwertsetzung eines Kulturerbes europäischer Industrie (Antike–21. Jahrhundert)*, edited by Alain Belmont and Fritz Mangartz, pp. 180–181. Römisch-Germanisches Zentralmuseum, Tagungen Band 2, Mainz, Germany.
Hunt, Patrick N., and Daffyd Griffiths 1992. The Suitability of Basalt, Andesite and Other Volcanic Stone for Querns, Millstones and Grinding Purposes. *Quern Study Group Newsletter* 2:4–6. England. [England]
Hurd and Burrows 1858. *Louisville City Directory and Business Mirror for 1858–9*. Hurd and Burrows Publishers, Louisville.
Ingle, Caroline J. 1982. A Petrological Study of Some Quernstones from the Bristol Region. Unpublished BA thesis, Department of Classics and Archaeology, Bristol University, Bristol, England. [England]
Ingle, Caroline J. 1984. A Petrological Study of Some Quernstones from the Bristol Region. *Bristol and Avon Archaeology* 3:8–12. [England]
Ingle, Caroline J. 1987. The Production and Distribution of Beehive Querns in Cumbria — Some Initial Considerations. *Transactions of the Cumberland and Westmorland Archaeological and Antiquarian Society* 87:11–17.
Ingle, Caroline J. 1989. Characterisation and Distribution of Beehive Querns in Eastern England. Ph.D. thesis, University of Southampton, Southampton.
Ingle, Caroline J. 1993. The Quernstones from Hunsbury Hillfort, Northamptonshire. *Northamptonshire Archaeology* 25:21–35.
Irwin, John Rice [Museum of Appalachia, Norris, Tennessee] 1993. Personal communication to Charles D. Hockensmith. [Millstone quarry in Union County, Tennessee.]
Ison, Cecil R. 2005. Occupational Hazards of Milling. *The Millstone* 4 (1):11–13. Kentucky Old Mill Association, Clay City, Kentucky.
Jäckle, Hans Werner 2002. Die Fabrik für Mühlstein-Schärfwerkzeuge J. C. Kupka, Schleuditz. Schriftenreihe, Heft, 76p. Sächsischer Landesverein für Mühlenerhaltung und Mühlenkunde e. V., Lehmann-Mühle, Klipphausen (Allemagne). [Discusses J. C. Kupka's role in

producing quality tools for dressing millstones at Saxony, Germany.]

Jäckle, Hans Werner 2003. Une Entreprise Spécialisée dans la Fabrication des Outils Destinés au Rhabillage des Meules: J. C. Kupa à Scheuditz près de Leipzig (Saxe, Allemagne). In *Meules à Grains: Actes du Colloque International de La Ferté-sous-Jouarre 16–19 Mai 2002*, edited by Mouette Barboff, François Sigaut, Cozette Griffin-Kremer, and Robert Kremer, pp. 368–379. Éditions Ibis Press and Éditions de la Maison des Sciences de l'Homme, Paris, France. [Discusses J. C. Kupka's role in producing quality tools for dressing millstones at Saxony, Germany.]

Jacobson, Maria 1998. *Upptäck Kvarnstensbrottet i Östra Utsjö, Malung*. Länsstyrelsen i Dalarnas län och Malungs kommun. [Millstone quarry at Östra Utsjö in Sweden.]

Jagailloux, Serge 2002. État de Santé des Meuliers, Accidents du Travail, Affections Professionnelles. In *Les Meuliers. Meules et Pierres Meulières* by Agapain, pp. 151–164. Presses du Village, Étrépilly, France. [France]

Jean, Marcel 1989. Une Tailleries de Meules de Moulin au Puget-sur-Argens. *Moulins de Provence Bulletin* 5:18–21.

Jecock, H. M. 1981. *The Production and Distribution of Prehistoric Rotary Querns in Wessex*. Unpublished B.A. dissertation, Southampton University, Archaeology Department, Southampton. [England]

Jecock, H. M. 1985. The Querns — Some Observations. In *The Prehistoric Settlement at Winnall Down, Winchester: Excavations of MARC 3 Site R 17 in 1976 and 1977* by P. J. Fasham, pp. 77–80. Hampshire Field Club and Archaeological Society Monograph 2.

Jegli, John B. 1845. *John B. Jegli's Louisville, New Albany, Jeffersonville, Shippingport Directory for 1845–1846*. The Louisville Journal, Louisville.

Jegli, John B. 1850. *The Louisville Business Register for 1850*. Printed by Brennan & Smith, Louisville.

Jegli, John B. 1851. *A Directory for 1851–1852, ... in the City of Louisville*. J. F. Brennan, Louisville.

Jobey, George 1981. Groups of Small Cairns and the Excavation of a Cairnfield on Millstone Hill, Northumberland. *Archaeologia Aeliana* 9:23–43. [England]

Jobey, George 1986. Millstones and Millstone Quarries in Northumberland. *Archaeologia Aeliana* 5th series, 14:49–80. [England]

Johnson, Bertrand L., and A. E. Davis 1936. Abrasive Materials. In *Minerals Yearbook, 1936*, pp. 877–894. Department of the Interior, U.S. Bureau of Mines, Government Printing Office, Washington, D.C.

Johnson, Bertrand L., and A. E. Davis 1937. Abrasive Materials. In *Minerals Yearbook, 1937*, pp. 1283–1300. Department of the Interior, U.S. Bureau of Mines, Government Printing Office, Washington, D.C.

Johnson, Bertrand L., and A. E. Davis 1938. Abrasive Materials. In *Minerals Yearbook, 1938*, pp. 1135–1150. Department of the Interior, U.S. Bureau of Mines, Government Printing Office, Washington, D.C.

Johnson, Bertrand L., and M. Schauble 1939. Abrasive Materials. In *Minerals Yearbook, 1939*, pp. 1225–1240. Department of the Interior, U.S. Bureau of Mines, Government Printing Office, Washington, D.C.

Johnson, Joseph Risk 1965. Transcribed version of William Risk's interview with John Shane (John Shane, Draper MSS. 11CC 86). Copy on file at the Kentucky Historical Society, Frankfort.

Jones, Joseph 1861. *Agricultural Resources of Georgia*. Address Before the Cotton Planters Convention of Georgia at Macon, December 13, 1860. Steam Press of Chronicle & Sentinel, Augusta. Electronic edition by University of North Carolina at Chapel Hill Libraries, Documenting the American South.

Jones, Walter B. 1926. *Index to the Mineral Resources of Alabama*. Bulletin No. 28. Geological Survey of Alabama, University, Alabama.

Joos, M. 1975. Eine permische Brekzie aus dem Südschwartzwald und ihre Verbreitung als Mühlstein im Spätlatène und in frührömischer Zeitz. *Archäologisches Korrespondenzblatt* 5:197–199.

Jottrand, M. G. 1895. L'Industrie de la Fabrication des Meules en Belgique avant et après la Conquête Romaine. *Bulletin de la Société Anthropogique de Bruxelles* 13:390–408. [Sandstone millstones made by the Romans in the L'Oise Valley on the border of France and Belgium.]

Judd, William [John's Island, South Carolina] 1999. Letter to Charles D. Hockensmith, December 14. On file at the Kentucky Heritage Council, Frankfort. [The use of Peak Millstones in the rice industry in South Carolina.]

Judd, William [John's Island, South Carolina] 2000. Letter to Charles D. Hockensmith, January 24. On file at the Kentucky Heritage Council, Frankfort. [The use of Peak Millstones in the rice industry in South Carolina.]

Kardulias, P. Nick, and Curtis Runnels 1995. The Lithic Artifacts: Flaked Stone and Other Nonflaked Lithics. In *Artifact and Assemblage: The Finds from a Regional Survey of the Southern Argolid, Greece*, edited by Curtis Runnels, Daniel J. Pullen, and Susan Langdon, pp. 74–139. Stanford University Press, Stanford. [Pages 109–139 deal with ground stone, polished stone, and other nonflaked artifacts including saddle querns, hopper querns, rotary querns, and rotary olive mills from Greece.]

Kars, H. 1980. Early-Medieval Dorestad, an Archaeo-Petrological Study. Part I: General Introduction. The Tephrite Querns. *Berichten van de Rijksdienst voor het Oudheidkundig Bodemonderzoek* 30: 393–422, Amersfoort. [England]

Kars, H. 1983. Un Centre de Production de Meules près de Mayen dans l'Eifel. *Rijksdienst voor het Oudheidkundig Bodemonderzoek Overdrakken Amersfoort* 194: 110–120. [Germany]

Katz, Frank J. 1913. Abrasive Materials. In *Mineral Resources of the United States, Calendar Year 1912*, pp. 819–831. Part II — Nonmetals. Department of the Interior, U.S. Bureau of Mines, Government Printing Office, Washington, D.C.

Katz, Frank J. 1914. Abrasive Materials. In *Mineral Resources of the United States, Calendar Year 1913*, p. 253–272. Part II — Nonmetals. Department of the Interior, U.S. Bureau of Mines, Government Printing Office, Washington, D.C.

Katz, Frank J. 1916. Abrasive Materials. In *Mineral Resources of the United States, 1914*, p. 549–568. Part II — Nonmetals. Department of the Interior, U.S. Bureau of Mines, Government Printing Office, Washington, D.C.

Katz, Frank J. 1917. Abrasive Materials. In *Mineral Resources of the United States, 1915*, p. 65–80. Part II — Nonmetals. Department of the Interior, U.S. Bureau of Mines, Government Printing Office, Washington, D.C.

Katz, Frank J. 1919. Abrasive Materials. In *Mineral Resources of the United States, 1916*, p. 197–212. Part II — Nonmetals. Department of the Interior, U.S. Bureau of Mines, Government Printing Office, Washington, D.C.

Katz, Frank J. 1920. Abrasive Materials. In *Mineral Resources of the United States, 1917*, pp. 213–232. Part II — Nonmetals. Department of the Interior, U.S. Bureau of Mines, Government Printing Office, Washington, D.C.

Katz, Frank J. 1921. Abrasive Materials. In *Mineral Resources of the United States, Calendar Year 1918*, pp. 1171–1187. Part II — Nonmetals. Department of the Interior, U.S. Bureau of Mines, Government Printing Office, Washington, D.C.

Katz, Frank J. 1926. Abrasive Materials. In *Mineral Resources of the United States, 1923*, pp. 327–337. Part II — Nonmetals. Department of the Interior, U.S. Bureau of Mines, Government Printing Office, Washington, D.C.

Katz, Frank J. 1927. Abrasive Materials. In *Mineral Resources of the United States, 1924*, pp. 241–252. Part II — Nonmetals. Department of the Interior, U.S. Bureau of Mines, Government Printing Office, Washington, D.C.

Katz, Frank J. 1928. Abrasive Materials. In *Mineral Resources of the United States, 1925*, pp. 171–174. Part II — Nonmetals. Department of the Interior, U.S. Bureau of Mines, Government Printing Office, Washington, D.C.

Kelleher, Tom 1990. The Kingsbury Mill: Rare New England Survivor (Medfield, Massachusetts). *Old Mill News* 18 (4):8.

Kelleher, Tom [Old Sturbridge Village, Sturbridge, Massachusetts] 1993. Letter to Charles D. Hockensmith, January 29. On file at the Kentucky Heritage Council, Frankfort. [Two millstone quarries in Norfolk County, Massachusetts.]

Keller, P. T. 1989b. Quern Production at Folkestone, South-East Kent: An Interim Note. *Britannia* 20:193–200.

Keller, Peter 1989a. The Folkestone Quernstones. *Current Archaeology* 114:236–237.

Kelly, Frank B. 1913. *Historical Guide to the City of New York*. F. A. Stokes Company, New York.

Kennedy, Joseph C. G. 1864. *Population of the United States in 1860; Compiled from the Original Returns of the Eighth Census*. United States Census Office, Washington, D.C.

Kentucky Gazette 1799. Ad for five pairs of Red River millstones at Cleveland's Landing. *Kentucky Gazette*, June 13, 1799, Lexington, Kentucky.

Kentucky Reporter 1818. Ad for Red River Millstones at quarry by Spencer Adams and James Daniel. *Kentucky Reporter*, April 8, 1818, Lexington, Kentucky.

Killebrew, J. B. 1874. *Introduction to the Resources of Tennessee*. First and Second Reports of the Bureau of Agriculture for the State of Tennessee. Travel, Eastman, and Howell, Nashville.

King, D. 1986. Petrology, Dating and Distribution in Querns and Millstones. The Results of Research in Bedfordshire, Buckinghamshire, Hertfordshire and Middlesex. University of London. *Institute of Archaeology, Bulletin* 23:65–126, London. [England]

Klausener, M. 2001. Châvannes-le-Chêne, Carrière Romain. Unpublished report prepared for the Canton de Vaud. [Châvannes-le-Chêne is a Roman era quern quarry in western Switzerland.]

Kleinmann, Dorothée 1984. La Fabrication de Meules à Cinq-Mars-la-Pile. *Bulletin de la Société des Amis du Vieux Chinon* 8 (8):1107–1116. [France]

Kling, Joern 2006. Die unterirdischen Mühlsteinbrüche von Niedermendig/Deutschland. Historische Kartographie und Detailinventarisierung ["The Millstone Mines of Niedermendig/Germany. Historical Cartography and Detailed Inventory"]. In *Mühlsteinbrüche. Erforschung, Schutz und Inwertsetzung eines Kulturerbes europäischer Industrie (Antike–21. Jahrhundert)*, edited by Alain Belmont and Fritz Mangartz, pp. 133–144. Römisch-Germanisches Zentralmuseum, Tagungen Band 2, Mainz, Germany.

Kling, Joern, and Charles D. Hockensmith, 2008. Millstones, Querns, and Millstone Studies in Germany: A Bibliography. Draft Manuscript.

Konschak, Alexander 1996. *Ein Mühlsteinbruch am Heideberg bei Zittau*, Bad Muskau, Verlag Gunter Oettel, 27 pages. [Germany]

Krämer, S. 1948. Preispolitik in der Basaltlavaindustrie. Dipl. Arbeit. Köln. [Germany]

Kranjc, A. 2005. Konglomeratni kras v Sloveniji: Zgodovina Raziskovanja in Poznavanja jam v Udin Borštu na Gorenjskem. *Acta Carsologica* 34 (2):521–532. [Cave in Gorenjsko, Slovenia where some conglomerate millstones were cut.]

Kremer, Bruno 2006. Inventar der unterirdischen Mühlsteinbrüche in der Eifel und in Luxemburg ["Inventory of the Millstone Mines of the Eifel and Luxembourg"]. In *Mühlsteinbrüche. Erforschung, Schutz und Inwertsetzung eines Kulturerbes europäischer Industrie (Antike–21. Jahrhundert)*, edited by Alain Belmont and Fritz Mangartz, pp. 145–154. Römisch-Germanisches Zentralmuseum, Tagungen Band 2, Mainz, Germany.

Kresten, Peter 1996a. Kvarnstenar i Roskilde Museum, Danmark. Geoarkeologiskt Laboratorium, Analysrapport nummer 24–1996. Riksanitikvarieämbetet. Avdelningen för arkeologiska underökningar. UV Uppsala. [Study of millstones from the collections at the Roskilde Museum in Denmark.]

Kresten, Peter 1996b. Kvarnstenar från Malung samt en inventering av kvarnberget. Dalarna, Malung Kvarnberget RAÄ 589, GAL analysrapport nr 1–1996. [Millstones from Malung, Sweden.]

Kresten, Peter 1996c. Analys av stenmaterial från Kyrkheddinge 2:19, Skåne, Kyrkheddinge sn. Geoarkeologiskt Laboratorium, Analysrapport nummer 8–1996. Riksanitikvarieämbetet. Avdelningen för arkeologiska underökningar. UV Uppsala. [Millstones recovered from excavations the Medieval village of Kyrkheddinge at Scania.]

Kresten, Peter 1996d. Kvarnstenar från Lund. Geoarkeologiskt Laboratorium, Analysrapport nummer 23–1996. Riksanitikvarieämbetet. Avdelningen för arkeologiska underökningar. UV Uppsala. [Geological analysis of millstones recovered from an excavation in Lund, Sweden.]

Kresten, Peter 1996e. Analys av stenmaterial från Visby, Gotland, Visby stad, RAÄ 107. Geoarkeologiskt Laboratorium, Analysrapport nummer 12–1996. Riksanitikvarieämbetet. Avdelningen för arkeologiska underökningar. UV Uppsala. [Geological analysis of quernstones from Ystad, Scania; Önneruo, Scania; and Visby, Gotland, Swedish island]

Kresten, Peter 1998. Malung eller Sala? Diskussion av kvarnstensbrottens betydelse från vikingatid till nyare tid, Geoarkeologiskt Laboratorium Forskningsrapport nr R1–1998. [Deals with millstone quarries from the Viking Age to modern times at Malung, Sweden or Sala.]

Kresten, Peter, and Magnus Elfwendahl 1994. Malung-en glömnd kvarnstensmetropol Skinnarebygd, sid 10–28. [Discusses the forgotten millstone making center at Malung, Sweden.]

Kresten, Peter, Magnus Elfwendahl, and Täpp John-Erik Pettersson 1996. Provenance of Quernstones, Grindstones, and Hones from Sweden. In *Proceedings from the 6th Nordic Conference on the Application of Scientific Methods in Archaeology, Esbjerg, 1993. Arkaeologiska Rapporter No. 1*, Esbjerg Museum.

Labs, William A. 1917. Survival of Corn Querns of an Ancient Pattern in the Southern United States. *A Collection of Papers Read Before the Bucks County Historical Society*, Volume 4, pp. 740–744. The Chemical Publishing Company, Easton, PA.

Lacombe, Claude 2000. Les Ouviers Meulières de la Plaine de Born, à Domme, à la fin du XIXe Siècle: Techniques et Conditions de Travail. *Bulletin de la Société d'Art et Hist. De Sarlat et du Périgord Noir* 81:65–73. [Millstones from the Born Plain in France.]

Lacombe, Claude 2003. Les Meulières de la Plaine de Born, à Domme (Dordogne), du XVIIIe au XXe Siècle: Histoire, Exploitation et Conditions de Travail. In *Meules à Grains: Actes du Colloque International de La Ferté-sous-Jouarre 16–19 Mai 2002*, edited by Mouette Barboff, François Sigaut, Cozette Griffin-Kremer, and Robert Kremer, pp. 312–333. Éditions Ibis Press and Éditions de la Maison des Sciences de l'Homme, Paris, France. [Millstones from the Born Plain at Dordogne, France.]

Lacroix, B. 1963. Les Moulins Domestiques du IVe Siècle aux Fontaines Salées (Commune de Saint-Père-sous-Vezelay, Yonne). *Revue Archéologique de l'Est et du Centre Est* 14:301–314. [Production centers of Gallo-Roman millstones within sedimentary rocks in France.]

Ladoo, Raymond B. 1925. *Non-Metallic Minerals: Occurrence-Preparation-Utilization*. McGraw-Hill Book Company, New York.

Ladoo, Raymond B. and W. M. Myers 1951. *Monometallic Minerals*. Second edition. McGraw-Hill Book Company, New York.

Laumanns, M. 1987. Üer Eis- und Mühlsteinhöhlen in der Eifel. *Karst und Höhle*. 87: 97–102, Munich. [Millstone caves in the Eifel area of Germany and ice formation.]

Laville, L. 1963. Déouverte d'une carrière Gallo-Romaine Spécialisée dans la Fabrications de Meules à Grain Domestiques à St. Christophe le Chaudy *Revue Archéologique du Centre* 2:146–151. [Production centers of Gallo-Roman millstones within sedimentary rocks (sandstone) in France.]

Laws, K. 1987. Quernstones. In *Hengistbury Head Dorset*, edited by B. Cunliffe, pp. 167–171. Oxford University for Archaeology, Oxford.

Lechevin, Jean-Michel 1975. Les Meules de la Carrière des Cabas à La Mouillère, Commune de Buxières les Mines. *Bulletin de la Société d'Emulation du Bourbonnais*, 3e trim., pp. 439–441. [France]

Leffler, Dankmar 2001. *Das Crawinkler Mühlsteingewerbe: Zur Geschichte eines der ältesten Gewerbe im Thüringer Wald*. Förderverein "Alte Mühle" e. V., Crawinkel. [Germany]

Leffler, Dankmar 2002. *Das Mühlsteingewerbe in Crawinkel/Thüringen. (Des Meules de Moulin de Crawinkel/Thüringen)*. Booklet prepared for Colloque International "Extraction, Façonnage, Commerce et Utilisation des Meules de Moulin — Une Industrie dans la Longue Durée," La Ferté-sous-Jouarre, France. [Germany]

Leffler, Dankmar 2003. *Das Mühlsteingewerbe in Crawinkel (Thüringen)*. In *Meules à Grains: Actes du Colloque International de La Ferté-sous-Jouarre 16–19 Mai 2002*, edited by Mouette Barboff, François Sigaut, Cozette Griffin-Kremer, and Robert Kremer, pp. 150–159. Éditions Ibis Press and Éditions de la Maison des Sciences de l'Homme, Paris, France. [The millstone industry at Crawinkel in Germany.]

Lemon, J. R. 1894. *Lemon's Hand Book of Marshall County Giving Its History, Advantages, Etc. and Biographical Sketches of Its Prominent Citizens*. J. R. Lemon, Benton, Kentucky.

Leung, Felicity L. 1997. *Grist and Flour Mills in Ontario: From Millstones to Rollers, 1780s–1880s*. Reprint of 1981 book by the National Historic Parks and Sites Branch, Parks Canada, Ottawa. Society for the Preservation of Old Mills, Newton, North Carolina.

Leveille, Alan, and Mary Lynne Rainey 2001. Intensive (Locational) Archaeological Survey, Proposed Saugus Residential Development, Saugus, Massachusetts. Technical Report No. 1166.01. Report submitted to Trammell Crow Residential, Fairfield, Connecticut, by PAL, Pawtucket, Rhode Island.

Liebgott, Niels-Knud 1989. *Dansk Middelalderarkæologi*. Copenhagen. [12th century quernstones from the magnate's farm of Sdr. Jernløse northwest of Sjælland, Denmark.]

Linney, W. M. 1884. *Report on the Geology of Montgomery County*. New Series, No. 2. Kentucky Geological Survey, Frankfort.

Linsley, Stafford M. 1990. Millstones from Brockholm Quarry, Northumberland. *Industrial Archaeology Review* 12 (2):178–184. [England]

Longepierre, Samuel 2004. Organisation, Aire de Diffusion et Évolution des Centres Producteurs de Meules de l'Époque Romaine au Moyen-Âge en Languedoc Oriental et en Provence: l'Exemple des Carrières de Saint-Quentin-la-Poterie (Gard). *Mémoire de DEA d'archéologie*. Université de Provence. Aix Marseille I, 123 pages. [France]

Longepierre, Samuel 2006. Aux Environs de Saint-Quentin-la-Poterie (Gard) durant L'Antiquité Tardive: Une Microrégion Très Impliquée dans L'Activité Meulière ["The Landscape Around Saint-Quentin-la-Poterie (Gard) during Late Antiquity: A Small Region with Important Millstone Production."]. In *Mühlsteinbrüche. Erforschung, Schutz und Inwertsetzung eines Kulturerbes europäischer Industrie (Antike–21. Jahrhundert)*, edited by Alain Belmont and Fritz Mangartz, pp. 47–51. Römisch-Germanisches Zentralmuseum, Tagungen Band 2, Mainz, Germany.

Lorenz, Claude, Michel Turland, and François Boitel 1992. L'Arkose de Blavozy (près du Puy-en-Velay, Haute-Loire) et Son Emploi Régional. *117e Congrès National des Sociétés Savantes, Clermont-Ferrand 1992, 2e coll. Carrières et Constructions II*, pp. 401–410. Paris, Éditions du CTHS. [France]

Lorenz, W. F., and P. J. Wolfram 2007. The Millstones of Barbegal. *Civil Engineering* 77 (6):62–67. [Millstones from Roman ruins in southern France.]

Lorenzoni, S., M. Pallara, D. Venturo, and E. Zanettin 1996. Archaometric Preliminary Study of Volcanic Millstones from Neolithic-Roman Archaeological Sites of the Altamura Area (Apulia, Southern Italy). *Sci. Tecnol. Cult. Hert.* 5:47–55.

Lorenzoni, S., M. Pallara, and E. Zanettin 2000a. Studio Archeometrico delle Macine in Rocce Vulcaniche della Puglia e Zone Limitrofe dall'Età Arcaica all' Età Romana. *Rassegna di Archeologia* 17:225–252.

Lorenzoni, S., M. Pallara, and E. Zanettin 2000b. Volcanic Rock Bronze Age Millstones of Apulia, Southern Italy: Lithology and Provenance. *European Journal of Mineralogy* 12:877–882.

Lorimer, Norma Octavia 1932. *Millstones*. Hutchinson & Company, London. [England]

Loubès, l'Abbé G. 1983. Les Meules de Laroque-Engalin.

Bulletin de la Société Archéologique et Historique du Gers, 84e année, 4e trim., pp. 421–427.

Loughridge, R. H. 1888. *Report on the Geological and Economic Features of the Jackson Purchase Region, Embracing the Counties of Ballard, Calloway, Fulton, Graves, Hickman, McCracken, and Marshall.* Geological Survey of Kentucky. Series 2, Volume F. John D. Woods, Frankfort.

Lovell, B., and J. Tubb 2006. Ancient Quarrying of Rare in Situ Palaeogene Hertfordshire Puddingstone. *Mercian Geologist* 16 (3):186–189. [Conglomerate quern quarries in England that were used during the Neolithic and Roman periods.]

Lundström, A., and I. Lindeberg 1964. Querns. In *Excavations at Helgö II*, pp. 238–239, Uppsala.

Lundström, P. 1961. Querns. In *Excavations at Helgö I*, p. 239, Uppsala.

Lyell, Charles 1853. A Manual of Elementary Geology; or, The Ancient Changes of the Earth and Its Inhabitants as Illustrated by Geological Monuments. Reprinted from the 4th revised edition. D. Appleton, New York.

Lynchburg Daily Virginian 1853. Ad by Israel Price and Company for the Brush Mountain Millstone Quarry in Montgomery County, Virginia. *Lynchburg Daily Virginian*, September 1853.

Lynn Historical Society 1949. *Records of Ye Towne Meetings of Lyn. Part 1691–1701/2.* Lynn Historical Society, Lynn, Massachusetts.

Lynn Historical Society 1956. *Records of Ye Towne Meetings of Lyn. Part 1701–1717.* Lynn Historical Society, Lynn, Massachusetts.

MacCabe, Julius P. Bolivar 1837. *A Directory of the Cities of Cleveland & Ohio, for the Years 1837–38...* Sanford & Lott, Cleveland.

MacKie, Euan W. 1971–2. Some New Quernstones from Brochs and Duns. *Proceedings of the Society of Antiquaries of Scotland* 104:137–146.

MacKie, Euan W. 1995. Three Iron Age Rotary Querns from Southern Scotland. *Glasgow Archaeological Journal* 19:107–109.

MacKie, Euan W. 2002. Two Querns from Appin. *Scottish Archaeological Journal* 24 (1): 85–92.

Madsen, H. J. 1967. Møllesten til Norden. *Skalk* nr 6/1967. [Quernstones in Scandinavia.]

Major, J. Kenneth 1982a. Eifel Millstone Production. *The International Molinological Society, Transactions of the Fifth Symposium*, pp.343–356, France. [Germany]

Major, J. Kenneth 1982b. The Manufacture of Millstones in the Eifel Region of Germany. *Industrial Archaeology Review* 6 (3):194–204.

Major, J. Kenneth 1985. The Tools of the Millstone Dresser. *Tools and Trades Historical Society Newsletter* 8:27–37.

Major, J. Kenneth 2005. Two Millstone Quarries Seen during the Congress in Grenoble, September 2005. *International Molinology* 71:31, Watford, England.

Malaws, B. A. 1990. A Quarry Rediscovered. *Melin* 6:41–42. [Millstone quarry at Conway, Wales.]

Mallet, Louis 1995. Les Carrières Souterraines de Carlus et d'Amarens (Tarn). *Subterranea, Bulletin de la Société Française d'Étude des Souterrains* 94:47–58. [France]

Mangartz, Fritz 1993. Mittelalterliche Mühlsteingewinnung am Hochstein. Der Hochstein-Führer zu einem Vulkan der Osteifel. Mendig, Schriften des Förderverein Kultur und Heimatmuseum, Band 1, pp.23–33. [Germany]

Mangartz, Fritz 1998. "Bernhard Keibs Lay." Neue Erkenntnisse zur spämittelalterlichen Mühlsteinproduktion im oberen Niedermendiger Lavastrom ["'Bernhard Keibs Lay.' New Facts on Late Medieval Millstone Production in the Upper Niedermendig Lava Flow in Germany"]. *Pellenz-Museum* 7:101–137.

Mangartz, Fritz 1999. Vom Napoleonshut bis Napoleon; über 2000 Jahre Steinabbau am Vulkan "Hohe Buche" bei Andernach-Namedy. In *Mayen-Koblenz: Heimatbuch*, pp. 62–65.

Mangartz, Fritz 2003a. Abbau und Produktion rheinischer Basaltlava-Mühlsteine vom Mittelalter bis 1900: archäologische Funde und technische Denkmäler ["Quarrying and Production of Rhineland Basaltic Lava Millstones in Germany from the Middle Ages to 1900"]. In *Meules à Grains: Actes du Colloque International de La Ferté-sous-Jouarre 16–19 Mai 2002*, edited by Mouette Barboff, François Sigaut, Cozette Griffin-Kremer, and Robert Kremer, pp. 160–168. Éditions Ibis Press and Éditions de la Maison des Sciences de l'Homme, Paris, France.

Mangartz, Fritz 2003b. Der Koloss von Oberlützingen. *Kreis Ahrweiler, 60 Jahrgang*, pp. 161–165. [Germany]

Mangartz, Fritz 2005. Römischer Basaltlava-Abbau zwischen Eifel und Rhein. Diss. Am Institut für Urind Frühgeschichte der Universität zu Köln bei Andreas Zimmermann. [Germany]

Mangartz, Fritz 2006a. Prehistoric to Medieval Quernstone Production in the Bellerberg Volcano Lava Stream Near Mayen, Germany. *Quern Study Group Newsletter* 7:10–13. Translated by Caroline Rann. Oxford, England.

Mangartz, Fritz 2006b. Vorgeschichtliche bis Mittelalterliche Mühlsteinproduktion in der Osteifel ["Millstone Production in the Eastern Eifel from Prehistory to the Middle Ages"]. *Mühlsteinbrüche. Erforschung, Schutz und Inwertsetzung eines Kulturerbes europäischer Industrie (Antike–21. Jahrhundert)*, edited by Alain Belmont and Fritz Mangartz, pp. 25–34. Römisch-Germanisches Zentralmuseum, Tagungen Band 2, Mainz, Germany.

Mangartz, Fritz 2006c. Introduction: Millstone Mines. In *Mühlsteinbrüche. Erforschung, Schutz und Inwertsetzung eines Kulturerbes europäischer Industrie (Antike–21. Jahrhundert)*, edited by Alain Belmont and Fritz Mangartz, p. 132. Römisch-Germanisches Zentralmuseum, Tagungen Band 2, Mainz, Germany.

Mangartz, Fritz, and Olaf Pung 2002. Die Holzkeilspaltung im alten Steinabbau. *Der Anschnitt. Zeitschrift für Kunst und Kultur im Bergbau*, 54e année, 6/2002, pp. 238–252. [Use of wooden wedges in stone quarrying in Germany.]

Mannino, Robert, Jr. [Medfield Historical Commission] 1988. Letter to Peter Mills, Massachusetts Historical Commission, Boston, July 21. Copy on file at the Kentucky Heritage Council, Frankfort. [Granite millstone quarry at Medfield.]

Manufacturer and Builder 1872. The Ventilation of Unwholesome Manufactories. *Manufacturer and Builder* 4 (4):81, New York.

Manufacturer and Builder 1876. Burr Stones. *Manufacturer and Builder* 8 (4):82, New York.

Manufacturer and Builder 1879. Ulster County Mill Stones. *Manufacturer and Builder* 11 (10):226–227, New York.

Manufacturer and Builder 1886. Mineral Production of the United States in 1885. *Manufacturer and Builder* 18 (11): 250–251, New York.

Manufacturer and Builder 1893. Rock Emergy Millstones. *Manufacturer and Builder* 25 (10):231, New York.

Marshall, C., H. J. James, R. E. Bevins, and J. Horák 2003. Quernstones. In *Roman Carmarthen: Excavations 1978–1993*, by Heather J. James, pp. 353–367. Britannia

Monograph Series No. 20, Society for the Promotion of Roman Studies, London.

Martel, Pierre 1973. Les Tailleries de Meules de Ganagobie, leur Intérêt pour l'Étude des Tailleries du Sud-Est de la France. *Le Monde Alpin et Rhodanien,* 1:77–96. [France]

Martin, Joseph, and William Henry Brockenbrough 1835. *A New and Comprehensive Gazetteer of Virginia, and the District of Columbia: Containing a Copious Collection of Geographical, Statistical, Political, Commercial, Religious, Moral and Miscellaneous Information.* J. Martin, Charlottesville.

Mary, Stéphane 2003. Les Meules Métalliques en France à la Fin du XIXe Siècle. In *Meules à Grains: Actes du Colloque International de La Ferté-sous-Jouarre 16–19 Mai 2002,* edited by Mouette Barboff, François Sigaut, Cozette Griffin-Kremer, and Robert Kremer, pp. 400–417. Éditions Ibis Press and Éditions de la Maison des Sciences de l'Homme, Paris, France.

Maryland Farmer 1882. Ad that mentions Moore County Grit Millstones from Moore County, North Carolina. *Maryland Farmer* 19:337.

Maryland Gazette 1753. Ad announcing that Herman Husbands was making millstones near the mouth of the Susquehanna. *Maryland Gazette,* October 25, 1753. Annapolis.

Massachusetts Office of the Secretary of State 1856. *Statistical Information Relating to Certain Branches of Industry in Massachusetts, for the Year Ending June 1, 1855.* Massachusetts Office of the Secretary of State, Boston.

Masson, L. 1950. Die Eis- und Mühlsteinhöhlen von Roth bei Gerolstein. *Die Eifel* 45 Jg., p. 79, Düren. [Germany]

Mather, Joseph H. 1847. *Geography of the State of New York. Embracing Its Physical Features, Climate, Geology, Mineralogy, Botany, Zoology, History, Pursuits of the People, Government, Education, Internal Improvements, &c. with Statistical Tables, and a Separate Description and Map of Each County.* J. H. Mather & Co. and M. H. Newman & Co., Hartford, New York.

Mather, William W. 1838a. *First Annual Report of the Geological Survey of Ohio.* Samuel Medary, Columbus.

Mather, William W. 1838b. *Second Annual Report of the Geological Survey of Ohio.* Samuel Medary, Columbus.

Mather, William W. 1839. Report on the Geological Reconnoissance of Kentucky, Made in 1838. *Journal of the Senate of the Commonwealth of Kentucky.* Reprinted by the Kentucky Geological Survey, Series 11, Reprint 25, 1988, Lexington.

Mather, William W. 1843. *Geology of New-York.* Part I: Comprising the Geology of the First District. Carroll and Cook, Albany.

Maume, Jack 1991. Millstones of Carrig-na-m Brónta. Mallow Archaeological and Historical Society. *Mallow Field Club Journal Number 9,* Mallow, County Cork, Ireland.

Maxwell, Hu 1968. *The History of Barbour County, West Virginia: From Its Earliest Exploration and Settlement to the Present Time.* McClain Printing Company, Parsons, West Virginia.

Mayer, Adele 1993. A Stone from Horwich, Bolton. *Quern Study Group Newsletter* 3:6. England. [England]

M'Cauley, I. H., J. L. Suesserott, and D. M. Kennedy 1878. *Historical Sketch of Franklin County, Pennsylvania: Prepared for the Centennial Celebration Held at Chambersburg, Penn'a, July 4th, 1876 and Subsequently Enlarged.* D. F. Pursel, Chambersburg, Pennsylvania.

McCalley, Henry 1886. *Geological Survey of Alabama.* Barrett and Company, Montgomery, Alabama.

McGee, Marty 2001. *Meadows Mills: The First Hundred Years.* Meadows Mills, Inc., Wilkesboro, North Carolina.

McGill, William M. 1936. *Outline of the Mineral Resources of Virginia.* Virginia Geological Survey, Bulletin 47, Educational Series No. 3.

McGrain, John W. 1982. "Good Bye Old Burr": The Roller Mill Revolution in Maryland, 1882. *Maryland Historical Magazine* 77 (2):154–171.

McGrain, John W. 1991. Fact cards on millstones from the personal files of John W. McGrain, Baltimore County Landmarks Commission, Towson, Maryland. Photocopies of cards shared with Charles D. Hockensmith on February 4.

McIlhaney, Calvert W. [Bristol, Virginia] 1992. Personal communication to Charles D. Hockensmith. [Conglomerate millstone quarries on Cloyd Mountain in Pulaski County, Virginia.]

McKechnie, Jean L. (editor) 1978. *Webster's New Twentieth Century Dictionary of the English Language.* Unabridged, second edition. Collins and World.

McKee, Harley J. 1971. Early Ways of Quarrying and Working Stone in the United States. *APT* 3 (1): 44–58. Bulletin of the Association for Preservation Technology.

McKee, Harley J. 1973. *Introduction to Early American Masonry: Stone, Brick, Mortar and Plaster.* National Trust for Historic Preservation and Columbia University.

McMurtrie, H. 1819. *Sketches of Louisville and Its Environs.* S. Penn, Louisville.

McWhirr, Alan D., L. Viner, C. Wells Witt, et al. 1982. *Cirencester Excavations II: Romano-British Cemeteries at Cirencester.* Cirencester Excavation Committee, Cirencester, England. [Includes a discussion of querns.]

Meadows, Larry G. 2002. Red River Millstone Quarry. *The Millstone* 1 (2):11–12. Kentucky Old Mill Association, Clay City, Kentucky.

Meadows, Larry G. [Clay City, Kentucky] 2006. Personal communication to Charles D. Hockensmith. [Concerning location of Cleveland's Landing in Kentucky.]

Menillet, F. 1985. Les Meulières et les Argiles à Meulières: Leurs Rapports Avec les Surfaces Néogènes à Quaternaire Ancien du Bassin de Paris. *Géol. de la France* 2:213–226.

Mercer, Henry C. 1917. Survival of Ancient Hand Corn Mills in the United States. *A Collection of Papers Read Before the Bucks County Historical Society,* Volume 4:729–735. The Chemical Publishing Company, Easton, PA.

Meredith, Rosamond 1981a. Millstone Making at Yarncliff in the Reign of Edward IV. *Derbyshire Archaeological Journal* 101:102–106. [England]

Meredith, Rosamond 1981b. Hathersage Affairs 1720–1735: Some Letters from Thomas Eyre of Thorp. *Transactions of the Hunter Archaeological Society* 11:14–27. [England]

Metcalf, Robert W. 1941. Abrasive Materials. In *Minerals Yearbook Review of 1940,* pp. 1239–1254. Department of the Interior, U.S. Bureau of Mines, Government Printing Office, Washington, D.C.

Metcalf, Robert W. 1943a. Abrasive Materials. In *Minerals Yearbook, 1941,* pp. 1339–1356. Department of the Interior, U.S. Bureau of Mines, Government Printing Office, Washington, D.C.

Metcalf, Robert W. 1943b. Abrasive Materials. In *Minerals Yearbook, 1942,* pp. 1331–1348. Department of the Interior, U.S. Bureau of Mines, Government Printing Office, Washington, D.C.

Metcalf, Robert W. 1949. Abrasive Materials. In *Minerals Yearbook, 1947,* pp. 97–113. Department of the Interior,

U.S. Bureau of Mines, Government Printing Office, Washington, D.C.

Metcalf, Robert W. 1950. Abrasive Materials. In *Minerals Yearbook, 1948*, pp. 98–115. Department of the Interior, U.S. Bureau of Mines, Government Printing Office, Washington, D.C.

Metcalf, Robert W. 1951. Abrasive Materials. In *Minerals Yearbook, 1949*, pp. 91–110. Department of the Interior, U.S. Bureau of Mines, Government Printing Office, Washington, D.C.

Metcalf, Robert W., and A. B. Cade 1945. Abrasive Materials. In *Minerals Yearbook, 1943*, pp. 1384–1399. Department of the Interior, U.S. Bureau of Mines, Government Printing Office, Washington, D.C.

Metcalf, Robert W., and A. B. Cade 1946. Abrasive Materials. In *Minerals Yearbook, 1944*, pp. 1341–1358. Department of the Interior, U.S. Bureau of Mines, Government Printing Office, Washington, D.C.

Metcalf, Robert W., and A. B. Holleman 1947. Abrasive Materials. In *Minerals Yearbook, 1945*, pp. 1357–1376. Department of the Interior, U.S. Bureau of Mines, Government Printing Office, Washington, D.C.

Metcalf, Robert W., and A. B. Holleman 1948. Abrasive Materials. In *Minerals Yearbook, 1946*, pp. 92–110. Department of the Interior, U.S. Bureau of Mines, Government Printing Office, Washington, D.C.

Michael, Ronald L. 1983. National Register of Historic Places Nomination form for Game Lands 51 Millstone Quarry, Fayette County, Pennsylvania. Prepared by California University, California, Pennsylvania.

Milburn, Mark 1992. Saharan Stone. *Quern Study Group Newsletter* 2:6–7, England. [Africa]

Milburn, Mark 2002. Further Thoughts on Saharan Querns and Allied Objects. *Quern Study Group Newsletter* 6:2, Reading, England.

Mills, William C. 1921. Flint Ridge. *Ohio State Archaeological and Historical Society Quarterly* 30:91–161.

Mitchell, J. E. 1855. Millstones and Burr-blocks of La Ferté-sous-Jouarre. In *The American Miller and Millwright's Assistant* by William C. Hughes, pp. 234–244. Henry Carey Baird, Philadelphia. New revised edition.

Mitchell, J. E. 1869. A History of the French Burr. In *The American Miller and Millwright's Assistant* by William C. Hughes, pp. 234–244. Henry Carey Baird, Philadelphia. New revised edition.

Montgomery County 1907. *Montgomery County, Virginia.* Jamestown Exposition Souvenir. Blacksburg, Virginia.

Moog, B. 1989. Alpenländische Wassermühlen, Teil 1. Der Mülstein, Periodikum für Mühlenkunde un Mühlenerhaltung, *Heft* 6:66–68, Minden. [Germany]

Moreau, Jean, and Yves Ruel 2003. Voyage au Coeur d'un Moulin à Meules ["Voyage to the Heart of a Millstone"]. In *Meules à Grains: Actes du Colloque International de La Ferté-sous-Jouarre 16–19 Mai 2002,* edited by Mouette Barboff, François Sigaut, Cozette Griffin-Kremer, and Robert Kremer, pp. 354–356. Éditions Ibis Press and Éditions de la Maison des Sciences de l'Homme, Paris, France.

Moritz, L., and C. Jones 1950. Experiments in Grinding Wheat with a Romano-British Quern. *Milling* 2–4. [England]

Müller, Jörg, and Manfred Lorenz 2002. *Eine Zeittafel zu den Mühlsteinbrüchen in Jonsdorf.* 64 page booklet, Jonsdorf, Germany. [Millstone quarry at Jonsdorf, Germany]

Müller, S. 1907. Omdreiende Kværn. *Aarbøger for Nordisk Oldkyndighed og Historie.* 11 Række 22. Bind. Copenhagen. [Discusses rotary querns in Scandinavia during the Iron Age.]

Mullin, David 1988. Some Millstone Quarry Locations in the Forest of Dean — Part I. *New Regard* 4:53–59. [England]

Mullin, David 1990. Some Millstone Quarry Locations in the Forest of Dean — Part II. *New Regard* 6:1–14. [England]

Munson Brothers 1860. *Munson Brothers, Proprietors and Manufacturers of French Burr Mill Stones ... Machine Finish for Mill Stones, Pat. Cast Iron Eyes ... Portable and Stationary Engines and Boilers ... and All Kinds of Mill Furnishings...* Catalogue produced by Munson Brothers, Utica, New York.

Munson Brothers 1875. *Illustrated Catalogue and Descriptive Pamphlet from Munson Brothers' Central New York Burr Mill Stone Manufactory, and Mill Furnishing Establishment...* Munson Brothers, Utica, New York.

Munson Brothers 1886. *Illustrated Catalogue of Munson Brothers, Mill Furnishers. Specialties: Munson's Patent Eyes and Spindles, Shafts, Gearing, Pulleys, and All Kinds of Mill Furnishings, at the Lowest Prices, Utica, N. Y.* L. C. Childs & Son, Printers, Utica, New York.

Murphy, James L. [Ohio State University, Columbus, Ohio] 2004. E-mail to Charles D. Hockensmith, April 9. [Flint Ridge millstones and Raccoon buhrstones of Ohio.]

Nason, F. L. 1894. *Economic Geology of Ulster County.* New York State Museum 47th Annual Report, Albany.

Nasz, Adolf 1950. *Zarna Wczesnodziejowe.* Studia Wczesnodziejowe, Seria Archeologiczna, T. 1. Archeogiczne, Polskie Tow, Warszawa. [Millstones in Poland; includes summary in French.]

Neri, Y. 1993. Archaeological and Geological Aspects of Beachrock Saddle-Quern Utilization in Antiquity. MA thesis, University of Haifa. [In Hebrew]

Newland, David H. 1907. Report of Operations and Productions during 1906. In *The Mining and Quarry Industry of New York State 1906.* New York State Museum Bulletin Number 305, Albany.

Newland, David H. 1908. *The Mining and Quarry Industry of New York State 1907.* New York State Museum Bulletin Number 426, Albany.

Newland, David H. 1909. *The Mining and Quarry Industry of New York State 1908.* New York State Museum Bulletin Number 451, Albany.

Newland, David H. 1910. *The Mining and Quarry Industry of New York State 1909.* New York State Museum Bulletin Number 476, Albany.

Newland, David H. 1911. *The Mining and Quarry Industry of New York State 1910.* New York State Museum Bulletin Number 496, Albany.

Newland, David H. 1916. *The Mining and Quarry Industry of New York State 1915.* New York State Museum Bulletin Number 190, Albany.

Newland, David H. 1921. *Mineral Resources of the State of New York for 1919.* New York State Museum Bulletins, Numbers 223 and 224, Albany.

Newland, David H., and C. A. Hartnagel 1932. *Mining and Quarry Industries of New York 1927–1929.* New York State Museum Bulletin Number 295, Albany.

Newland, David H., and C. A. Hartnagel 1936. *Mining and Quarry Industries of New York 1930–1933.* New York State Museum Bulletin Number 305, Albany.

Newland, David H., and C. A. Hartnagel 1939. *Mining and Quarry Industries of New York 1934–1936.* New York State Museum Bulletin Number 319, Albany.

New York 1912. *Preliminary Report of the Factory Investigating Commission, 1912...* The Argus Company, Albany.

New York Chamber of Commerce 1867. *Colonial Records of the New York Chamber of Commerce, 1768–1784.* J. F. Trow & Co., New York.

New-York Historical Society 1868. *The John Watts DePeyster Publication Fund Series,* Volume 38. New-York Historical Society, New York.

New York State Museum 1918. Millstone Producers in New York—1918, Ulster County. Typed list from the open file with the Geological Survey, New York State Museum, Albany.

New York State Museum 1934. Millstones in 1934, New York, Ulster County. Typed list from the open file with the Geological Survey, New York State Museum, Albany.

New York State Secretary's Office 1857. *Census of the State of New York, for 1855.* New York State Secretary's Office, Albany.

Niles Register 1812. Report of Stone Suitable for Making Burr Stones in Georgia. *Niles Register* 1:418, February 8, 1812.

North Carolina Land Company 1869. *A Statistical and Descriptive Account of the Several Counties of North Carolina, United States of America.* Published by the North Carolina Land Company. Nichols & Gorman, Raleigh.

North Carolina Millstone Co. n.d. North Carolina Millstone Co. Brochure describing the "Moore County Grit" stone and price lists for mills and millstones, Parkewood, North Carolina. University of North Carolina Library, Chapel Hill.

North Carolina Millstone Co. 1885. "Moore County Grit" the Best Stone for Grinding Corn Meal for Table Use in the World. North Carolina Millstone Co. Catalog, Parkewood, North Carolina. University of Delaware Library.

North Carolina State Geologist 1875. *Report of the Geological Survey of North Carolina.* J. Turner, State Printer, Raleigh.

Nuttall, Thomas 1821. Journal of Travels into the Arkansa Territory, during the Year 1819. Reprinted in *Early Western Travels 1748–1846*, edited by Reuben Gold Thwaites (1905), Volume 13, pp. 19–366. Arthur H. Clark Company, Cleveland.

Oakes, R. A. 1905. Genealogical and Family History of the County Jefferson, New York: A Record of the Achievements of Her People and Phenomenal Growth of Her Agricultural and Mechanical Industries. Jefferson County Historical Society, Watertown, New York.

O'Callaghan, E. B. 1855. History of New Netherland. Volume I. D. Appleton and Company, New York.

O'Callaghan, E. B., and Dorthy C. Barck 1929. *Calendar of New York Colonial Commissions, 1680–1770.* New-York Historical Society, New York.

Ogden, George W. 1823. *Letters from the West, Comprising a Tour Through the Western Country, and a Residence of Two Summers in the States of Ohio and Kentucky: Originally Written in Letters to a Brother.* Melcher & Rogers, New Bedford. Reprinted in *Early Western Travels 1748–1846,* edited by Reuben Gold Thwaites (1905). Arthur H. Clark Company, Cleveland.

Ohio Valley Publishing Company 1873. *Kentucky State Directory and Shipper's Guide for 1873–4.* Ohio Valley Publishing Company, Louisville.

Oliva, Priscia, Didier Béziat, Claude Domergue, Catherine Jarrier, François Martin, Bernard Pieraggi, and Francis Tollon 1999. Geologic Source and Use of Rotary Millstones from the Roman Iron-Making Site of Les Martys (Montagne Noire, France). *European Journal of Mineralogy* 11 (4):757–762.

Ortuno, Joëlle 1999. Une Famille Dauphinoise d'Artisans Ruraux: Les Sébelin (Meuliers à Vinay, XVIIe-XVIIIe s.), Mémoire de Maîtrise sous la dir. d'A. Belmont, Université P. Mendès France-Grenoble. 2 vols. 276 p. [France]

Owen, David Dale 1858. *First Report of a Geological Reconnoissance of the Northern Counties of Arkansas, Made during the Years 1857 and 1858.* Arkansas Geological Survey. Johnson & Yerkes, Little Rock.

Owen, David Dale 1859. *Report of a Geological Reconnoissance of the State of Indiana Made in the Year 1837, in Conformity to an Order of the Legislature.* J. C. Walker, Indianapolis.

Owen, David Dale 1860. *Second Report of a Geological Reconnoissance of the Middle and Southern Counties of Arkansas, Made during the Years 1859 and 1860.* Arkansas Geological Survey. C. Sherman & Son, Philadelphia.

Owen, David Dale 1861. *Fourth Report of the Geological Survey in Kentucky Made during the Years 1858 and 1859.* Printed at the Yeoman Office, Frankfort.

Palausi, G. 1965. Les Tailleries de Meules Anciennes dans la Région de l'Esterel et leur Relation avec la Géologie. *Bulletin Philologique et Historique du Comité des Travaux Scientifiques et Historiques,* 44:707–714. [France]

Parker, Edward W. 1893a. Abrasive Materials. In *Mineral Resources of the United States, Calendar Year 1891,* pp. 552–554. Part II—Nonmetals. U.S. Bureau of Mines, U.S. Geological Survey, Government Printing Office, Washington, D.C.

Parker, Edward W. 1893b. Abrasive Materials. In *Mineral Resources of the United States, Calendar Year 1892,* pp. 748–750. Part II—Nonmetals. U.S. Bureau of Mines, U.S. Geological Survey, Government Printing Office, Washington, D.C.

Parker, Edward W. 1894. Abrasive Materials. In *Mineral Resources of the United States, Calendar Year 1893,* pp. 670–679. Part II—Nonmetals. U.S. Bureau of Mines, U.S. Geological Survey, Government Printing Office, Washington, D.C.

Parker, Edward W. 1895. Abrasive Materials. In *Sixteenth Annual Report of the United States Geological Survey to the Secretary of the Interior 1894–1895,* by Charles D. Walcott, pp. 927–950. Part III—Mineral Resources of the United States. U.S. Geological Survey, Government Printing Office, Washington, D.C.

Parker, Edward W. 1896. Abrasive Materials. In *Seventeenth Annual Report of the United States Geological Survey to the Secretary of the Interior 1895–1896,* by Charles D. Walcott, pp. 586–587. Part IV—Mineral Resources of the United States. Washington, D.C.

Parker, Edward W. 1897. Abrasive Materials. In *Eighteenth Annual Report of the United States Geological Survey to the Secretary of the Interior 1896–1897,* by Charles D. Walcott, pp. 1219–1231. Part V—Mineral Resources of the United States. Washington, D.C.

Parker, Edward W. 1898. Abrasive Materials. In *Nineteenth Annual Report of the United States Geological Survey to the Secretary of the Interior 1897–1898,* by Charles D. Walcott, pp. 515–533. Part VI—Mineral Resources of the United States. Washington, D.C.

Parker, Jenny Marsh 1884. *Rochester: A Story Historical.* Scranton, Wetmore & Company, Rochester, New York.

Parker, Steve [Lancaster, Ohio] 2005. Letter to Charles D. Hockensmith, April 25. On file at the Kentucky Heritage Council, Frankfort. [Raccoon Creek Buhr millstone quarry at McArthur, Vinton County, Ohio; also, photographs, brief report, and sketch map prepared by Steve Parker.]

Parkhouse, Jonathan 1976. The Dorestad Quernstones. *Berichten van de Rijksdienst voor het Oudheidkundig Bodemonderzoek* 26, pp. 181–188, Amersfoort. [England]

Parkhouse, Jonathan 1977. Early Medieval Basalt Lava Quernstones and Their Use as an Indicator of Trade. Unpublished M. A. thesis, University of Manchester, Manchester. [Germany and England]

Parkhouse, Jonathan 1991. An Assemblage of Lava Querns from the Thames Exchange Site, City of London. Unpublished archive report for the Museum of London, London. [Germany and England]

Parkhouse, Jonathan 1997. The Distribution and Exchange of Mayen Lava Quernstones in Early Medieval Northwestern Europe. In *Exchange and Trade in Medieval Europe, Volume 3*, edited by Guy de Boe and Frans Verhaeghe, pp. 97–106. Institute for the Archaeological Heritage, Rapporten 3, Zellik. [Germany]

Parkhouse, Jonathan 1998. Evidence for a Y-Shaped Rynd on a Quern from Western Underwood, Buckinghamshire. *Quern Study Group Newsletter* 5:2, Reading, England.

Pascal Mayoral, Pilar, and Pedro Garcia Ruiz 2001. Canteras y Tecnología Molinar en el Río Jubera (La Rioja) ["Quarries and Technology of Millstones in the Jubera River (the Rioja)"]. *Revista Murciana de Antroplogía* 7:237–266. [Spain]

Pascal Mayoral, Pilar, and Pedro Garcia Ruiz 2002. Nuevas Canteras de Piedras de Molino y Trujal, Valle del Cidacos (Arnedillo, La Rioja) ["New Quarries of Millstones and Presses, Valley of Cidacos (Arnedillo, La Rioja)"]. *Kalakorikos* 7:209–219. [Spain]

Pascal Mayoral, Pilar, and Pedro Garcia Ruiz 2003a. Canteras de Molino y Trujal. Cuenca del Río Linares: Muro de Aguas, Villarroya y Grávalos ["Millstone Quarries and Presses, Basin of Linares River: Wall of Water, Villaroya and Gravalos"]. *Graccurris* 14:221–241. [Spain]

Pascal Mayoral, Pilar, and Pedro Garcia Ruiz 2003b. Nuevo Hallazgo de Canteras de Piedras de Molino, Río Iregua ["New Discovery of Millstone Quarries, Iregua River"]. *El Serradero* 76:16–18. [Spain]

Pascal Mayoral, Pilar, and Pedro Garcia Ruiz 2003c. Las Canteras de Piedra de Molino: Una Industria Riojana Desconocida ["The Millstone Quarries: An Unknown Industry of Rioja"]. *Altza* VII: 135–146, San Sebastian. [Spain]

Pascal Mayoral, Pilar, and Francisco Moreno Arrastio 1980. Prensas de Aceite Romanas en La Rioja ["Roman Oil Presses in La Rioja"]. *Archivo Español de Arqueología* 53 (141–142):199–211. [Spain]

Passemard, E., É Laoust, and J. Bourrilly 1923. Mode d'Extraction des Pierres Meulières au Maroc. Association Française pour l'avancement des Sciences, 46e Session, pp. 497–482. Monpellier, Paris.

Patapsco Land Company of Baltimore City 1874. *Curtis' Bay*. Patapsco Land Company of Baltimore City.

Peacock, D. P. S. 1980a. The Roman Millstone Trade: A Petrological Sketch. *World Archaeology* 12:43–53, London.

Peacock, D. P. S. 1980b. The Mills of Pompeii. *Antiquity* 63 (239):205–214.

Peacock, D. P. S. 1986. The Production of Roman Millstones Near Orvieto, Umbria, Italy. *The Antiquaries Journal* 66 (1):45–51, London.

Peacock, D. P. S. 1987. Iron Age and Roman Quern Production at Lodsworth, West Sussex. *The Antiquaries Journal* 67 (1):61–85, London. [England]

Peacock, D. P. S. 2004. Quern Quarries and Iron-Working Near Châbles. *Journal of Roman Archaeology* 17:650–651. [Review of Anderson et al. book *Des Artisans à la Campagne: Carrière de Meules, Forge et voie Gallo-Romaines à Châbles (FR.)]*

Peacock, T. B. 1860. On French Millstone-Makers' Phthisis. *British & Foreign Medico-Chirugical Review* XXV, January–April 1860, pp. 214–224.

Pearson, T. 2000. *Quern Manufacturing at Wharncliffe Rocks, Sheffield South Yorkshire*. English Heritage, National Monuments Record, York, England.

Peck, J. M. 1840. A Gazetteer of Illinois. *The North American Review* 51 (108):92–141, Cedar Falls, Iowa.

Peters, J. T., and H. B. Carden 1926. *History of Fayette County, West Virginia*. Jarrett Printing Company, Charleston, West Virginia.

Pettersson, Täpp John-Erik 1973. Kvarnstenbrytningen i Malung. Uppsats i Etnologi, Uppsala Universitet HT, 1973, Grundkurs B1, 2 betyg. [A report 29 pages long dealing with millstone quarries at Malung, Sweden.]

Pettersson, Täpp John-Erik 1977a. Kvarnstenshuggning. En inledande översikt på skandinavisk grund med utgångspunkt från Malung socken. Uppsala Universitet, Uppsats i Etnolo-Gi, påbyggnadskurs C, pp. 35–43, 71. [A review of Scandinavia beginning with Malung, Sweden.]

Pettersson, Täpp John-Erik 1977b. Kvarnstenshuggning Malung-ur en sockens historia, pp. 103–136. [Millstone quarries at Malung, Sweden.]

Pettersson, Täpp John-Erik 1981. Malungsfolkets kvarnstenshuggning i äldre tid, Kulturdagarna i Bonäs bygdegård 23–24 juni 1980, Uppsala 1981, pp. 85–90. [Millstone quarries at Malung, Sweden.]

Phalen, W. C. 1908. Abrasive Materials. In *Mineral Resources of the United States, Calendar Year 1907*, pp. 607–626. Part II — Nonmetals. Department of the Interior, U.S. Bureau of Mines, Government Printing Office, Washington, D.C.

Phalen, W. C. 1909. Abrasive Materials. In *Mineral Resources of the United States, Calendar Year 1908*, pp. 581–598. Part II — Nonmetals. Department of the Interior, U.S. Bureau of Mines, Government Printing Office, Washington, D.C.

Phalen, W. C. 1910. Abrasive Materials. In *Mineral Resources of the United States, Calendar Year 1909*, pp. 609–627. Part II — Nonmetals. Department of the Interior, U.S. Bureau of Mines, Government Printing Office, Washington, D.C.

Phalen, W. C. 1911. Abrasive Materials. In *Mineral Resources of the United States, Calendar Year 1910*, pp. 683–690. Part II — Nonmetals. Department of the Interior, U.S. Bureau of Mines, Government Printing Office, Washington, D.C.

Phalen, W. C. 1912. Abrasive Materials. In *Mineral Resources of the United States, Calendar Year 1911*, pp. 835–854. Part II — Nonmetals. Department of the Interior, U.S. Bureau of Mines, Government Printing Office, Washington, D.C.

Philip, G., and Olwen Williams-Thorpe 1993. A Provenance Study of Jordanian Basalt Vessels of the Chalcolithic and Early Bronze Age I Periods. *Paleorient* 19 (2): 51–63.

Philip, G., and Olwen Williams-Thorpe 2000. The Production and Distribution of Ground Stone Artefacts in the Southern Levant during the Fifth to Fourth Millenia B C: Some Implications of Geochemical and Petrographic Analysis. In *Proceedings of the 1st International Congress on the Archaeology of the Ancient Near East, Rome, May 18–23rd 1998*, edited by P. Matthiae, A. Enea, L. Peyronal, and F. Pinnock, pp. 1379–1396. Rome.

Philip, G., and Olwen Williams-Thorpe 2001. The Production and Consumption of Basalt Artefacts in the Southern Levant during the 5th-4th Millenia B C: A Geochemical and Petrographic Investigation. In *Proceedings of Archaeological Sciences 1997*, edited by A. R. Millard, pp. 11–30. BAR International Series, Volume 939, Archeopress, Oxford.

Philips, Judith T. 1950. A Survey of the Distribution of Querns of Hunsbury or Allied Types. In "Excavations at Breedon-on-the-Hill, 1946," edited by K. Kenyon, pp. 75–82. *Transactions of the Leicestershire Archaeological Society* XXVI.

Piot, Auguste 1871. *Traité Historique et Pratique sur la Meulière et la Meunerie*. Second edition. Grenoble. [France]

Pitt-Rivers, A. 1884. *Report on Excavations in the Pen Pits, Near Penselwood, Somerset*. Harrison & Sons, London.

Polak, Jill P. 1987. The Production and Distribution of Peak Millstones from the Sixteenth to the Eighteenth Centuries. *Derbyshire Archaeological Journal* 107:55–72. [England]

Polk, R. L., Company and A. C. Danser 1876. *Kentucky State Gazetteer and Business Directory, for 1876–77*. R. L. Polk Company and A. C. Danser, Detroit and Louisville.

Polk, R. L. Company, and A. C. Danser 1879. *Kentucky State Gazetteer and Business Directory, for 1879–80*. R. L. Polk Company and A. C. Danser, Detroit and Louisville.

Polk, R. L., Company and A. C. Danser 1881. *Kentucky State Gazetteer and Business Directory, for 1881–82*. R. L. Polk Company and A. C. Danser, Detroit and Louisville.

Poor, N. Peabody (compiler) 1844. *Haldeman's Picture of Louisville Directory and Business Advertiser, for 1844–1845*. W. N. Haldeman, Louisville.

Power, Patrick 1939. A Decorated Quern-Stone and Its Symbolism. *Proceedings of the Royal Irish Academy* 45 (4):25–30. [Religious aspects of quern-stone in Ireland.]

Pratt, Joseph H. 1901. Abrasive Materials. In *Mineral Resources of the United States, Calendar Year 1900*, pp. 787–801. Part II — Nonmetals. Department of the Interior, U.S. Bureau of Mines, Government Printing Office, Washington, D.C.

Pratt, Joseph H. 1902. Abrasive Materials. In *Mineral Resources of the United States, Calendar Year 1901*, pp. 781–809. Part II — Nonmetals. Department of the Interior, U.S. Bureau of Mines, Government Printing Office, Washington, D.C.

Pratt, Joseph H. 1904a. Abrasive Materials. In *Mineral Resources of the United States, Calendar Year 1902*, pp. 873–890. Part II — Nonmetals. Department of the Interior, U.S. Bureau of Mines, Government Printing Office, Washington, D.C.

Pratt, Joseph H. 1904b. Abrasive Materials. In *Mineral Resources of the United States, Calendar Year 1903*, pp. 989–1015. Part II — Nonmetals. Department of the Interior, U.S. Bureau of Mines, Government Printing Office, Washington, D.C.

Pratt, Joseph H. 1905. Abrasive Materials. In *Mineral Resources of the United States, Calendar Year 1904*, pp. 995–1015. Part II — Nonmetals. Department of the Interior, U.S. Bureau of Mines, Government Printing Office, Washington, D.C.

Pratt, Joseph H. 1906. Abrasive Materials. In *Mineral Resources of the United States, Calendar Year 1905*, pp. 1069–1085. Part II — Nonmetals. Department of the Interior, U.S. Bureau of Mines, Government Printing Office, Washington, D.C.

Prevot, J. 1975. *Les Meulières du sud de la Région Parisienne*. Laboratoire central des ponts à Chaussées, Paris. Rapport de Recherche LPC; No. 51. 128 pages. [France]

Procopiou, H., E. Jautee, R. Vargiolu, and H. Zahouani 1996. Petrographic and Use Wear Analysis of a Quern from Syvritos Kephala. *Proceedings of the 13th Congress–International Union of Prehistoric and Protohistoric Sciences 1996*. The British Library Document Supply Centre, Paper CN034043887.

Py, Michel 1992. Meules d'Époque Protohistoriques et Romaines Provenant de Lattes. *Lattara* 5:183–232. [Querns recovered from archaeological excavations on the Mediterranean shore of France dating to prehistoric and Roman periods.]

Quinnell, Henrietta 2004. Querns and Mortars. In *Trethurgy: Excavations at Trethurgy Round, St Austell: Community and Status in Roman and Post Roman Cornwall*, by Henrietta Quinnell, pp. 145–152. Cornwall County Council, Cornwall.

Raborg, William A. 1887. Buhrstones. In *Mineral Resources of the United States, Calendar Years 1886*, pp. 581–582. Part II — Nonmetals. U.S. Bureau of Mines, U.S. Geological Survey, Government Printing Office, Washington, D.C.

Radley, Jeffrey 1964. A Millstone Makers Smithy on Gardom's Edge, Baslow. *Derbyshire Archaeological Journal* 84:123–127. [England]

Radley, Jeffrey 1966. Peak Millstones and Hallamshire Grindstones. *The Newcomen Society for the Study of the History of Engineering and Technology Transactions* 36:165–173, London. [England]

Rahtz, Philip, and Robert Meeson 1992. *An Anglo-Saxon Watermill at Tamworth*. Research Report 83. Council for British Archaeology. [Chapter 3 includes information on millstones, pp. 70–78.]

Raistrick, Arthur 1979. *Industrial Archaeology: An Historical Survey*. Granada Publishing, London. Reprint of 1973 edition.

Rawlinson, R. 1954. The Abandoned Millstones of Hathersage. *Derbyshire Countryside* 20 (3):60–61. [England]

Rawnsley, W. F. 1925. Roman Nether Millstone from Bramley. *Surrey Archaeological Collections* xxxvii:241–242.

Rawson, Marion, N. 1935. *Little Old Mills*. E. P. Dutton and Company, New York.

Raymo, Chet 2005. Millstones. *http://www.sciencemusings.com/blog/2005/12/millstones.html*. [Millstone quarry at the Stonehill College campus at Easton, Massachusetts].

Read, M. C. 1883. Berea Grit. In *Mineral Resources of the United States for 1882*, U.S. Bureau of Mines, Washington, D.C.

Reichstein, J. 1987. Mühlsteine aus Basaltlava. In G. Kossack u. A (Hrsg.). Archsum auf Sylt. Teil 2. *Römisch-Germanische Forschungen* 44, pp. 133–136, Mainz. [Germany]

Reigniez, Pascal 2003. Les Étapes de la Fabrication des Meules de Moulins d'Après les Matériaux de la Géorgie

Caucasienne. In *Meules à Grains: Actes du Colloque International de La Ferté-sous-Jouarre 16–19 Mai 2002*, edited by Mouette Barboff, François Sigaut, Cozette Griffin-Kremer, and Robert Kremer, pp. 67–85. Éditions Ibis Press and Éditions de la Maison des Sciences de l'Homme, Paris, France. [Manufacturing stages of millstones in Caucasian Georgia.]

Reinemund, John A. 1955. *Geology of the Deep River Coal Field, North Carolina*. Geological Survey Professional Paper 246. U.S. Geological Survey, Washington, D.C.

Renouf, J. T. 1993. The Rotary Quern. In *A Gallo-Roman Trading Vessel from Guernsey: The Excavation and Recovery of a Third Century Ship Wreck*, by Margaret Rule and Jason Monaghan, pp. 103–106. Guernsey Museum Monograph 5. Guernsey Museum and Galleries, Candie Gardens.

Renzulli, Alberto, Patrizia Santi, Giovanni Nappi, Mario Luni, and Daniele Vitali 2002. Provenance and Trade of Volcanic Rock Millstones from Etruscan-Celtic and Roman Archaeological Sites in Central Italy. *European Journal of Mineralogy* 14 (1):175–183.

Reyes Mesa, José Miguel 2006. Los Molinos Hidráulicos Harineros de la Provincia de Granada ["The Hydraulic Flour Mills of the Province of Granada"]. Colección Molinológica, Granada. [Spain]

Ridenour, George L. 1977. *Early Times in Meade County, Kentucky*. Ancestral Trails Historical Society, Vine Grove, Kentucky. Reprint of 1929 edition published by Western Recorder, Louisville, Kentucky.

Riedl, Norbert F., Donald B. Ball, and Anthony P. Cavender 1976. *A Survey of Traditional Architecture and Related Material; Folk Culture Pattern in the Normandy Reservoir, Coffee County, Tennessee*. Report of Investigations 17. Department of Anthropology, University of Tennessee, Knoxville.

Robertiello, Barbara 1994. Vincient Dunn Remembers a Pataukunk Boyhood and More. *The Accordian* 8 (4):1–7, Rochester, New York.

Robertiello, Barbara 1995. An Odyssey, a Dream, and a Machine Shop: Otto Paul Tolski Tells His Story. *The Accordian* 9 (3):7–10, Rochester, New York.

Roberts, Ellwood 1904. Biographical Annuals of Montgomery County, Pennsylvania: Containing Genealogical Records of Representative Families, Including Many Early Settlers and Biographical Sketches of Prominent Citizens. T. S. Benham, New York.

Roberts, Joseph K. 1942. *Annotated Geological Bibliography of Virginia*. The Dietz Press, Richmond, Virginia.

Roberts, Niall 1989. Some Chinese Millstones. *The International Molinological Society Transaction of the 7th Symposium on Molinology*, pp. 329–342. Flensburg, Germany.

Roberts, Niall 1993. Rotation and Dressing in Cornmill Stones. *Proceedings of the Ninth and Tenth Mill Research Conferences*, edited by D. Breckels, pp. 39–46. Mills Research Group, Manningtree, Essex.

Roberts, Niall 1994. Lutyens' and Jekyll's Garden Millstones. Proceedings of the Twelfth Research Conference, pp. 51–64. The Mills Research Group, Manningtree, England. [England]

Robinson, Rosemary 1981. Millstone Working on Brent Moor. *Devon and Cornwall Notes and Queries* XXXIV: 333–334.

Rockwell, Peter 1993. *The Art of Stoneworking: A Reference Guide*. Cambridge University Press, New York.

Röder, J. 1955. Zur Lavaindustrie von Mayen und Volvic (Auvergne). *Germania* 31:24–27. [Germany]

Röder, J. 1956. Das Werden der Besitzverhaltniesse in Mayener Basaltgebiet. *Germania* 34. [Germany]

Röder, J. 1970. Die Mineralischen der Romischen Zeit in Rheinland. *Bonner Universitätsblatte*. [Germany]

Röder, J. 1972. Die Mühlsteinbrücke von Mayen. Geländedenkmäler einer vor- und frühgeschichtlichen Großindustrie. *Bonner Universitätsblätter*, pp. 35–46. [Germany]

Rodríguez Rodríguez, Amelia C., and V. Barrosa Cruz 2001. Labrar la Piedra para Moler el Grano. La Explotación Prehistórica de las Canteras de Molinos de Toba en la Isla de Gran Canaria ["Working the Stone to Grind the Grain. The Prehistoric Exploitation of the Mill Quarries of Tufa in the Grand Canary Islands"]. *El Pajar* 10:4–9. [Spain]

Rodríguez Rodríguez, Amelia C., Ernesto Martín Rodríguez, J. Mangas Viñuela, Mac González Marrero and J. Buxeda-Garrigós 2006. La Explotación de los Recursos Líticos en la Isla de Gran Canaria. Hacia la Reconstrucción de las Relaciones Sociales de Producción en Época Preeuropea y Colonial ["The Exploitation of the Lithic Resources in the Grand Canary Islands. Towards the Reconstruction of the Social Relationships of Production in the Pre–European and Colonial Epochs"]. In *Sociedades Prehistóricas, Recursos Abióticos y Territorio, Granada*, edited by Gabriel Martínez, A. Morgado, and J. A. Afonso, pp. 367–391. Fundación Ubn al-Jatib de Estudios de Cooperación Cultural, Loja, Granada. [Spain]

Roe, Fiona 1991. A Decorated Mortar or Quern Fragment from Alvaston. *Quern Study Group Newsletter* 1:4, England. [England]

Roe, Fiona 2001. Querns and Millstones. In *Excavation of a Romano-British Roadside Settlement in Somerset* by P. Leach with J. C. Evans, pp. 235–237. Fosse Lane Shepton Mallet Britannia Monograph Series 18.

Roe, Fiona 2004. Worked Stone. In *Yarnton: Saxon and Medieval Settlement and Landscape*, by Gill Hey, pp. 293–295, 304–305. Oxford Archaeology, Thames Valley Landscapes Monograph No. 20. [Provides information on Saxon querns discovered at sites along the Thames River.]

Rogers, N. 1988. Quern Queries. *Interim* 13 (4):35–38.

Rogers, Nicolas S. H. 1993. Rotary Querns. *Anglian and Other Finds from 46–54 Fishergate. The Archaeology of York. The Small Finds*, by Nicolas S. H. Rogers, pp. 1321–1329. Volume 17 (Fasc. 9). Published for York Archaeological Trust by Council for British Archaeology, London.

Rogers, Steve [Tennessee Historical Commission, Nashville] 1989. Letter to Charles D. Hockensmith. Copy on file at the Kentucky Heritage Council, Frankfort. [Millstone quarry in Williamson County, Tennessee.]

Rogers, William Barton 1884. A Reprint of the Annual Reports and Other Papers, on the Geology of the Virginias. Original papers from 1835 to 1841. D. Appleton and Company, New York.

Rolseth, Ingeniør P. O. 1947. *Kvernfjellet*. Selbu og Tydals Historielag, Oslo. [Includes quarries, quarrying, and millstones in Norway.]

Rolseth, Ingeniør P. O., and E. Alsvik 2000. *Selbu og Tydals Historie Kvernfjellet*. Selbu, 204 pages. Reprint of 1947 book. [Includes quarries, quarrying, and millstones in Norway.]

Rønneseth, Ottar 1968. Das Zentrum der Ältesten Mühlsteinindustrie in Norwegen. In *Studien zur europäischen Vor- und Frühgeschitchte*, edited by M. Claus, W. Haarnagel, and K. Raddatz, Festschrift Herbert Jankuhn, pp. 241–252, Neumünster. [Millstone industry at Rönneset, Norway.]

Rønneseth, Ottar 1977. Kvernsteinsbrota ved Åfjiorden. *Sogeskrift for Hyllestad,* hefte 2. [Millstone industry at Hyllestad, Norway.]

Rosen, Steven A. 2003. Early Multi-Resource Nomadism: Excavations at the Camel Site in the Central Negev. *Antiquity* 77:749–760.

Rosen, Steven A., and J. Schneider 2001. Early Bronze Age Milling Stone Production and Exchange in the Negev: Preliminary Conclusions. *Journal of the Israel Prehistoric Society* 31:201–212.

Rotenizer, David E. [Blacksburg, Virginia] 1989. Letter to Charles D. Hockensmith. Copy on file at the Kentucky Heritage Council, Frankfort. [Brush Mountain millstone quarry, Montgomery County, Virginia.]

Ruben, Walter 1954. Die Mühlsteinbrüche bei Johnsdorf. Unser kleines Wanderheft. Heft 31, Veb Bibliographisches Institut, Leipzig. 47 pages. [Millstone industry of Johnsdorf in East Germany.]

Ruben, Walter 1961. Die Mühlsteinbrüche bei Johnsdorf. Unser kleines Wanderheft. Heft 31, Veb Bibliographisches Institut, Leipzig. Second edition. 48 pages. [Millstone industry of Johnsdorf in East Germany.]

Ruben, Walter 1967. Die Mühlsteinbrüche bei Johnsdorf. Unser Kleines Wanderheft, Heft 31, Veb Brockhaus, Leipzig. Fourth edition. 51 pages. [Millstone industry of Johnsdorf in East Germany.]

Ruben, Walter 1973. Die Mühlsteinbrüche bei Johnsdorf. Unser kleines Wanderheft. Heft 31, Veb Brockhaus, Leipzig. Fifth edition. 45 pages. [Millstone industry of Johsdorf in East Germany.]

Rucker, B. H. [Missouri Department of Natural Resources, Jefferson City] 1996. Letter to Charles D. Hockensmith. Copy on file at the Kentucky Heritage Council, Frankfort. [Grindstone and millstone quarries in Missouri.]

Rudge, E. A. 1966. Interim Report on the Distribution of the Puddingstone Quern. *Transactions of the Essex Archaeological Association* 1 (4):247–249.

Rule, William, George F. Mellen, and J. Wooldridge 1900. *Standard History of Knoxville, Tennessee: With Full Outline of the Natural Advantages, Early Settlement, Territorial Government, Indian Troubles, and General and Particular History of the City Down to the Present Time.* Lewis Publishing Company, Chicago.

Runnels, Curtis N. 1981. A Diachronic Study and Economic Analysis of Millstones from the Argold, Greece. Ph.D. thesis, Indiana University, Bloomington.

Runnels, Curtis N. 1985. Trade and the Demand for Millstones in Southern Greece in the Neolithic and the Early Bronze Age. In *Prehistoric Production and Exchange. The Aegean and Eastern Mediterranean*, edited by A. Bernard Knapp and Tamara Stech, pp. 30–43. Monograph XXV, Institute of Archaeology, U. C. L. A., Los Angeles.

Runnels, Curtis N. 1988. The Rotary Querns. In "The Glass Wreck: An 11th-Century Merchantman." *Institute of Nautical Archaeology Newsletter* 15 (3):30–31.

Runnels, Curtis N. 1990. Rotary Querns in Greece. *Journal of Roman Archaeology* 3:147–154.

Runnels, Curtis N. 1992. The Millstones. In "Groundstone" with contributions by D. Evely and C. Runnels. *Well Built Mycenae* series, Fascicule 27, pp. 35–38. Oxbow Books, Oxford, England. [Greek millstones.]

Runnels, Curtis N. 1994. On Lithic Studies in Greece. In *Beyond the Site: Regional Studies in the Aegean Area*, edited by P. Nick Kardulias, pp.161–172. University Press of America, Lanham, Maryland.

Runnels, Curtis N. 2004. The Querns. In *Serce Limani: An Eleventh-Century Shipwreck, Volume I: The Ship and Its Anchorage, Crew, and Passangers*, edited by George F. Bass, Sheila D. Matthews, J. Richard Steffy, and Frederick H. van Doorminck, pp. 255–262. Texas A & M University Press, College Station.

Runnels, Curtis N., and Priscilla M. Murray 1983. Milling in Ancient Greece. *Archaeology* 36 (6):62–63, 75.

Russell, John 1949. Millstones in Wind and Water Mills. *The Newcomen Society for the Study of the History of Engineering and Technology Transactions* 24:55–64, London.

Ryan, John. 1972. Rotary Querns. *Old Kilkenny Review* 24:68. [Ireland]

Safford, James M. 1869. *Geology of Tennessee.* S. C. Mercer, Nashville.

St. John, Samuel 1854. *Elements of Geology, Intended for the Use of Students.* G. P. Putnam, New York.

Salac, V. 1993. Production and Exchange during the La Tene Period in Bohemia. *Journal of European Archaeology* 1 (2):73–99. [This study includes the manufacture and trade of quern-stones during the La Tene period in the Czech Republic.]

Salley, A. S. 1911. *Narratives of Early Carolina, 1650–1708.* C. Scribner's Sons, New York.

Samuel, Delwin 1993. A Comprehensive Surface Survey of Saddle Querns at Amarna, Egypt. *Quern Study Group Newsletter* 3:7–9, England. [Egypt]

Sanchez Navarro, Joaquin 2001. Estudi de les Pedres de Molins Manuals i de les Seves Zones d'Extracció a Menorca. *Treballs de la Secció d Estudis, Publicacions des Born* 10:49–179, Ciutadella. [Spanish Island of Menorca]

Sanchez Navarro, Joaquin 2005. Estudi de les Pedres de Molins Manuals i de les Zones d'Extracció a Menorca. In *Els Barrancs Trancats. L'orde Pagès al Sud de Menorca en Època Andalusina (Segles X-XIII)* edited by M. Barceló and F. Retamero, pp. 235–267, Barcelona. [Spanish Island of Menorca]

Sanchez Navarro, Joaquin 2006. Esta Actual de Investigacions Sobre les Pedres de Molins Manuals i de les Seves Zones d'Extracció a Menorca. I Jornades de Recerca Històtica de Menorca. La Manurqa de Sa'id Ibn Hakam, Un País Islàmic a Occident, Publicacions des Born 15–16:154–196. Ciutadella. [Spanish Island of Menorca]

San Juan, Guy, Michel Gasnier, and Xavier Savary 1999. Histoire et Archéologie des Meules en Granite depuis les Origines. *In L'Exploitation Ancienne des Roches dans le Calvados: Histoire et Archéologie,* edited by Guy San Juan and Jack Maneuvrier, pp. 333–339. Caen, Service Départemental d'Archéologie. [France]

Sankalia, H. 1959. Rotary Querns from India. *Antiquity* 33:128–130.

Sass, Jon A. 1982. A Virginia Millstone Quarry. *Old Mill News* 10 (4):6–7.

Sass, Jon A. 1984. *The Versatile Millstone: Workhorse of Many Industries.* Society for the Preservation of Old Mills, Knoxville.

Sauldubois, Bernard 2003. Fabrication d'Unité de Mouture à Meules dans l'Ouest de la France ["Manufacturing Milling Units with Millstones in the West of France"]. In *Meules à Grains: Actes du Colloque International de La Ferté-sous-Jouarre 16–19 Mai 2002*, edited by Mouette Barboff, François Sigaut, Cozette Griffin-Kremer, and Robert Kremer, pp. 418–422. Éditions Ibis Press and Édi-

tions de la Maison des Sciences de l'Homme, Paris, France.
Saunders, Ruth 1997. Quernstones from Ashton Keynes, Wiltshire. *Quern Study Group Newsletter* 4:4–6, Reading, England. [Quernstones and fragments from the Iron Age and Roman hamlet of Ashton Keynes south of Cirencester, England.]
Saunders, Ruth 1998. The Use of Old Red Sandstone in Roman Britain: A Petrographical and Archaeological Study. *Quern Study Group Newsletter* 5: 7, Oxford, England. [Includes querns.]
Savage, James 1853. *The History of New England from 1630 to 1649 by John Winthrop ... from His Original Manuscripts with Notes....*Volume I. Little, Brown and Company, Boston.
Schaaff, Holger 2006. Der Vulkanpark Osteifel—Wissenschaft und Tourismus in einem Alten Steinbruch- und Bergwerksreier ["The Volcanic Park of Eastern Eifel—Science and Tourism in an Old Quarrying and Mining District"]. In *Mühlsteinbrüche. Erforschung, Schutz und Inwertsetzung eines Kulturerbes europäischer Industrie (Antike–21. Jahrhundert)*, edited by Alain Belmont and Fritz Mangartz, pp. 215–224. Römisch-Germanisches Zentralmuseum, Tagungen Band 2, Mainz, Germany. [Germany]
Schmandt, J. 1930. Die Historsche Entwicklung der Rheinischen Basalt- und Basaltlavaindustrie. Dissertation, Köln. [Germany]
Schön, Volkmar 1989. Betrachtungen zum Handel des Mittelalters am Beispiel von Mühlsteinfunden us Schleswig-Holstein. *Hammaburg* 9:185–190. [Germany]
Schön, Volkmar 1995. Die Mühlsteine von Haithabu und Schleswig. Ein Beitrag zur Entwicklungsgeschichte des mittelalterlichen Mühlenwesens in Nordwesteuopa. *Berichte über die Ausgrabungen in Haithabu,* 0525–5791; Bericht 31. Wachholtz, Neumünster. [Germany]
Schoonhoven, J. 1977. Grinding with Stones. *The International Molinological Society Transaction of the 4th Symposium on Molinology*, pp. 269–283, Matlock, England.
Schrader, Frank C., Ralph W. Stone, and Samuel Sanford 1917. *Useful Minerals of the United States.* Bulletin 624. U.S. Geological Survey, Washington, D.C.
Schüller, Hans 2006. Die historischen Mühlsteinbrüche Mayens im Nutzungskonflikt zwischen Steinindustrie, Siedlungsentwicklung, Kulturlandschaftspflege, Naturschultz und Tourismus ["The Historical Millstone Quarries of Mayen in the Conflict Between the Stoneworking Industry, Settlement Growth, Cultural Landscape Heritage, Nature Conservation and Tourism"]. In *Mühlsteinbrüche. Erforschung, Schutz und Inwertsetzung eines Kulturerbes europäischer Industrie (Antike–21. Jahrhundert)*, edited by Alain Belmont and Fritz Mangartz, pp. 207–214. Römisch-Germanisches Zentralmuseum, Tagungen Band 2, Mainz, Germany. [Germany]
Schulze 1828. Die Mühlsteinbrüche zwischen Mayen und dem Laacher See. *Archiv f. Bergbau und Hüttenwesen*, Bd. 17:386–432, Berlin. [Germany]
Scientific American 1849a. Georgia Burr Mill Stones. *Scientific American*, Volume 4, Issue 48, August 18, 1849, page 380, New York.
Scientific American 1849b. Georgia Burr Mill Stones. *Scientific American*, Volume 5, Issue 14, December 22, 1849, page 106, New York.
Scientific American 1850a. Comments on the Georgia Burr Mill Stones. *Scientific American*, Volume 5, Issue 20, February 2, 1850, page 154, New York.
Scientific American 1850b. Georgia Burr Versus French Burr Stones. *Scientific American*, Volume 5, Issue 32, April 27, 1850, page 253, New York.
Scientific American 1850c. Wilson's Patent Stone Dressing Machine. *Scientific American*, Volume 5, Issue 36, May 25, 1850, page 284, New York.
Scientific American 1857. Burr Stone. *Scientific American*, Volume 12, Issue 34, May 2, 1857, page 268, New York.
Scientific American 1868. Answers to Correspondents. *Scientific American*, Volume 19, Issue 10, September 2, 1868, page 151, New York.
Seeley, Thaddeus D. 1912. *History of Oakland County, Michigan: A Narrative Account of Its Historical Progress, Its People, and Its Principal Interest.* Lewis Publishing Company, Chicago.
Selkirk, Thomas Douglas 1958. *Lord Selkirk's Diary, 1803–1804.* Edited by C. T. White. The Champlain Society, Toronto.
Selmeczi Kovács, Attila 1989. Hand-Mills in the Carpathian Basin. In *Estudos em Homenagem a Ernesto Veiga de Oliveria*, pp. 339–354, Lisbon. [Carpathian Basin in Hungary.]
Selmeczi Kovács, Attila 1990. Querns. Historical Layers—Technical Regression. *Ethnologia Europaea* 20:35–46, Lingby.
Selmeczi Kovács, Attila 1999. *Kézimalmok a Kàrpàtmedencében: eszköztörténeti monográfia*. Agroinform, Budapest. [Includes millstones, mills, and mill work in Carpathian Mountains region of Hungary; summary in English.]
Servelle, Christian 2006. Les Meulières Antiques de la Marèze (Saint-Martin-Laguépie et le Riols, Tarn, France): Géologie, Géomorphologie, Techniques d'Exploitation et de Façonnage ["The Antique Millstone Quarries of Marèze (Saint-Martin-Laguépie and Riols-Tarn, France): Geology, Geomorphology, Techniques of Exploitation and Shaping."]. In *Mühlsteinbrüche. Erforschung, Schutz und Inwertsetzung eines Kulturerbes europäischer Industrie (Antike–21. Jahrhundert)*, edited by Alain Belmont and Fritz Mangartz, pp. 61–69. Römisch-Germanisches Zentralmuseum, Tagungen Band 2, Mainz, Germany.
Shaffrey, Ruth 2003. The Rotary Querns from the Society of Antiquaries' Excavations at Silchester, 1890–1909. *Britannia* 34:143–174.
Shaffrey, Ruth 2006a. *Grinding and Milling: A Study of Romano-British Rotary Querns and Millstones Made from Old Red Sandstone.* BAR British Series. Archaeopress, Oxford.
Shaffrey, Ruth 2006b. Review of Recent Finds in Southern England. *Quern Study Group Newsletter* 7:5–6. Oxford, England. [This brief review presents a listing of beehive querns from 12 counties and information on 17 querns from Fairfield Park, an early to middle Iron Age site, in Bedfordshire.]
Shear, Hazel M. 1960. *The Willing Story, 1795–1850.* H. M. Shear, Wellsville, New York.
Sheldon, Hezekiah Spencer 1879. *Documentary History of Suffield: In the Colony and Province of Massachusetts Bay in New England, 1660–1749.* C. W. Bryan Company, Springfield, Massachusetts.
Shepard, Charles Upham 1837. *Report on the Geological Survey of Connecticut.* B. L. Hamlen, New Haven.
Siegert, J. 1921. Die Basaltlavaindustrie von Mayen und Umgebung. Dissertation, Würzburg. [Germany]
Sigaut, François 2002. Sortir de l'oubli. In *Les Meuliers. Meules et Pierres Meulières* by Agapain, pp. 9–20. Presses du Village, Étrépilly, France. [France]

Sitte, Josef 1954. *Der Junge Vulkanismus der Mühlsteinbrüche von Jonsdorf bei Zittau.* Ein Erdgeschichtlicher Führer Durch das Naturschutzgebiet. Jena, Urania. [Millstone industry at Johnsdorf, Germany.]

Skinner, Raymond J. 1996. Pen Pits and Sir Arthur Bliss. *Wiltshire Archaeology Magazine* 89:139–143.

Small, Alsstair M., and Robert J. Bruce 1994. V. Volterra [millstones]. In *The Excavations of San Giovanni di Ruoti*, edited by Alsstair M. Small and Robert J. Bruce. University of Toronto Press, Toronto.

Smerdel, Inja 2003. Notre Mère, Donnez-nous Aujourd'hui Notre Pain Quotiden!: De la Fabrication, de la Vente et de l'Emploi du Moulin à Bras dans les Régions Reculées de la Slovénie. In *Meules à Grains: Actes du Colloque International de La Ferté-sous-Jouarre 16–19 Mai 2002*, edited by Mouette Barboff, François Sigaut, Cozette Griffin-Kremer, and Robert Kremer, pp. 125–148. Éditions Ibis Press and Éditions de la Maison des Sciences de l'Homme, Paris, France. [Study of making, selling, and using hand querns in remotes areas of Slovenia.]

Smerdel, Inja 2006. Les Pierres, les Hommes et les Boeufs: Mémoires Sur le Travail dans Certaines Carrières de Pierres à Moulins en Slovénie ["Stones, Men and Oxen. Memories on the Work in Some Millstone Quarries in Slovenia."]. In *Mühlsteinbrüche. Erforschung, Schutz und Inwertsetzung eines Kulturerbes europäischer Industrie (Antike–21. Jahrhundert)*, edited by Alain Belmont and Fritz Mangartz, pp. 115–127. Römisch-Germanisches Zentralmuseum, Tagungen Band 2, Mainz, Germany.

Smerdel, Inja [Slovene Ethnographic Museum, Ljubljana, Slovenia] 2007. E-mail to Charles D. Hockensmith, December 20. [Millstone quarries in Slovenia.]

Smith, D. 1978. The Lava Querns. In "The Graveney Boat" edited by V. Fenwick. *British Archaeological Reports* 53: 131–132, Oxford. [Germany]

Smith, Dwight L., and Ray Swick (editors) 1997. *A Journey Through the West: Thomas Rodney's 1803 Journal from Delaware to the Mississippi Territory.* Ohio University Press, Athens. Excerpts reprinted with permission of Ohio University Press, Athens, Ohio (www.ohioswallow.com).

Smith, Eugene A. 1924. Abrasives. In *Statistics of the Mineral Production of Alabama for 1922* by Walter S. Ernest, Geological Survey of Alabama, Bulletin 27.

Smith, H. P. 1884. *History of the City of Buffalo and Erie County: With Illustrations and Biographical Sketches of Some of Its Prominent Men and Pioneers.* D. Mason and Company, Syracuse, New York.

Smith, R. F. 1868. *Doniphan County, Kansas, History and Directory for 1868–9: Containing the State Constitution, a Concise History of Kansas, Also of Doniphan County, and Sketches of Each Village in the County, Citizens' and Business Directory, Revenue and Postal Information, Legal Forms, etc., etc.* Smith, Vaughn & Co.

Smith, William, Jr. 1972. *The History of the Province of New-York from the First Discovery to the Year 1732.* Volume One, edited by Michael Kammen. The Belknap Press of Harvard University Press, Cambridge, Massachusetts.

Snyder, Bradley 1981. *The Shawangunk Mountains: A History of Nature and Man.* Mohonk Press, New Paltz, New York.

Sopko, Joseph 1991. Memorandum to Charles Florance, January 28, 1991. New York State Office of Parks, Recreation, and Historic Preservation, Waterford, New York. Copy on file at Kentucky Heritage Council, Frankfort. [Ulster County millstone industry.]

Southern Publishing Company 1868. *Edward's Fourth Annual Directory of the Inhabitants, Institutions, Incorporated Companies, Manufacturing Establishments, Business, Business Firms, etc., etc. in the City of Louisville for 1868–9.* Southern Publishing Company, Louisville.

Southern Publishing Company 1869. *Edward's Fifth Annual Directory of the Inhabitants, Institutions, Incorporated Companies, Manufacturing Establishments, Business, Business Firms, etc., etc. in the City of Louisville for 1869.* Southern Publishing Company, Louisville.

Spafford, Horatio Gates 1813. *A Gazetteer of the State of New-York;....* H. C. Southwick, Albany.

Spain, R. J. 1986. Millstones from Barton Court Farm. In *Archaeology at Barton Court Farm, Abingdon, Oxon: An Investigation of Late Neolithic, Iron Age, Romano-British and Saxon Settlements* by David Miles. Oxford Archaeological Unit and the Council for British Archaeology, Oxford.

Spence, Sheila 1988. The Quarryman in Sandwick. *Old Orkney Trades,* edited by Sheila Spence, pp. 33–39. Orkney Press, Kirkwall. [Millstone making on Orkney.]

Spiteri, Danielle 2002. Exploitants, Fabricants et Marchands. In *Les Meuliers. Meules et Pierres Meulières* by Agapain, pp. 179–184. Presses du Village, Étrépilly, France. [France]

Spofford, Jeremiah 1828. *A Gazetteer of Massachusetts: Containing a General View of the State, with an Historical Sketch of the Principal Events from Its Settlement to the Present Time, and Notices of Several Towns Alphabetically Arranged.* C. Whipple, Newburyport, Massachusetts.

Sprague, Mary Gabrielle 1979. The Life Cycle of a Quern: An Analysis of the Grindstones from the Neolithic Site of Selevac, Yugoslavia. Senior honors thesis, Department of Anthropology, Harvard University, Cambridge, Massachusetts.

Stanley, Philip R. 1975. The Querns from South Cadbury. Undergraduate dissertation, Department of Archaeology, University College, Cardiff, Wales. [Wales]

Sterrett, Douglas B. 1907. Abrasive Materials. In *Mineral Resources of the United States, Calendar Year 1906*, pp. 1043–1054. Part II — Nonmetals. Department of the Interior, U.S. Bureau of Mines, Government Printing Office, Washington, D.C.

Stewart, I. J. 1994. Querns in a Romano-British Villa Estate at Mantles Green, Amersham, Buckinghamshire. *Records of Buckinghamshire* 34:172.

Still, Bayrd 1948. *Milwaukee: The History of a City.* State Historical Society of Wisconsin, Madison.

Stirling, M. 1958. An Upper Quernstone from Perthshire, Near Bridge of Allan. *Proceedings of the Society of Antiquaries of Scotland* 91:187–188.

Stonehill College n.d. *The King Philip History Trail.* Outer Area Restoration Project. Stonehill College, 1995–1996, Easton, Massachusetts.

Stoner, Jacob 1947. Old Millstones, May 31, 1934. In *Historical Papers, Franklin County and the Cumberland Valley, Pennsylvania* by Jacob H. Stoner and Lu Cole Stoner, pp. 411–430. Craft Press, Chambersburg, Pennsylvania.

Storck, John, and Walter D. Teague 1952. *Flour for Man's Bread: A History of Milling.* University of Minnesota Press, Minneapolis.

Stoyel, A. 1967. Millstones in the North-East. *Industrial Archaeology Group for North-East*, No. 2. [England]

Stoyel, Alan 1992. Medieval Millstones — The Clockwise Theory and Its Applications. *Wind & Watermill Section Newsletter* 50:18–24.

Strawhacker, William 2004. E-mails to Charles D. Hockensmith, January 19, January 30, February 7. [Involvement of the Strohecker family of Berks County, Pennsylvania, in millstone making.]

Sullivan, George M. 1891. *Report on the Geology of Parts of Jackson and Rockcastle Counties.* Kentucky Geological Survey, Series 2, 20 pages, Frankfort.

Summer, William M. 1976. A Typology of Ancient Middle Eastern Saddle Querns. Unpublished M. A. thesis, Department of Anthropology, University of Pennsylvania, Philadelphia.

Sussenbach, Tom 1990. Personal communication to Charles D. Hockensmith, December 7. [Concerning report of millstone quarry on Indian Creek in McCreary County, Kentucky.]

Sussenbach, Tom 1991. Personal communication to Charles D. Hockensmith. [Concerning report of millstone quarry near Cumberland Falls in Whitley County, Kentucky.]

Swain, Frank K. 1917. Hand Corn Mill at Georgetown, South Carolina. *A Collection of Papers Read Before the Bucks County Historical Society,* Volume 4:735–739. The Chemical Publishing Company, Easton, PA.

Swallow, G. C. 1855. *The First and Second Annual Reports of the Geological Survey of Missouri.* By order of the Legislature. James Lusk, Jefferson City.

Swift, Michael 1988. The Millstone Industry. *The Accordian* 2 (2):6–7, Rochester, New York.

Swisher, Jacob A. 1940. *Iowa, Land of Many Mills.* The State Historical Society of Iowa, Iowa City.

Taché, J. C. 1855. *Sketch of Canada, Its Industrial Condition and Resources.* Hector Bossange and Sons, Paris.

Takaoğlu, Turan 2005. Coşkuntepe: An Early Neolithic Quern Production Site in NW Turkey. *Journal of Field Archaeology* 30 (4):419–433.

Talbert, Charles G. 1962. *Benjamin Logan, Kentucky Frontiersman.* University of Kentucky Press, Lexington.

Tchilakadze, M. 1985. Les Meules à Main, Atlas Historico Ethnographique de la Géorgie, éd. Sciences, Tbilissi. pp. 113–122.

Telford, Thomas 1938. On Mills. *The Newcomen Society for the Study of the History of Engineering and Technology, Transactions,* Volume 17, 1936–1937, London.

The Argus of Western America 1812. Ad for Samuel Taylor's millstone quarry in Rockcastle County, Kentucky. *The Argus of Western America,* January 8, Frankfort, Kentucky.

The Argus of Western America 1813. Ad for Charles Colyer's millstone quarry in Rockcastle County, Kentucky. *The Argus of Western America,* October 9, Frankfort, Kentucky.

The Argus of Western America 1821a. Ad for Miller, Railsback & Miller's millstone quarry in Franklin County, Kentucky. *The Argus of Western America,* August 9, Frankfort, Kentucky.

The Argus of Western America 1821b. Ad for Jeremiah Buckley's millstone quarry in Franklin County, Kentucky. *The Argus of Western America,* November 8, Frankfort, Kentucky.

The Augusta Chronicle and State Messenger 1792. Ad for John Murray's Millstone Quarry in Burke County, Georgia. *Chronicle and State Messenger,* August 15, Augusta, Georgia.

The Eagle 1825. Ad for Laurel-Hill Millstones from Pennsylvania. *The Eagle,* April 13, Maysville, Kentucky.

The Maysville Eagle 1828. Ad for Morris & Egenton French Burr Millstone Manufactory of Baltimore for a branch office in Cincinnati, Ohio. *The Maysville Eagle,* January 23, Maysville, Kentucky.

The Maysville Eagle 1834. Ad for Raccoon Burr Millstones for sale. *The Maysville Eagle,* November 20, Maysville, Kentucky.

The Miller 1878. French Millstones — A Warm Dispute. *The Miller* 4:646.

The Morning Star 1887. Comment Concerning the Moore County, North Carolina, millstone industry. *The Morning Star,* April 29, Wilmington, North Carolina.

The New York Times 1883. Was There Foul Play! *The New York Times,* August 13, page 8, New York. On ProQuest website. [Concerning the death of Peter Nelson, a employee of the millstone quarry at New London, Connecticut.]

The New York Times 1900a. Strikers Held in Check by Troops, Calvalry and Infantry Control Situation at Cornell Dam.... *The New York Times,* April 18, page 1, New York. On ProQuest website. [Concerning National Guard's protection of old Croton Dam and the millstone quarries near Peekskill, New York.]

The New York Times 1900b. Contractors May End Big Strike. Announce Their Willingness to Increase Hand-Drillers' Pay. Strikers' View of Offer.... *The New York Times,* April 19, page 2, New York. On ProQuest website. [Concerning National Guard's protection of old Croton Dam and the millstone quarries near Peekskill, New York.]

The New York Times 1900c. Troops Back from Croton. Only a Few Cavalrymen Left to Guard Cornell Dam. No More Trouble Expected... *The New York Times,* April 25, page 9, New York. On ProQuest website. [Concerning National Guard protecting old Croton Dam and the millstone quarries near Peekskill, New York.]

The North American Review 1839. Missouri. *The North American Review* 48 (103):514–527, Cedar Falls, Iowa.

The Palladium 1800. Ad for John Tanner's millstone quarry in Woodford County, Kentucky. *The Palladium,* February 27, Frankfort, Kentucky.

The Public Advisor 1827. Ad for sale of French Burr Millstones in Louisville. *The Public Advisor,* May 15, Louisville, Kentucky.

Thue, Johs. B. 2000. *Livets Steinar. Produksjon og Eksport av Kvernstein frå Hyllestad i Mellomalderen.* Leikander, SKALD AS, 80 pages. [Millstone quarry near Hyllestad, Norway.]

Tomalin, D. J. 1977. The Quernstones. In "Brooklands, Weybridge: The Excavation of an Iron Age and Medieval Site 1964–65 and 1970–71" by R. Hansworth and D. J. Tomalin. *Surrey Archaeological Society Collections* 4:81–85.

Tomlinson, Tom D. 1981. *Querns, Millstones, and Grindstones Made in Hathersage & District.* Hathersage Parochial Church Council, Sheffield. 20 pages. [England]

Trask, John B. 1854. *Report on the Geology of the Coast Mountains, and Part of the Sierra Nevada: Embracing Their Industrial Resources in Agriculture and Mining.* Geological Survey of California. B. B. Redding, Sacramento.

Triboulet, C., L. Langouet, and C. Bizien-Jaglin 1996. Recherches sur les Origins de Meules Gallo-Romaines dans le Nord de la Haute-Bretagne. *Les Dossiers du Ce. R. A. A.* 24: 39–47.

Truax, J. W. 1896. *The Eagle Mill Pick.* J. W. Truax Firm, Essex Junction, Vermont.

Tucci, P, E. Azzaro, P. Morbidelli, G. Agostini, and V. Misiti n.d. Pompeii (Naples, Italty): The Nature and Provenance of the Lavas Used for Making Flour Millstones. The British Library Document Supply Centre, Paper CN054354846.

Tucker, D. Gordon 1971. Millstone Making at Penallt, Monmouthshire. *Industrial Archaeology* 8:229–239. [Wales]

Tucker, D. Gordon 1973. Millstone Making in Gloucestershire: Wm. Gardner's Gloucester Millstone Manufactory with a Note on Hudsons of Penallt and Redbrook. *Gloucestershire Society for Industrial Archaeology Journal*, pp. 6–16. [Wales]

Tucker, D. Gordon 1977. Millstones, Quarries and Millstone Makers. *Post-Medieval Archaeology* 11:1–21. [England]

Tucker, D. Gordon 1980a. Millstone Making in Anglesey. *Wind and Water Mills* 1:16–23. [Wales]

Tucker, D. Gordon 1980b. Millstones. *S.P.A.B. Wind & Water Mills Section Newsletter*, Volume 5, October. [England]

Tucker, D. Gordon 1982a. Millstones North and South of the Scottish Border. *Industrial Archaeology Review* 6 (3):186–193. [Scotland]

Tucker, D. Gordon 1982b. Millstone Making in France: When Éperson Produced Millstones. *Wind and Water Mills* 3, 32 pages. [Translation of article published in French]

Tucker, D. Gordon 1984a. The Dressing of Millstones: English Practice as Described by Bryan Corcoran in 1882. *Wind and Water Mills* 5:24–26. [England]

Tucker, D. Gordon 1984b. Millstone Making in Scotland. *Proceedings of the Society of Antiquaries of Scotland* 114:539–556. [Scotland]

Tucker, D. Gordon 1985. Millstone Making in the Peak District of Derbyshire: The Quarries and the Technology. *Industrial Archaeology Review* 8 (1):42–58. [England]

Tucker, D. Gordon 1987. Millstone Making in England. *Industrial Archaeology Review* 9 (2):167–187. [England]

Tucker, D. Gordon 1988. A 14th Century Millstone Transaction. *Wind and Water Mills* 8:18–20. [England]

Tuomey, M. 1848. *Report on the Geology of South Carolina.* A. S. Johnston, Columbia, South Carolina.

Tuttle, Charles R. 1875. *An Illustrated History of the State of Wisconsin.* B. B. Russell & Co., Madison, Wisconsin.

Utter, William T. 1942. *The Frontier State, 1803–1825.* Ohio Historical Society, Columbus, Ohio.

Valero, J. F. 1983. Enquéte sur les Meules en Grès de Facture Gallo-Romaine. *Archéologia* 175:60. [France]

Valero, J. F. 1984. Note Préliminaire sur l'Antelier de Taille de Meules Antiques de "La Marèze" à Saint-Martin-Laguépie (Tarn). *Bulletin de la Société des Sciences, Arts et Belles-Lettres du Tarn*, N^elle^ série, années 1981–1982, 1984, No. 37:729–736. [Millstone quarries in southern France near Tarn.]

Vincent, Francis 1870. *A History of the State of Delaware, from Its First Settlement Until the Present Time, Containing a Full Account of the First Dutch and Swedish Settlements, with a Description of Its Geography and Geology.* John Campbell, Philadelphia.

Volterra, V., and R. G. V. Hancock 1984. Provenancing of Ancient Roman Millstones. *Journal of Radioanalytical and Nuclear Chemistry* 180 (1):37–44. [Study of millstones from two Roman villa sites in southern Italy.]

Vrettou-Souli, Margarita 2002. *H Milopetra tis Milou. Apo tin exorixi stin emporiki a diakinisi.* S&B Industrial Minerals S.A., Athens. ["*The Millstone of Milos. From Mining to Commercial Circulation*" at Milos, Greece]

Vrettou-Souli, Margarita 2007. E-mail to Charles D. Hockensmith, August 12. [Detailed English summary of her study of the Greek millstone industry on the Island of Milos.]

Wailes, B. L. C. 1854. *Report on the Agriculture and Geology of Mississippi, Embracing a Sketch of the Social and Natural History of the State.* Philadelphia.

Wallcut, Thomas 1879. *Journal of Thomas Wallcut, in 1790.* University Press. J. Wilson and Son, Cambridge, Massachusetts.

Wallis, F. S. 1988. Petrology of Nineteenth Century Millstones in the Wells Area, Somerset. *1987/1988 Annual Report, Wells Natural History and Archaeological Society*, pp. 9–11.

Walton, James 1953. Pestles, Mullers and Querns from the Orange Free State and Basutoland. *South African Archaeological Bulletin* 8 (3):32–39. [In Lesotho in southeast Africa.]

Walton, James 1997. Old Yorkshire Stone Crafts. *Folk Life* 35:78–90. [Millstone production]

Ward, Owen H. 1982a. Millstones from La Ferté-sous-Jouarre, France. *Industrial Archaeology Review* 6 (3):205–210. [France]

Ward, Owen H. 1982b. French Millstones. *Wind and Water Mills* 3:36–43. [France]

Ward, Owen H. 1984a. The Making and Dressing of French-Burr Millstones in France in 1903. *Wind and Water Mills* 5:27–32. [France]

Ward, Owen H. 1984b. The Slaughter of the French Millstone-makers. *BIAS Journal* 17:30–31. [France]

Ward, Owen H. 1985. British Burrstones, 1799–1821 *Melin* 1:33–47. [England]

Ward, Owen H. 1986a. French Burrstones for America and the Siege of Paris. *Old Mill News* 14 (12):12–14. [France]

Ward, Owen H. 1986b. The French Millstone Story: The Emergence of the Société Générale Meulière. *Proceedings of the Third & Fourth Research Conferences*, pp. 39–41. The Mills Research Group, Manningtree, England. [France]

Ward, Owen H. 1888. Millstone Extraction in Normandy, France. *Bulletin Subterranea Britannica* 24:17–18. [France]

Ward, Owen H. 1990a. Welsh Millstones. *Melin* 6:15–40. [Wales]

Ward, Owen H. 1990b. English Millstones, the Seven Years War, and the Society for the Encouragement of Arts, Manufactures and Commerce. *Proceedings of the 5th and 6th Conferences*, pp. 3–5. The Mills Research Group, Manningtree, England. [England]

Ward, Owen H. 1992a. Millstones of La Ferté-sous-Jouarre: The Dechan Report of 1796. *Proceedings of the Seventh & Eighth Research Conferences*, pp. 8–12. The Mills Research Group, Manningtree, England. [France]

Ward, Owen H. 1992b. The Millstones of Sacrewell. *Proceedings of the Seventh & Eighth Research Conferences*, pp. 50–51. The Mills Research Group, Manningtree, England. [England]

Ward, Owen H. 1992c. Millstone Makers in London. *Proceedings of the Ninth & Tenth Research Conferences*, pp. 29–31. The Mills Research Group, Manningtree, England. [England]

Ward, Owen H. 1993a. *French Millstones: Notes on the Millstone Industry at La Ferté-sous-Jouarre.* The International Molinological Society. [France]

Ward, Owen H. 1993b. Millstones at Willsbridge Mill. *Bristol Industrial Archaeological Society Journal* 25:28–32. [Avon County, England]

Ward, Owen H. 1994. The Brits in Paris 1878: British Millstone Makers at the Paris Universal Exhibition. *Proceedings of the Twelfth Research Conference*, pp. 1–6. The Mills Research Group, Manningtree, England. [England]

Ward, Owen H. 1995a. The Survey of Flour Mills in France 31 Sec 1808: "Where Do You Get Your Millstones?" Some Problems of Interpretation. *The International Molinological Society 8th Symposium Transactions*, pp. 75–80. [France]

Ward, Owen H. 1995b. Balancing Burrstones. *Melin* 11:8–19. [England]

Ward, Owen H. 1996a. French Millstones for King's Lynn. *International Molinology* 53:13–17. [France]

Ward, Owen H. 1996b. The King's Lynn Millstone Accounts. *Proceedings of the Fourteenth Research Conference*, pp. 1–11. The Mills Research Group, Manningtree, England. [France]

Ward, Owen H. 1997. The 1808/9 Survey of Corn Mills for the Department of Indre-et-Loire (37). *International Molinology* 55:15–17. [France]

Ward, Owen H. 1998. French Millstones Made-to-Order. *Proceedings of the Sixteenth Research Conference*, pp. 59–63. The Mills Research Group, Manningtree, England. [France]

Ward, Owen H. 2002a. The Millstone Industry of Houlbec-Cocherel (Eure). *International Molinology* 64:23–25, Holland. [France]

Ward, Owen H. 2002b. French Millstones (les Meules Françaises). In *Les Meuliers. Meules et Pierres Meulières* by Agapain, pp. 21–34. Presses du Village, Étrépilly, France. [France]

Ward, Owen H. 2003. Millstones from France: The Survey of 1808/1809. In *Meules à Grains: Actes du Colloque International de La Ferté-sous-Jouarre 16–19 Mai 2002*, edited by Mouette Barboff, François Sigaut, Cozette Griffin-Kremer, and Robert Kremer, pp. 267–280. Éditions Ibis Press and Éditions de la Maison des Sciences de l'Homme, Paris, France. [France]

Ward, Owen H. 2005. Review of "Carriers et Meuliers de Région d'Epernon" by Jean-Paul Duc. *International Molinology* 71:43, England.

Ward, Owen H. 2007a. Millstone Matters — Catching Up. *International Molinology* 75:52–54, England. [Provides commentary on sources of millstones from archival sources in France.]

Ward, Owen H. 2007b. Review of "Millstone Quarries. Research, Protection and Valorization of a European Heritage" by A. Belmont and F. Mangartz. *International Molinology* 75:56–57, England.

Watson, Thomas 1907. *Mineral Resources of Virginia*. J. P. Bell, Lynchburg, Virginia.

Watts, Martin 1986. The Tools of the Millstone Dresser. *Tools and Trades Historical Society Newsletter* 12: 9 [Comments on Major 1985.]

Watts, Martin 2002. *The Archaeology of Mills and Milling*. Tempus, Stroud, Gloucestershire [Includes sections on querns and millstones from Roman to Post-Medieval periods in England.]

Watts, Sue 1996. The Rotary Quern in Wales (Part 1). *Melin* 12:26–35. [Wales]

Watts, Sue 1997. The Rotary Quern in Wales (Part II). *Melin* 13:47–54. The Welsh Mills Society, Aberystwyth.

Watts, Sue 2000. Grinding Stones. In *The South Manor Area. Wharram. A Study of Settlement on the Yorkshire Wolds VIII*, by Paul A. Stamper and R. A. Croft, pp. 111–115. York University Archaeological Publications 10, York.

Watts, Sue 2003a. *The Form and Function of Querns and Mortars in Iron Age and Roman Cornwall*. Unpublished undergraduate dissertation, Department of Archaeology, University of Exeter.

Watts, Sue 2003b. The Longis Querns Uncovered. *Alderney Society Bulletin* XXXVIII:67–75. [Querns from Alderney, the Channel Island in the English Channel.]

Watts, Sue 2006. The Longis Querns Uncovered. *Quern Study Group Newsletter* 7:6–8. Oxford, England. [Describes the discovery of 14 broken and unfinished rotary querns on Longis beach on Alderney, the Channel Island in the English Channel.]

Weaver, Valerie 1995. The Lawrence Brothers Share 80+ Years of Saint Josen Memories. *The Accordian* 9 (2):1–5, Rochester, New York.

Webb, William S. 1933. The Millstone as an Antique. *Kentucky School Journal*, March, pp. 30–34.

Webb, William S. 1935. Old Millstones of Kentucky. *The Filson Club History Quarterly* 9 (4):209–221, Louisville.

Wefers, Stefanie 2006. Latènezeitliche Handdrehmühlen im Nordmainischen Hessen. In *Mühlsteinbrüche. Erforschung, Schutz und Inwertsetzung eines Kulturerbes europäischer Industrie (Antike–21. Jahrhundert)*, edited by Alain Belmont and Fritz Mangartz, pp. 15–24. Römisch-Germanisches Zentralmuseum, Tagungen Band 2, Mainz, Germany. ["La Tène Period Rotary Querns from Hessen North of the River Main" in Germany.]

Weidmann, D. 2002. Chronique Archéologique 2001, Châvannes-le-Chêne. *Revue Historique Vaudoise* 2002, pp. 115–116. [Châvannes-le-Chêne is a Roman era quern quarry in western Switzerland.]

Weise, Arthur James 1891. *Troy's One Hundred Years: 1789–1889*. W. H. Young, Troy, New York.

Welfare, A. 1981. The Milling Stones. In *Whitton, an Iron Age and Roman Farmstead in South Glamorgan*, edited by Michael G. Jarrett and Stuart Wrathmell. University of Wales, Cardiff. [Wales]

Welfare, A. T. 1985. The Milling Stones. In *The Roman Fort of Vindolanda at Chesterholm, Northumberland*, by P.T. Bidwell, pp. 154–164. Historic Buildings and Monuments Commission for England, London.

Wetmore, Alphonso 1837. *Gazetteer of the State of Missouri: With a Map of the State, from the Office of the Surveyor-General, Including the Latest Additions and Surveys: To Which is Added an Appendix, Containing Frontier Sketches, and Illustrations of Indian Character*. C. Keemle, St. Louis.

White, D. 1963. A Survey of Millstones from Morgantina. *American Journal of Archaeology* 67:199–206. [Discusses lava millstones from Morgantina, Austria, dating between the Greek and Roman periods.]

Wilkie, Aitken 1874. The Only True and Practical Way to Make Edge Tools: Adapted for Edge Tool-Makers, Millers for Mill-Picks, Blacksmiths, and All Kinds of Tool-Dressers and Steelworkers. Blade Printing and Paper Company, Toledo, Ohio.

Willcox, G. 2002. Charred Plant Remains from a 10th Millennium B. P. Kitchen at Jerf el Ahmar (Syria). *Vegetation History and Archaeobotany* 11 (1–2):55–60. [Includes a discussion of saddle querns.]

Williams, Albert, Jr. 1885. Abrasive Materials. In *Mineral Resources of the United States, Calendar Years 1883–1884*, pp. 581–594. Part II — Nonmetals. U.S. Bureau of Mines, U.S. Geological Survey, Government Printing Office, Washington, D.C.

Williams, David, and David Peacock 2006. Roman Querns and Mills in the Red Sea Area. In *Mühlsteinbrüche. Er-*

forschung, Schutz und Inwertsetzung eines Kulturerbes europäischer Industrie (Antike–21. Jahrhundert), edited by Alain Belmont and Fritz Mangartz, pp. 35–40. Römisch-Germanisches Zentralmuseum, Tagungen Band 2, Mainz, Germany.

Williams, Joseph S. 1873. *Old Times in West Tennessee*. W. G. Cheeney, Memphis, Tennessee.

Williams & Company 1861. *Williams' Covington Directory*. Williams & Company, Cincinnati, Ohio.

Williams-Thorpe, Olwen 1986. Exploitation of Lithic Resources — Quernstones in the Roman Mediterranean. 14 pages plus figures. In *The Social and Economic Contexts of Technological Change*, edited by S. van der Leeuw and R. Torrence. The World Archaeological Congress, Southampton and London, in association with Allen and Unwin (London).

Williams-Thorpe, Olwen 1988. Provenancing and Archaeology of Roman Millstones from the Mediterranean Area. *Journal of Archaeological Science* 15:253–305.

Williams-Thorpe, Olwen 1994. Geological Provenancing of Millstone MJ 89–40. In *Deep Water Archaeology: A Late-Roman Ship from Carthage and an Ancient Trade Route Near Skerki Bank off Northwest Sicily*, edited by A. M. McCann and J. Freed, pp. 122–126. *Journal of Roman Archaeology*, Supplementary Series Number 13, Ann Arbor, Michigan.

Williams-Thorpe, Olwen 1995. Comments in 'Objects of Stone.' In *Tattenhoe and Westbury* by R. Ivens, P. Busby, and N. Sheperd, pp. 312–313. The Buckinghamshire Archaeological Society Monograph Series No. 8, Buckinghamshire Archaeological Society, Aylesbury, UK.

Williams-Thorpe, Olwen, and David Peacock 1995. The Quernstone Remains. In *The Biferno Valley Survey*, edited by G. Barker, pp. 141–142. Leicester University Press, Leicester.

Williams-Thorpe, Olwen, and Richard S. Thorpe 1987. Els Origens Geologics dels Molins Roman de Pedra del Nord-Est de Catalunya il'us de la Reglo Volcanica d'Olot. *Vitrina* 2:49–58.

Williams-Thorpe, Olwen, and Richard S. Thorpe 1988. The Provenance of Donkey Mills from Roman Britain. *Archaeometry* 30:275–289. [England]

Williams-Thorpe, Olwen, and Richard S. Thorpe 1989. Provenancing and Archaeology of Roman Millstones from Sardinia (Italy). *Oxford Journal of Archaeology* 8:89–117.

Williams-Thorpe, Olwen, and Richard S. Thorpe 1990. Millstone Provenancing Used in Tracing the Route of a 4th Century BC Greek Merchant Ship. *Archaeometry* 32:115–137.

Williams-Thorpe, Olwen, and Richard S. Thorpe 1991. Millstones that Mapped the Mediterranean. *New Scientist* 129 (1757):42–45. [Transportation of millstones by the Romans from production centers in Spain, France, North Africa, Italy, and Levant.]

Williams-Thorpe, Olwen, and Richard S. Thorpe 1991. The Import of Roman Millstones to Mallorca. *Journal of Roman Archaeology* 4:152–159.

Williams-Thorpe, Olwen, and Richard S. Thorpe 1993. Geochemistry and Trade of Eastern Mediterranean Millstones from the Neolithic to Roman Periods. *Journal of Archaeological Science* 20 (3):263–320.

Williams-Thorpe, Olwen, and Richard S. Thorpe 2004. Appendix I: Comparison of Millstone Samples from the Serce Limani Shipwreck with Possible Source Rocks from Rema. In *Serce Limani: An Eleventh-Century Shipwreck, Volume I: The Ship and Its Anchorage, Crew, and Passangers*, edited by George F. Bass, Sheila D. Matthews, J. Richard Steffy, and Frederick H. van Doorminck, pp. 256–261. Texas A & M University Press, College Station.

Williams-Thorpe, Olwen, Richard S. Thorpe, C. Elliott, and C. Xenophontos 1991. Archeology, Geochemistry and Trade of Igneous Rock Millstones in Cyprus during the Late Bronze Age to Roman Periods. *Geoarchaeology* 6:27–60.

Williams-Thorpe, Olwen, J. S. Watson, and P. C. Webb 1992. Report on the Geological Provenance of Rotary Quern 33 from Totternhoe Roman Villa. In *A Roman Villa at Totternhoe* by C. L. Matthews, J. Schneider, and B. Horne. *Bedfordshire Archaeology* 20:93–94.

Wilson, Catherine 2003. An Unusual Millstone at Heckington Windmill, Lincolnshire. In *Meules à Grains: Actes du Colloque International de La Ferté-sous-Jouarre 16–19 Mai 2002*, edited by Mouette Barboff, François Sigaut, Cozette Griffin-Kremer, and Robert Kremer, pp. 334–336. Éditions Ibis Press and Éditions de la Maison des Sciences de l'Homme, Paris, France. [England]

Winwood, H. H. 1884. The Results of Further Excavations at Pen Pits. *Proceedings of the Somerset Archaeological and Natural History Society* 30:149–152.

Wisdom, L. 1981. A Petrological Typological and Distribution Study of Quernstones Quarried from the Quartz Conglomerate of the Forest of Dean. Unpublished BA thesis, England. [England]

Worsham, Gibson 1986a. *Montgomery County Historic Sites Survey, Volume 1*. Unpublished report submitted to the Montgomery County Planning Commission and the Virginia Division of Historic Landmarks, Richmond.

Worsham, Gibson 1986b. *Montgomery County Historic Sites Survey, Volume 2*. Unpublished report submitted to the Montgomery County Planning Commission and the Virginia Division of Historic Landmarks, Richmond.

Worth, Henry B. 1902. *Nantucket Lands and Land Owners*. Nantucket Historical Association, Nantucket, Massachusetts.

Wright, M. Elizabeth 1988. Beehive Quern Manufacture in the South-East Pennines. *Scottish Archaeological Review* 5:65–77

Wright, M. Elizabeth 2002. Querns and Millstones. In *Cataractonium: Roman Catterick and Its Hinterland. Excavations and Research, 1958 1997, II*, by P. R. Wilson, pp. 267–280. Council for British Archaeology Research Report 129, York.

Young, Alexander 1846. *Chronicles of the First Planters of the Colony of Massachusetts Bay, from 1623 to 1636*. C. C. Little and J. Brown, Boston.

Young, Andrew W. 1875. *History of Chautauqua County, New York, from Its First Settlement to the Present Time; with Numerous Biographical and Family Sketches*. Matthews & Warren, Buffalo.

Zanesville, Ohio, Board of Trade 1874. *The Agricultural and Mineral Resources of Muskingum County*. Zanesville, Ohio, Board of Trade, Zanesville.

Zerfass, Samuel G. 1921. *Souvenir Book of the Ephrata Cloister: Complete History from Its Settlement in 1728 to the Present Time: Included is the Organization of Ephrata Borough and Other Information of Ephrata Connected with the Cloister*. John G. Zook, Lititz, Pennsylvania.

Zirkl, E. J. 1954. Mikroskopische Untersuchungen über die Herkunft der Basaltischen Rohstoffe einiger latènezeitlicher Mühlstein aus Wien. *Mikroskopie* 9: 95–109. [Millstones in Austria.]

Zirkl, E. J. 1955. Zur Herkunft der Rohstoffe einiger latènezeitlicher Handmühlen. *Archeologica Austriaca* 18: 90–92. [Millstones in Austria.]

Zirkl, E. J. 1963. Über die Herkunft eines römischen Mühlsteines von Magdalensburg in Kärnten. *Carinthia I* 153:287–290. [Petrological study of Roman millstones from Magdalensburg in Carinthia, Austria.]

Index

Abbeville, South Carolina 60
Accord, New York x, xiii, 36, 38, 39, 41, 102, 103
Accord Station 41
The Accordian x, xiii
Ackworth, New Hampshire 79
Adams, Joseph 67
Adams, Spencer 24–25, 109
Addis, W. C. Stone Company 103
Addis, Wilson C. 102
Addis Stone Co., Wilson C. 103
Addison, Jon 217
Aegen Sea 167
Aegina Island Quarry 166, 169, 197
Åfjorden, Norway 173
Agapain 7, 217
Agard, Patrick 217
Agostini, G. 218, 248
Agustoni, Clara 218
Aigina, Greece 167
Akron, Ohio 82
Alabama 19, 20, 21, 54, 60, 61–62, 65, 77–78, 80, 86, 101, 109, 110–111, 115
Alaimo, R. 228
Alaska viii
Albania 4, 125, 212
Albany, New York ix, xiii, 32, 33, 102, 112
Albany, Oregon viii
Alderney 162
Aldsworth, F. G. 217
Alfolden Site 154
Allabert Blandine 217, 221
Allegany County, New York 85
Allegheny Mountains 47, 48
Allen, A. B. 65
Allen, John W. 217
Allen, Joseph 84
Allen, Z. G. 92
Alligerville, New York 36, 38, 39, 41, 102, 103
Allis, E. P. 99
Almarza de Cameros Quarry 187
Alonso, Natalià 182
Alos, Didier 225
Alps 135
Alricks, J. 11
Alsvik, E. 217
Alvaston 162
Alvis, Gayle viii
Ambrose, Paul M. 217
American Geological Institute xiii
American Miller 208, 218
American millstone industry decline 117–118
American millstone industry 1, 4, 5, 9–118
Amiran, R. 218
Amouric, Henry 218
Ancient Rome 4
Andernach, Germany 196
Andernach, Hohen Buche Quarry 150
Andernach-Eich, Sattelberg Quarry 156
Anderson, Timothy xi, xii, 7, 150, 182, 194, 196, 218, 223
andesite 6, 165, 196
Andrews, E. B. 218
Andrieu, Gaby 218
Anglesey 162, 164
Anglo-Norse period 161
Anglo-Saxon period 157, 159
Angola 197
Annapolis, Maryland viii
Anonymous 218–219
Antonelli, Fabrizio 219
Antwerp, New York 15, 16, 78, 79, 105
Apling, Harry 219
Apollonia Site 125
Appalachian Hills 15, 32
Appalachian Mountains 20
Apperson, Richard 219
apple cider 6
Applegate, Elisha L. 87
Aquia Creek, Virginia 86
Archaeological Society of Virginia viii
archaeology 28–32
Archaeology Magazine xiii
Argentina 91
Argold, Greece 165, 166
The Argus of Western America 16, 17, 25, 26, 66, 67, 109, 248
Aris, R. 219
Arizona viii
Arkansas 19, 20, 54, 55, 60–62, 80, 86
Arlington, Virginia viii
Arndt, Karl J. R. 219
Arnedillo Quarry 187
Arnow, Harriette S. 28, 219
artifacts 32
artificial millstones 199–201
Ashton Keynes, Wiltshire 151, 154
Asia 42
Athens County, Ohio 73–75
Atlantic Ocean 1
Attica, Greece 165
Atwater, Caleb 73, 219
The Augusta Chronicle and State Messenger 65, 248
Australia 42, 91, 212
Austria 4, 125–126
Aydolett, Zadock 83
Ayrshire, Scotland 163
Azéma, Jean-Pierre Henri 219
Azzaro, E. 248

Bad-Bertrich- Kennfuß Quarry 150
Bad-Bertrich, Seenflürchen Quarry 150
Bad-Godesberg-Mehlem, Roddenburg Quarry 150
Bailly-Maître, Marie-Christine 219
Baines Cragg, Quarry 152, 161
Baird, Joseph 73, 109
Bajulaz, Lucien 219
Baker, Ira O. 219
Baker, John L. 84
Baker Millstone Quarry 24, 30, 32
Balearic Islands, Spain 187
Ball, Donald B. viii, 219, 233, 244
Ballard, John 11
Baltimore, Maryland viii, 58, 84, 88, 90, 91, 94, 102
Baltimore American 44, 219
Baltimore County, Maryland viii
Baltzell, George 67
Baltzell, John 67
Bancroft, Thomas 10, 11
Bankhead National Forest 21
Barbera Miralles, Benjami 220
Barboff, Mouette xi, 7, 220

Barbour County, West Virginia 46, 47
Barck, Dorthy C. 241
Barford, P. M. 220
Barkley, William 9
barley 167, 200, 201; pearling 6, 82
Barnes & Pearsol 220
Barnwell, South Carolina 75
Barrosa Cruz, V. 244
Barsinghausen Millstone Quarry 148
Barton Court Farm, Abingdon 156, 157
barytes 6, 21, 37, 38
basalt 6, 119, 140, 148, 150, 160, 162, 170, 171, 182, 187, 197
basaltic andesites 167, 169, 171, 197
Bates Mill 69
Bath, England viii, xii
Battelle, Col. 13
Bauer, Herbert 220
Baug, Irene xii, xiii, 220
Beach, L. M. 220
Beadle, H. L. 220
Beaman, Lela 220
Beauvois, Jacques xi, 220
Beaver Creek 47
Beck, Lewis C. 220
Beckwith, H. W. 220
Bedford County, Pennsylvania 48
Bedfordshire, England 162
bedstone 5, 213
Bedukadze, Sarah 220
beehive querns 151, 161, 162, 165
Beiron, Ingemar xii, xiii, xiv, 191, 192, 220
Beitzell, Edwin W. 220
Belgian millstones 119
Belgium 119, 139, 200
Bell, Edward L. viii, xii, 200
Belmont, Alain xi, xii, xiii, xiv, 1, 2, 7, 124, 131–137, 140, 150 169, 170, 172, 173, 177, 181, 193, 194, 195, 220, 221
Bennett, Pamela J. xiii
Bennett, Richard 221
Beranová, Magdalena 221
Berdorf, Hohllay Millstone Quarry 171
Berea grit 81
Berg, Thomas N. 48, 221
Bergen County, New Jersey 85
Berks County, Pennsylvania 47, 48, 97, 108
Bern, Switzerland 194
Bernardini, F. 219
Bevins, R. E. 238
Béziat, Didier 241
Biddle, M. 221
Biferno Valley 162
Big Barren River 84
Big Creek 62
Big Hatchie 86
Big Sandy Creek 66
Big Spring, Tennessee 76
Birk Bank Fell Quarry 155
Birreshorn Quarry 150
Bishop, John Leander 221
Bishop's Castle, Shropshire 159
Bizien-Jaglin, C. 248
Black, Amos 56
Black Bank at Quernmore Quarry 158, 161
Black Creek 25
Black Fell Quarry 156
Black River 55
Black Water Lake 9
Blackburn, Glen A. 221
Blacksburg, Virginia ix, 20, 49, 52, 105
blacksmiths 39, 40
Bledsoe Creek 84
Bloch Jorgensen, A. 221
blocking hammers 202, 213
Blossom, Benjamin 13
Blount County, Alabama 20
blue basalt 140
blue granite 58
Blue Ridge 42, 51
blue stones 121, 122, 140
Bluegrass Region 29
Bohemia 72, 126, 148
Bois, Michèle 221
Boisvert, Richard viii, 57, 221, 222
Boitel, François 237
bone mills 6, 117, 118, 145
Bonn, Germany xiv
Bonneff, Léon 222
Bonneff, Maurice 222
Bonson, Tony xiii, 222
Borgone Susa, Arco di Maometto 171
Borgone Susa, Molere 171
Borgone Susa, Roccafurà Quarry 171
Borken Quarry 150
Bosset, L. 222
Bost, Gerald 222
Boston, Massachusetts viii, xi, 56, 83
Boston University xi
Bottin, C. 222
boulders 16, 20, 31, 86–87, 213
Bourbon County, Kentucky 17, 26
Bourrilly, J. 242
Bowles, Oliver 222
Bowmanstown, Pennsylvania 48
Boyd, Andrew 222
Boyer, François 222
Bradford, James & Co. 95
Bradford, T. and Company 95
Bradford, Thomas and Company 96
Bradford County, Pennsylvania 47, 108
Bradley, P. 222
Bragg, M. M. 222
Brainard, Jehu 222
Brandywine, Pennsylvania 65
Branson, George 222
Braun, John & Company 89
Breedon-on-the-Hill 159
Bregadze, Nelli 222
Briggs, S. 222
Bristol, Virginia viii
Bristol region, England 161
Brittany, France 138, 139
Broadhead, G. C. 222
Broadhead, Garland C. 222
Brockenbrough, William Henry 239
Brockholm Quarry 159, 162
Bronze Age 163, 166, 196, 197, 212
Brose, David S. 224
Brouchard family 13
Broughton, John G. 231
Brown, Bruce B. viii, 222
Brown, James B. 222
Brown, Lisa 222
Brown, Richard x
Brown, Robin 222
Browne, Tony 222
Bruce, Robert J. 247
Bruggeman, Jean 222
Brundage, Larry 223
Brush Creek 25
Brush Mountain Millstone Quarry ix, 20, 37, 49
Brush Mountain stone 50–53, 101, 105, 213
Buchanan, Mungo 223
Buchsenschutz, Olivier 222, 223
Buckey's Ferry 67
Buckinghamshire, England 154
Buckley, D. G. 223
Buckley, Jeremiah 17, 66, 67, 68, 109
Buckman, J. 223
Bucks conglomerate 159
buckwheat, hulling 6
Buentke, Holger xi, xii, 190, 223
Buffalo, New York 91, 92, 105
Buffone, Luigi 223
Buford, John 67
bull point 38
Bulleid, Arthur 223
Bulliot, J. G. 223
Bullock, James M. 219
Bullock County, Georgia 83
bullrigging 213
bullset 202, 203, 213
Buor, Major 55
Burdett, Charles 223
Burges, Robert 11, 18
Burgomaster and Schepens 10
Burke County, Georgia 63, 65, 83, 109
Burnes, Dan and Son 97
Burnham, Barry 223
Burnham, Helen 223
burr millstone companies 15
burrhstone 213
burrs 213
Bury Hill 161
Bush, Frend 103
bush hammers 30, 202, 203, 213
bushing hammer 205
Butcher, L. H. 223
Butler, Brian M. 223
Butrint Site 125
Butt, John 223
Buxeda-Garrigós, J. 244
Bwlch Gwyn Quarry 164
Byzantine era 167

Cabezuelo, Ulysse 223
Cade, A. B. 240
Cahawba Valley 20
Caldwell County, Kentucky 67
California viii, 1, 19, 60–63
calipers 203, 213
Cambria, Virginia 53, 105
Camel Site 198
Campbell, E. 223
Campbell, Marius 223
Canada 4, 87, 92, 126, 212
Canary Islands, Spain 187, 190
Cannon County, Tennessee 48
Canos-La Cuerda Millstone Quarry 185
Cantera de Cuatro Puertas Quarry 187
Cantera de El Queso Quarry 187
Cantera de La Calera (La Suerte-Las Piletas) Quarry 187
Cantera de Riquiánez Quarry 187
Cantet, Jean-Pierre 223
Canton, Connecticut 78
Cap d'Ail Millstone Quarry 131, 132, 140
Cape Fear River 58
Capedri, S. 219, 223
Carbon County, Pennsylvania 47, 48
Carcassone, France 139
Carcauzon, C. 223
Carden, H. B. 242
Carelli, P. 223
Carinthia, Austria 126
Caron, C. K. 223
Carpathian Basin 169
Carpathian Mountains 169
Carr Crags, Newbiggin-in Teasdale 159
Carreman, Jan 10, 18
Carrig-na-m Brónta 162
Carrio, Elisabeth 228
Carrol County, Missouri 69, 85
Carroll County, New Hampshire 109
Carter County, Tennessee 60
Casseday, Ben 223
Casson, Edward 9
cast-iron millstones 201
Castella, D. 218, 223
Castle, Henry 16, 71, 109
Catahoula formation 81
Cataractonium at the Roman Catterick 154
Cattani, M. 224
Caufield, John J. 224
Caulfield, Séamas 224
Caulkins, Frances Manwaring 224
Cavender, Anthony P. 244
Cayahoga River 82
Cecil County, Maryland 109
Cedar County, Missouri 70
celtic querns 139
cement 6, 38, 117
Cesari, Bernard xiii
Châbles Quern Quarry 7, 194, 196
chaffing wheat 6
Chalkousaki, Elena 224
Chamberlain, Joseph 57
Chambers, Benjamin 12
Chambers County, Alabama 54
Chambersburg, Pennsylvania 15, 97
Chambon, R. 224
Chandler, Henry P. 224
Channel Island, England 162
Chapman, Hilary S. 224
charcoal 6, 117
Charlottesville, Virginia x
Charlton, Robert 9
chasers 6, 21, 37, 39, 112, 213
Chatham County, North Carolina 42
Chautauqua County, New York 87
Châvannes-le-Chêne Quarry 194–196
Chazin, Daniel 224
Cheat Mountain 86
Cherau, Suzanne G. 226
Cherry's Mill 81
Chesapeake Bay area 86
Cheshire, England xiiii
Chestnut Ridge, Pennsylvania Quarry 46
Chevillot, C. 224
Child, Hamilton 224
Childe, V. G. 224
China 197, 198
chipping hammers 202, 203, 213
chisels 203, 213
chocolate industry 6
Christian, Patricia x
Christiansburg, Virginia 105
Chunkey Creek 81
cider mills 164
Cinadr, Thomas J. 224
Cincinnati, Ohio 46, 65, 94, 95–97
Cinq-Mars-la-Pile 139
Cisneros Cunchillos, M. 224
Cist, Charles 224
Civil War 58, 77, 108
Claiborne County, Mississippi 81
Clairbourne County, Tennessee 76 77, 98
Claracq, Paul 224
Clark, Kate viii
Clark, Victor S. 224
Clark County, Kentucky xii, 109, 224–225
Clark Pilot knob 25
Clarke, Robin 225
Clarke County, Mississippi 81
classical period 165, 166
Clay, R. Berle 225
Clay, Sidney Payne 17, 26–27
cleaning clover seeds 6
Clear Creek Limestone 66
Clear Creek 67
Cleare, John 225
Cleary, Andrew 15, 97, 108
Cleveland, Abel 87
Cleveland, and Cochran 96
Cleveland, Ohio 96
Cleveland's Landing 23–24
Clifford, J. D. 48, 225
Clinch Mountain,Tennessee 16, 48
Clinton, Ohio 61
Close-Brooks, J. 225
Clougha east of the Pike 156, 161
Clougha Scar 161
Clouse, Jerry A. viii, 225
clover seeds, cleaning 6
Cloyd Mountain 53
Cocalico Millstone Quarry 43
Cocalico Millstones 20, 37, 90, 91, 101, 102, 108, 117, 213
Cocalico Township 43, 97
Cochran, Robert 96
Coddington, E. D. 103
Coddington, Frank 103
Coddington, George 103
Coddington, Joachim 40, 103
Coddington, Joseph 102
Coddington, Lavoid 40
Coddington, Lester 103
Coddington, Oscar 103
Coddington, Q. D. 103
Coddington, Simon 102
Coddington, Willfred 103
Coffee County, Tennessee 48, 86
Coffeen, David 78, 105
Cohen, Ronald 225
Colardelle, Renée 231
Colen, Germany 16, 36
Cole's Pit Site 157
Collins, Gabriel 225
Collins, Lewis 28, 225
Collins, Lewis Dale viii, ix
Collins, Richard H. 225
Cologne stones 90, 91, 102, 121, 146
colonial period 19
color mills 6, 32, 145
Colorado viii
Columbian Foundry and Burr Mill-Stone Manufactory 91
Colyer, Charles, Jr. 17, 25, 26, 109, 225
composite millstones 6, 74, 127, 160, 213
Concord, Ohio 96
conglomerate 3, 6, 18, 19–54, 138, 151, 157, 160, 162–164, 171–172, 179, 182, 183, 200, 207, 212, 213
Conley, James F. 225
Connally, Patrick B. 65
Connecticut 19, 20, 54, 60, 67, 77–78, 83, 88
Connecticut Historical Commission viii
Connier, Yves 223
Conococheague Creek 12
Constantinople 168
Converse, Brigham 56
Conway, Wales 165
Cookson, Mildred M. 225
Coons, A. T. 220, 225
Cooper, James D. 225
Coosa County, Alabama 54
Coquebert-Montbret 225
Cordier, Louis 225
cork mills 6

Corley, Lewis 47
corn 6, 34, 38, 41, 42, 54, 67, 75, 79, 80, 83, 84, 86, 106, 145, 170
corn mills 159, 164
Cornell Dam 85
Cornell University 4
Cornwall, England 151
Cors Goch Quarry 164
Coşkuntepe Site 196
Cossons, Neil 225
Côtes d'Armour 139
Cothren, William 225
Countryman, William 102, 103
Court Farm, England 151
Coustet, Robert 225
Couston, Jean-Claude 225
Coutard, André 225
Cove of Wachitta, Arkansas 62
Cox, Deborah C. xiii
Coxe, Tench 225
Coy, Fred E., Jr. viii, ix, x, xiv, 41, 44, 45, 48, 49–53, 233
cracking picks 205
Cragg Wood Quarry 161
Craigmaddie Muir, Stirlingshire 163
Craik, David 225
Crawford, H. S. 225
Crawford, O. G. S. 225, 226
Crawinkel, Germany 148
Crawinkel millstones 148
Črnilec, M. 226
Croft Mills 84
Cromer, Edward 106
Cromer, Thomas 106, 209
Cromer, Willard 106
Cross, Alice xiii, 226
cross grain 213
Croton Dam 85
Croudance, I. W. 226
crow bars 203
Cruse, John 226
crushers 6
Cruz, Alberto Machdo 226
Cullin stones 119, 121, 122, 140, 213
Culver, Oliver 13
Cumberland Falls 27–28
Cumberland River 81
Cumberland Valley 12
Cumbria, England 151
Cuming, F. 226
Cumming, David A. 226
Curran, J. 226
Curwen, E. Cecil 226
cutting hammers 202, 203, 213
Cypress 197
Czech Republic 4, 124–126, 141, 212
Czechoslovakia 197

D & H Canal 39
Dailey, John 81
Dalarna sandstone 193
Dalbeatties, Scotland 163
Dals-Rostock Quarry 143
damsel 93, 95
Dana, James D. 226
Daniel, James 24, 25
Danser, A. C. 243
Danvers, Massachusetts 83
Darton, N. H. 226
Daun, Germany 140
Dauphin County, Pennsylvania 97, 108
Dauphine 135
Davenport, Ira 102
Davenport, W. 103
Davidson County, North Carolina 70
Davies, David Llewelyn xii, 226
Davis, A. E. 222, 226, 231, 235
Davis, Devotion 58, 107
Davis, Jack L. 229
Davis, Robert S., Jr. 226
Day, David T. 226
Dayton, Ohio 96
Dean, J. O. 81
Dean, Lewis S. ix, 226
Deaton, Jo A. 226
Decker, C. S. & Company 96
Decker and Seville 99
Decker, Floyd 103
Decker, Miles 102, 103
Dedrick, B. W. 226
Deep River 41, 42
Deep River Coal Field 411
Deering, Richard 226
Deeter, John 47, 48, 108
Deffontaines, Benoît 226
Delaware 11, 18, 83
Delmas, Jean 226
Dendaletche, Claude 226
Denmark xi, 197
dental pathologies 139
Depuy, B. H. 102
Depuy, J. S. 102, 103
Derbyshire, England 120, 159, 161
derrick 214
Des Moines, Iowa viii
Deschamps, Marise 226
Desirat, Guy 227
Dever, Garland R. x
Devon County Council Archaeology Service xii
DeWitt, Martin 25
DeWitt, Peter, Sr 109
DeWitt, Peter 24
De Witt County, Illinois 87
D'Hinoyossa 11, 18
Dickinson, David 87
Dietsch, Craig 229
Dillingham, William 89
Dinwiddie County, Virginia 60
Dodekones Islands, Greece 169
Doe River 79
Dohm, B. 227
Dolaucothi Area 165
dolomite 20, 82–83
Domergue, Claude 241
Donack Gora Quarry 179
Donipan County, Kansas 88
donkey mills 140, 157, 159
Donnelly, William 58, 107
Dordogne, France 139
Doswald, C. 218
Doswald, Cornel 150, 227
Douglass, Harry S. 234
Dresden, Germany 141
drill 38, 52, 105, 214
drill holes 31, 214
drugs 6
Drumm, Adam 73, 109
Drumm, Samuel 71, 73, 75, 109
Dubech, Pascal 227
Duc, Jean-Paul 227
Ducassé, Élie 227
Duffy, Gordon 40
Dufrénoy, Ours Pierre Armand 227
Duhard, Jean-Pierre 227
Dunn, Vincent 40
Dunn Mountain granite 106
Durán, H. M. 227
Durán-Garcia, H. M. 227
Durand-Vaugaron, L. 227
Durham, Emma 227
Durlach, Pennsylvania x, 108
Dutch 13, 122
Dutch Colony 10
Dutton, Alabama 21, 80, 109
Duvauchelle, Anika 218
Dworakowska, A. 227
dye mills 6

The Eagle 46, 248
Early Byzantine 125
Early County, Georgia 83
Easton, Massachusetts xiii, 56, 80
Ebert, Milford 227
Eck, T. 227
economic geology 3
Edge, Henry 48, 108
edge-runners 6, 145, 158, 198, 214
Éditions Ibis Press xiii
Edwards, Greenough & Deved 227
Edwards, James 101
Edwards, Richard 227
Egenton, William 95
Egle, William H. 227
Egleston, Thomas 227
Egoumenidou, Euphrosyne 227
Egypt 168, 197
Eibensteiner, F. K. 227
Eiffel Region, West Germany 140, 141, 145, 148, 150, 157
Ekroll, Ø. 227
El Alaoui, Narjys 227, 228
Elfwendahl, Magnus 228, 236
Elizabethtown, Tennessee 79
Elk, Ohio 61
Ellard, Benjamin 55
Ellenville, New York x
Ellenville Public Library and Museum x
Elliott, C. 228, 251
Elliott, Daniel T. 228
Elliott, Jack D. viii
Elton, John 221, 228
Emerson, Edgar C. 228

emery rock millstones 117
Emmons, Ebenezer 228
Empire Stone Works 64
Enfield, New Hampshire 57
England ix, 7, 9, 17, 75, 92, 119, 120, 126, 145, 151–162, 196, 212
Englert, A. 220
English 1, 3, 7, 11, 12, 16–18, 19, 63, 74, 88, 92, 120, 125, 135, 148, 167, 169, 190, 194, 197, 209
Enoch, Harry xii
Epernon, France 129, 134, 135
Ephrata, Pennsylvania x, 43, 97
Erb, Edward E. xiii
Erie County, New York 91, 94, 105
Eritrea 197
Erpelding, Émile 228
Esopus Millstone Company 32, 103
Esopus millstones 16, 32–37, 39, 41, 54, 90, 91, 101, 102, 117, 214
Essex County, Massachusetts 83
Etna Quarry 171
Europe 1, 3, 4, 39, 42, 71, 119, 124–198, 212
European 6, 7
Evans, J. 228
Evans, Joshua & Company 71, 109
Evans, Oliver 63, 65
Evans and Morton 65
Eversen, Wessel 10
Everton Manor 160
Ewen Millstone Quarry 27–28, 30, 31
excavations, quarry 31–32
Exeter University xii
eyes (millstone) 30, 38, 90, 96, 214

Fabre, Denis 228
face 214
face-grinder millstones 6, 214
facing hammer 202, 203
Fahmy, A. G. 228
Fair Dealing 23
Fair Haven, Vermont 54, 109
Fairfield County, Connecticut 88
Faith, North Carolina 106
Falcone, R. 224
Falkenstein, Franz 228
Falkirk, Scotland 163
Falkner, James, Jr. 97
Farmer, D. O. 230
Farmer, David L. 228
Farmers Chase River 9
Fassbind-Bacq, Andrée 228
Faujas de Saint-Fond, Barthélemy 228
Faulkner, James, Jr. 108
Fayette County, Kentucky 24–25, 67, 228
Fayette County, Pennsylvania xi, 46, 108
Fayette County, West Virginia 53
Feachem, R. W. 228
feathers 32
feldspar 6, 21, 37, 38, 112, 118, 191
Felin Ly corn Mill xii
Fell End Beck Quarry 161
Fell End Farm Quarry 152
Fell Sandstones 161
Ferla, P. 228
fertilizers 6, 117
Fess, Simeon D. 228
The Filson Historical Society 17, 26
Finland xi
Fisher, Bob 106
Fisher, Enos 106
Fisher, George 75
Fisher, John 106
Fitzsimons, Christopher 65
Flanders 139
flat boats 46
Flat Swamp Mountain 70
flax mills 6
Flick, Alexander C. 228
flint 6, 15, 16, 17, 19, 20, 60–77, 138
Flint, James 228
flint grinding mills 6
Flint Ridge, Ohio 16–17, 73–75, 214
Flint Ridge Millstones xi, 71
Florida viii
Flory, Paul B. 229
flour 6, 12, 42, 68, 74, 75, 77, 93
Flourney, Matthew 67
Floyd, Virginia viii, 77
Folkestone formation 157
Folkestone querns 161–162
Fontaine, William M. 229
Forest County, Pennsylvania 86
Forest of Dean 151, 162
Fort Payne Formation 77
Foster, Edward H. 231
Foster, John W. 229
Fourche Cove 54
Fowler & Company 97
Fox, H. S. A. 229
Fox's Clough Quarry 161
Frame, Robert M. III viii
France x, xiv, 4, 6, 7, 19, 39, 40, 60, 63, 73, 75, 88, 90, 92, 93, 95, 98, 119–121, 124–140, 164, 196, 197, 207, 212
Francillon, John G. 164
François, Léon 229
Frankel, R. 229
Frankfort, Kentucky 25, 67
Franklin County, Kentucky 16, 17, 66, 109
Franklin County, New York 200
Franklin County, Pennsylvania 12, 15, 16–17, 47, 97, 108
Frazee, Mr. 89
French viii, xi, 1, 4, 7, 88, 90, 95–98, 124–125, 172, 177, 179, 182, 190, 194, 196, 209, 210
French Alps 135
French Burr Stones 12, 13, 34–36, 41, 42, 60–63, 64–67, 68, 69, 71, 73–74, 76–78, 78, 81, 85, 86, 88, 93
French National Centre for Scientific Research 2, 124
fresh water quartz 6
Freshwater, Tom 229
The Friends of Historic Rochester, Inc. xiii
Fries, Robert 229
Frontier Mills 92
Fuentelárbol Millstone Quarry 183, 184
Fuentelárbol 2 Millstone Quarry 183, 184
Fuliman, R. 96
Fuller, Tom C. viii
furrowing picks 205
furrows 5, 203, 204, 214

Gaborit, Thibaut 227
Galer, Greg xii, 80, 229
Galetti, Paola 229
Gallia County, Ohio 73
Gallo-Roman period 135, 138, 139
Gandilhon, René 229
Gannett, Henry 229
Gap Springs 62
Garber, D. W. 229
Garcia Ruiz, Pedro xiv, 182–186, 242
Gardener, New York 102
Gardner, Charles H. ix
Gardner, William 164
Gardner Brothers Granite Works 59, 107
Gardner Granite Works 107
Gardom's Edge 162
garnet mica schist 6
garnet schist 6
garnet slate 170
Gary, Margaret 229
Gasconade River 69, 81
Gasnier, Michel 245
Gast, Marceau 229
Gates, Janet ix
Gaucheron, André 229
Gauthier, Fabrice 223
Gehagan, George 78, 107
Geist, Henri 229
Gembe, Ruth Baer x
Geological Survey, New York State Museum ix, 102
Geological Survey of Alabama ix
Georgetown, South Carolina 86
Georgia 19, 60–61, 63–66, 70, 83, 86, 106, 109
Georgia Burr millstone 64, 65, 214
Georgian Bay 126
Gerke, Tammie 229
German 4, 6, 7, 17, 92, 119, 121–122, 124–125, 139, 140–151, 172, 196, 200
Germany xi, xiii, xiv, 1, 4, 7, 19, 71, 124–125, 169, 1978, 200, 212
Gerolstein, Germany 140
Gettel, Phil viii
Gibbings, Chris 230
Gibbons, Harry 230
Gibert, Louis-François 230
Gifford, Alan 230
Giles, Jacob 84
Gilpin, Joshua 230

Gislovshammar Quarry 193
glass block millstones 199–201
Glastonbury Lake Village 159
Gleisberg, H. 230
Glenstocken, Scotland 163
Gloucester Millstone Manufactory 164
Gloucestershire, Wales 164
glucose 6
gneiss 6, 15, 16, 17, 20, 78–79
Goffin, R. 230
Gold Hill, North Carolina 70
gold mining 154, 165
Gonzalez-Galvan, E. J. 227
González Marrero, Mac 244
Goose Creek Mill Stones 84, 214
Göpelwerk 146
Gordon, Kentucky 22
Gordon, William 24
Gorsuch, Dickinson 84, 85, 09
Gorsuch, John M. 84, 109
Gradie, Robert R. III 230
Graham, Alan H. 230
Grand Gulph hills 55
granite xi, 6, 16, 20, 35, 54–60, 86, 87, 100, 109, 126, 138, 160, 163, 190, 212
Granite, New York 37, 38, 102, 103
Graniteville, Missouri 57
Grassi, Robert 204, 230, 151, 154
Grávalos Quarry 187
Gravelly Guy (Oxfordshire) 151, 154
Gravesend, New York 10
Gray, W. D. 99
Great Britain 3, 4, 6, 19, 71, 120, 124, 151–165, 197
Greece xiii, 4, 7, 124, 125, 165–169, 199, 212
Greek xi, xii, 7, 124–125, 166, 167, 196, 212
Green family 56
Greencastle, Pennsylvania viii
Greensand formation 151, 152
Grenoble, France xi, 1, 2, 7, 135, 139
grey millstones 120
Griffin-Kremer, Cozette xi, xiv, 7, 220
Griffiths, Daffyd 234,
Grillo, Paolo 230
Grimm, Louis 58, 107
Grimshaw, Robert 230,
grinding bone 6
grinding phosphate rock 6
grinding quartz 6
grinding surface 5
grindstones 12, 25, 28, 82
Grypari, Maria 230
Guérin, Hubert 230
Guettard, Jean-Etienne 230
Gulden's mustard 39
Gullbekk, S. 220
Gun, Thomas 11
gun powder mills 6
Gustafsson, Hans xii, 223
Gwilt, Adam 230
gypsum grinding mills 6
Haberbosch, P. 230
Hackshall, Bro. & Co. 64
Haddock, John A. 230
Hagley Museum and Library x, 43
hailstone grit 25, 27
Haimley, Mr. 45, 108
Halcomb, Clarence 22, 28, 230
Haliburton, Thomas C. 230
Hall, James 230
Hallison, North Carolina 43
Hamilton County, Ohio 96
Hamon, Caroline 222
Hancock, R. G. V. 249
hand drills 202, 203
hand-mills (querns) 179, 194
Hands, Edmund C. 230
Hange-Persson, Berit 190, 230
Hanks, Absolom 25, 109
Hansen, A. Marøy 220
Hansen, P. B. 230
Hansen, Steffen Stummann 230
Hanson, J. W. 230
Harless, John Phillip 105
Harms, Eduard 148, 230
Harper, L. 230
Harpeth River 77
Harris, John 47
Harrisburg, Pennsylvania viii, 97
Harrison, Emma 230
Harrison, Glenn viii, 230, 231
Harrison County, Indiana 66
Hart, Alexander B. 92, 105
Hart, Martin 105
Hart & Munson 93
Hartford, Connecticut 109
Hartford County, Connecticut viii
Hartnagel, C. A. 231
Hartsville, Tennessee 48, 76
Harverson, Michael 231
Harvey, H. H. 231
Hasbrouch, J. 32–33
Hasbrouck, Bruyn 102, 103
Hasbrough, A. 34
Hastings, Mr. 55
Hathersage, England 162
Hatmaker, Paul 231
Haug, James K. viii
Hawes, George W. 231
Hawkes, John 60
Hawksley, Jeff 231
Hayes, R. H. 231
Hayward, Elisha 92
Hayward, Nelson 92
Hayward & Noye 92
Haywood, John 231
Hazard, Samuel 231
Hazen, Theodore R. 204, 231
Hazlett, William 75
Heace, Will 27
health hazards 4, 168, 207–211
Heckington Windmill 162
Hedges, Henry P. 231
Hege, Robert, III xiii
Hehnly & Wike x, 108, 231
Heldal, T. 220
Helgö I 197
Helgö II 197
Heller, William Jacob 231
Hemingway, J. E. 231
hemp mills 6, 84
Hemphill, Robert 231
Hendaye Millstone Quarry 133, 134, 140
Henderson, A. H. 231
Hendrickson, John 102
Hengistbury Head Dorset 162
Henry, M. 48
Henslee, S. 73, 109
Herbert, G. W. & Co. 89
Herbert & Wright 89
Herrscher, Estelle 231
Hertfordshire Puddingstone 154, 162
Heslop, D. H. 231
Heslop, David 230
Hesselgessel, James 48, 108
Heverly, Clement F. 231
Hewitt, Elmer 56
Higbee, John 25
High Falls, New York 20, 32, 38, 39, 41, 102, 103
Higham, Arthur P. 231
Hildreth, Samuel P. 231
Hirsch, Steve 41, 231
historians 1
Historical Society of Pennsylvania xiii
The Historical Society of the Cocalico Valley, Ephrata, Pennsylvania x
history 3
Hitchcock, Edward 231
Hochstein Quarry 150
Hockensmith, Charles D. xiv, 1, 2, 22–29, 33–36, 121, 127–131, 219, 221, 231, 232, 233, 236
Hockensmith, Susie xii
Hocking County, Ohio 73
Hodgman & Co. 233
Hoewich, Bolton 162
Hohenfels-Hensigen, Mühlenberg Quarry 150
Holeman, Joseph 15
Holford, Henry 233
Holland 13
Holland millstones 91, 122, 140, 200
Holleman, A. B. 240
Holloway, William R. 233
Holmans, Mr. 15
Holmstrong, John G. 233
Holtmeyer-Wild, Vera 233
Homans, I. Smith 233
Hoornbeek, Charlotte 40
Hoornbeek, George A. 40
Hopewell, M. 227
Hopewell, Ohio 61
hopper querns 166
Horák, J. 238
Hörmann, P. K. 151, 233
horse mill 83
Hörter, F., Jr. 233
Hörter, F., Sr. 233, 234
Hörter, Fridolin xi, 148, 234

Hörter, P. 234
Horton, John T 234
Horvat, A. 234
Hough, Franklin B. 234
Howard, George W. 234
Howe, Henry 234
Howell, Charles 54, 234
Hoyt, H. 65, 109
Hoyt, S. 65, 109
Hudson, Philip John xii, xiv, 152–156, 158, 161, 234
Hudson River 15, 32, 35, 37, 85
Huff, Valentine 24
Hughes, William C. 234
hulling buckwheat 6
hulling of rice 7
Hume, Abraham 234
Hungary 4, 169, 197, 212
Hunold, Angelika 234
Hunsbury Hill 161
Hunsbury Hillfort 161, 234
Hunt, Patrick N. 234
Huntsville, Alabama 21
Hurd and Burrows 234
Hurricane Creek 54
Husbands, Herman 84, 109
Huybert, Sergeant 10, 18
Hyllestad, Norway xii, xiii, xiv, 1, 172
Hyllestad Millstone Quarry 172–178

Iblem Fields 125
Iceland 72
Igea Quarry 187
Illes Balears, Spain xiv
Illinois 16, 19, 66, 83, 87, 88
Illinois River 66
Independence County, Arkansas 21
India 197, 198
Indian Creek 84, 86
Indiana viii, 13, 16, 17, 19, 60, 66, 86
Indiana Historical Bureau xiii
Indianapolis, Indiana xiii, 92
industrial archaeology 3, 124
Industrial Revolution 7, 135
Ingle, Caroline J. 234
Ingles conglomerate 49, 52, 53
injuries, tool related 207–208
International Molinology xiii
The International Molinology Society xiii
The International Molinology Society of America xii
Interstate Millstone Company 105
Invernessshire, Scotland 163
Iowa viii, 87
Ireland 151, 162–163
Iron Age 151, 164, 165, 174
Iron County, Missouri x, 57
Ironbridge Gorge Museums Trust viii
Irondequoit Creek 13
Ironton, Missouri 57
Ironton, Ohio 85
Irwin, John Rice viii, 49, 234
Island of Gran Canaria, Spain 187
Island of Menorca, Spain xiv, 187
Island of Nesis 170
Island of Ustica Quarry 171
Ison, Cecil R. 234
Israel 197, 198, 212
Italy ix, 4, 124, 125, 166, 168, 170–171, 196, 212
Izard County, Arkansas 62

Jaccottey, Luc 222
Jäckle, Hans Werner 234, 235
Jackson, Claude V. viii
Jackson, Mrs. Philomena ix
Jackson County, Alabama 21, 61, 80, 109
Jackson County, Ohio 73, 75
Jackson Purchase region 22
Jacobson, Maria 235
Jagailloux, Serge 207, 235
Jama Quarry 181
James, H. J. 238
January & Huston 75
January & Sutherland 46
Jarrier, Catherine 241
Jasper County, Missouri 70
Jasper County, Texas 82
Jasper Stone Company 109
Jaspery Alunite 169
Jautee, E. 243
Jean, Marcel 235
Jecock, H. M. 235
Jefferson County, Alabama 20
Jefferson County, Georgia 64, 83
Jefferson County, New York 14, 16, 78, 105
Jefferson County, Pennsylvania 64, 83
Jefferson County, Tennessee 76
Jegli, John B. 235
Jennings County, Indiana 66
Jobey, George 235
Johnson, Bertrand L. 235
Johnson, John G. 47
Johnson, Joseph Risk 235
Johnson, Martin 25, 109
Johnson, Robert 68
Johnson, Sir William 32–33
Johnson County, Tennessee 79
Johnston, M. 66
Joliet City, Illinois 87
Jonathan Creek 23
Jonathan Mills Dis Machine 201
Jones, Amanda C. x
Jones, C. 240
Jones, Hugh 76
Jones, Joseph 235
Jones, Samuel 46, 108
Jones, Walter B. 235
Jones County, North Carolina 70
Jones Mill 62
Jonsdorf Quarries 141, 145, 158, 150
Joos, M. 235
Jordan 197
Jottrand, M. G. 235
Jubera (Santa Engracia de Jubera) Quarry 187
Jubera 2 (Santa Engracia de Jubera) Quarry 187
Judd, William 235

Kaim Hill, Ayrshire 163
Kansas 88
kaolinized gneiss 190
Karapidakis, Louisa 230
Kardulias, P. Nick 235
Karman, Mr. 10
Kars, H. 235
Kash's Knob 25
Katz, Frank J. 235, 236
Keith, John R. xiii
Kelder, Franklin 40
Kelleher, Tom viii, 56, 236
Keller, Allan 234
Keller, P. T. 236
Keller, Peter 236
Kelly, Frank B. 236
Kelly, William ix
Kennedy, D. M. 239
Kennedy, Joseph C. G. 236
Kennedy, P. S. 220
Kent, James R. 105
Kenton County, Kentucky 89, 90
Kentucky 19–20, 24, 66–68, 83, 84, 86, 88–90, 109, 121, 203
Kentucky Department of Libraries and Archives vii, viii
Kentucky Gazette 23, 236
Kentucky Geological Survey x
Kentucky Heritage Council vii
Kentucky Historical Society ix
Kentucky Reporter 24, 236
Kentucky River 24
Kepner, Samuel 97
Kepner, Sara 97
Kepner, William H. 97, 108
Kerhonkson, New York 37, 38, 39, 102, 103
Kerman, Mr. 10
Kidville, Kentucky 23
Kilger, C. 220
Killafady, Ireland 94
Killebrew, J. B. 236
Kimolos, Mprovarma Quarry 169
Kindig, Steve 48
King, D. 236
Kingsbury Mill 56
Kingston, New York x, 32, 37, 38, 102
Kirk, Jarvis 9
Kirker, Jabez G. 89
Kit Point 29
Klausener, M. 236
Kleinmann, Dorothée 236
Kling, Joern xiv, 141–147, 149, 151, 236
knobs 23, 29
Knox County, Tennessee 76
Knox Dolomite 76, 83
Knoxville, Tennessee viii, 16, 76, 83
Koerner, U. 228

Kokelj, E. Montagnari 219
Konigmacher, William 45, 108
Konschak, Alexander 236
Korinththia, Greece 165
Kottenheim, Germany 140
Krämer, S. 236
Kranjc, A. 236
Kremer, Bruno 236
Kremer, Robert 7
Kresten, Peter 173, 223, 228, 236
Kupla, J. C. 206
Kvarnberger Quarries 193
Kvennberget, Norway 172
Kvernsteinsparken Quarry 177, 178
kyanite-garnet-muscovite-mica schist 175
Kyserike, New York 36, 38, 39, 102

La Ferté-sous-Jouarre, France xi, 1, 7, 127, 128, 134, 135, 209
La Narne River 127
La Tene Period 126, 148, 150, 212
Labs, William A. 237
Laclede County, Missouri 81
Lacombe, Claude 237
Lacroix, B. 237
Lacy, Henry de 159
Ladoo, Raymond B. 237
LaFayette Burr Mill Stone Company 63, 65, 109
LaFayette County, Georgis 64
LaForest, Michael J. viii
Lagadec, Jean-Paul 222
Laguna Seca, California 62
Lake Erie 95
Lake Simcoe 26
Lalm Millzstone Quarry 172, 173
Lambrugo Quarry 171
Lancaster, Ohio 241
Lancaster County, Pennsylvania x, xii, xiii, 20, 37, 43, 48, 97, 108
Lancaster County Historical Society x, xiii, 48
Lancastershire, England 159
Landmarks Preservation Commission viii
Lange Brothers 103
Langouet, L. 248
language barriers 4
Lantz, Aaron 75, 109
Laoust, É. 242
Lapuente Mercadal, María Pilar 182, 224
Larsen, Anne-Christine 230
Latin 1
Lauderdale County, Mississippi 81
Laumanns, M. 237
Laurel Hill Millstone Quarry 45, 47, 214
Laurel River 78
lava 6, 119, 157, 196
Laville, L. 237
Lawrence, Gilbert 36
Lawrence, Harry 102, 103
Lawrence, Henry 103
Lawrence, Russell 102
Lawrence, Vincent x, 39, 40, 41, 202
Lawrence, Wally x, 39, 40, 41, 202
Lawrence County, Arkansas 62
Lawrence County, Ohio 85
Lawrence Hill, New York 40, 41
Lawrence Hill Millstone Quarry 33–36
Lawrenceburg, Kentucky 67
Laws, Kathy 222, 237
Lazzarini, L. 224
Lazzarini, Lorenzo 219
Leahy, P. Patrick xiii
Lechevin, Jean-Michel 237
Lee & Johnson 50
Lee County, Alabama 78
Leffler, Dankmar 148, 237
Lehigh Valley Railroad Company 92
Leicestershire, England 89, 159
Lemon, J. R 237
Letcher County, Kentucky 22
leucitite rock 171, 196
Leung, Felicity L. 237
Leveille, Alan 237
leveling crosses 30, 214
Lewis County, New York 78, 79
Lexington, Kentucky x
Libya 125
Lichty, Mr. 45
Licking County, Ohio 16, 71–74, 109
Liebgott, Niels-Knud 237
limestone 6, 20, 68, 69, 77, 82, 127, 138, 160, 182, 193, 194, 212
Lincoln, Pennsylvania 108
Lindeberg, I. 238
Line Creek 42
Linn County Historical Society viii
Linney, W. M. 237
Linsay, Mr. 13
Linsley, Stafford M. 237
Litchfield County, Connecticut 83
Little Doe River 79
Liverpool, England 159, 161
Ljubljana, Slovenia xiv, 182
Llevant-ses Anglades–Cap d'en Font Quarry 187, 188
Lodsick & Co. 95
Lodsworth Quarries 12, 157, 162
Logan, Benjamin 23
Logan, W. E. 126
Logan County, Kentucky 84
Logroñ, Spain xiv
L'Oise Valley 135
London, England 12, 157, 162
Long, Donald 106
Long, Jack 106
Long, Ted 106
Long Island City, New York xiii
Longepierre,Samuel 237
Longis querns, Channel Island 162
Lorenz, Claude 237
Lorenz, Manfred 240
Lorenz, W. F. 237
Lorenzoni, S. 237
Lorenzoni, Serjio 223
Lorimer, Norma Octavia 237
Lost Creek 54
Loubès, l'Abbé G. 237, 238
Loughridge, R. H. 238
Louisa County, Virginia 60
Louisiana 16, 19, 55
Louisville, Georgia 65, 66
Louisville, Kentucky viii, ix, 26, 89
Louisville District, Corps of Engineers viii
Lounsbury, James 102
Lovell, B. 238
Low, Isaac 12
Lowell, Massachusetts 83
Luezas Quarry 187
Lugnås, Sweden xiii, 191
Lugnås Millstone Quarry 191–193
Lulbegrud Creek 25
Lundström, A. 238
Lundström, P. 238
Luni, Mario 219, 244
Luxemburg 4, 124, 125, 148, 171–172, 212
Lyell, Charles 238
Lynchburg, Virginia 50
Lynchburg Daily Virginian 50, 238
Lynchburg Press 49
Lynn, Massachusetts 11, 12, 18, 56
Lynn Historical Society 16, 238
Lyons, Pam vii

Maaskel, Mr. 65
MacCabe, Julius P. Bolivar 238
MacCulloch, John 161
Macedonia 197
MacKie, Euan W. 238
Madison County, Alabama 21
Madison County, Kentucky 22
Madison County, Missouri 69
Madison County, North Carolina 78, 107
Madsen, H. J. 238
Magallón Botaya, M. A. 224
Maher, N. A. 88
Main Elkhorn Creek 67
Maine vii, 19, 54, 60, 101, 110
Mainz, Germany xiii, xiv
Major, J. Kenneth xi, xii, 238
making apple cider 6
Making of America Books 4
Making of America periodicals 4
Malacour, Claudine 228
Malaws, B. A. 238
Mallet, Louis 238
Mallow Archaeological and Historical Society of Ireland 168
Malpas, J. G. 228
Malthouse 157
Malung Quarries 193
Manchester, Tennessee 48
Mangartz, Fritz xi, xii, xiii, xiv, 7, 141, 148, 169, 171, 182, 221, 230, 238
Mangas Viñuela, J. 244
Manhattan Island 9
Mannino, Robert, Jr. 56, 238
Mansfield Gazette 71
Mantle Green Site 154

Manufacturer and Builder 36, 64, 77, 96, 117, 199, 200, 209, 238
Marbletown, New York 102
Maries County, Missouri 69
Mariestad, Sweden xiv, 191, 193
Mariette Mill Company 73
Marion, Mississippi 81
Marion County, Alabama 21
Marks, Annie L. 224
Marquet, Cynthia x
Marshall, C. 238
Marshall County, Kentucky 22, 23
Martel, Pierre 239
Martin, François 241
Martin, Joseph 239
Martín Rodríguez, Ernesto 244
Marty, R. 217
Marty, Robert 221
Mary, Stéphane 239
Maryland 11, 18, 19, 54, 60, 84, 88, 90, 101, 109, 110
Maryland Farmer 58, 239
Maryland Gazette 84, 109, 239
Maryland Historical Trust viii
Maryland Millstone Manufactory 109
Massachusetts xi, 11, 12, 13, 15, 16, 17, 19, 20, 54, 56–57, 68, 78, 80, 81, 83, 88, 91, 109
Massachusetts Bay Colony 9, 15, 16
Massachusetts Historical Commission viii, xi
Masson, L. 239
Matadamas, P. O. 227
Matchedash River 126
Mather, Joseph H. 35, 239
Mather, William W. 34, 68, 71, 72, 239
Maume, Jack 239
Maxwell, Hu 239
Mayen, Germany xiv, 1, 7, 121, 140, 141, 148, 197
Mayen Millstone Quarries 141–145, 150, 157
Mayer, Adele 239
Maynard, J. Barry 229
Maysville, Kentucky 46
The Maysville Eagle 75, 94, 248
McAlfee, Robert, Jr. 67, 229
McArthur, Ohio xi, 75–76
M'Cauley, I. H. 97, 239
McCalley, Henry 239
McClain Printing Company xiii
McClung, Alexander 94
McCoum and Kennedy 67
McCreary County, Kentucky 84
McDaniel, Hurt & Preston 50
McDougal, Richard 75, 109
McGee, Marty 239
McGill, William M. 239
McGrain, John W. vii, 84, 239
McGuire Millstone Quarry 22, 23, 24, 30, 32
McIlhaney, Calvert W. viii, 53, 239
McKain, H. B. 13
McKechnie, Jean L. 239
McKee, Harley J. 239
Mckeen, Mr. 13–14
McKenzie, Mr. 200
McLean County, Illinois 87
McLennon's Creek 239
McMurtrie, H. 239
McWhirr, Alan D. 239
Meade County, Kentucky 46
Meadows, Larry G. 233, 239
Meadows Mills Company 100
Meadows Mills, Inc. xiii, 53, 59
Medfield, Massachusetts 56
medieval period 1, 150, 151, 157, 161, 162, 165–167, 170, 182, 183, 191, 196
Medina sandstone 35
Mediterranean 166, 197
Meek, F. B. 222
Meeson, Robert 243
Mellen, George F. 245
Melos, Greece 166, 167
Mendig, Germany 140, 150
Menillet, F. 239
Mercer, Henry C. 239
Meredith, Rosamond 239
Merovingian period 177
Mesa, Reyes 182
mescal manufacturing 198
Mesolithic period 165, 166
Metcalf, Robert W. 239, 240
Meucci, Roger 219
Meulières 7, 214
Mexico 198
Michael, Ronald L. 46, 240
Michels, F. X. 234
Michigan viii, 87
Middle Ages 7, 126, 135, 138, 139, 141, 149, 150, 167, 169, 171, 173, 177, 178, 193, 212,
Middle Creek 45
Middle Saxon period 161
Middlewood, Esther xii, xiii
Milburn, Mark 240
The Mill Monitor xii
mill pick 93, 98, 204, 214
Mill Stone Swamp 9, 10, 15
Miller, Henry 67
Miller, Jeremiah 65, 109
Miller, Marcia M. viii
Miller, Railsback & Miller 17, 66, 67, 109
Miller, Tamara G. xiii
The Miller 248
Milleville, Annabelle 222
Mills, William C. 74, 240
millstone cutter 214
Millstone Edge 161
millstone grit 21, 25, 27, 214
Millstone Hill 23, 56, 161
Millstone Knob 83
Millstone Mountain 21, 61–62, 86
Millstone Point 55
Millstone Ridge 22
Millstone Run 47
millstonequarries.eu website 7, 124, 140, 148, 150, 163, 170–172, 177, 182, 183, 187, 193, 212
Milos, Greece xii, xiii, 7, 167–169
Milos, Rema 1 Quarry 169
Milos, Rema 2 Quarry 169
Milroy, Samuel 13–15
Milwaukee, Wisconsin 98
mineral paint 38
Mineral Resources of the United States xiii, 104, 106, 109, 110–115, 117–118, 123, 214
Minerals Yearbook xiii, 214
mines and minerals 3
Minnesota viii, 19, 54, 60, 101, 109, 110
Minnewaska Mountains 40
Mirrer, Louise xiii
Misiti, V. 248
Mississippi 20, 58, 81
Mississippi Department of Archives and History viii
Mississippi River 66, 75
Missouri 19, 20, 57, 66, 69–70, 81, 83, 85–87, 88, 91
Missouri Department of Natural Resources x
Mitchell, J. E. 98, 240
Mohonk Mountain House 40
Molyneux, R. 160
Moniteau County, Missouri 69
Monmouthshire, England 161, 164
monolithic millstones 6, 29, 59, 127, 141, 160, 163, 164, 172, 214
Montagne-Saint-Emilion Millstone Quarry 136, 137, 140
Montana viii
Monte Vulture Quarry 171
Montgomery Buhr 77
Montgomery County, Indiana 16, 86
Montgomery County, Kentucky 23, 25, 240
Montgomery County, Missouri 69, 85
Montgomery County, Ohio 96
Montgomery County, Pennsylvania 97, 108
Montgomery County, Virginia 20, 37, 50, 51, 53, 105, 108, 202
Monti Iblei Quarry 171
Montmirail Millstone Quarry 130, 131, 140
Moog, B. 240
Moore County, North Carolina 41, 42, 94, 106, 107
Moore County grit 42, 214
Moore County Grit millstones 58, 59
Moose Mountain 57
Morbidelli, P. 240
Moreau, Jean 248
Moreno Arrastio, Francisco 248
Morgan, David L. vii
Morgantina, Austria 126, 196
Morgantown, West Virginia 109
Moritz, L. 240
The Morning Star 42, 248

Morocco 187, 197
Morris, M. 21
Morris & Egerton 90, 94, 95
Morris & Egenton Company 95
Morris & Trimble Company 90, 102
Morris Mountain 25
Morro Llevant-Ses Anglades Millstone Quarry 187, 188, 189
Mt. Etna 125
Mt. Oros 166
Mount Sterling, Kentucky 23
Mount Tom,Connecticut 54
Mount Vouan Millstone Quarry 128, 129, 140
Mowcop-Hill 161
Mühlenstein 214
Mulargia Quarry 171
Mullenax-McKinnie, Michelle L. xiii
Müller, Jörg 240
Müller, S. 240
Mullin, David 240
Münden stone 148
Munson, Alfred 16, 92, 105
Munson, Edmund 92, 93, 105
Munson & Co. 102
Munson & Hart 92, 93
Munson Brothers 92, 93, 240
Munson's Patent Eyes 93
Murist Millstone Quarry 196
Muro de Aguas Quarry 187
Murphy, James L. viii, xi, 75, 240
Murray, John 65, 66, 109
Murray, Kentucky xii
Murray, Priscilla M. 245
Murray State University xii
Muscatatuk River 66
Muscovy 161
Museum of Anthropology, University of Kentucky xiv
Museum of Appalachia viii, 49
Muskingum County, Ohio 16, 61, 71–74, 109
Muskingum River 13, 18
Musselman, Mr. 74, 109
mustard 6, 38
Myers, W. M. 237
Myrianthefs, Diomedes 227

Nagle, William D. 43, 108, 209
Naklo, Slovenia 129
Naklo Quarry 181
Namurian Gritstone 161
Nantucket, Massachusetts 15, 16, 109
Naples, Italy 170
Nappi, G. 219
Nappi, Giovanni 244
Nason, F. L. 240
Nasz, Adolf 240
National Historic Preservation Act of 1966 vii
Natural Bridge 21
Naud, Georges 219
Neal, Robert 67
Neary, Donna M. vii
Nebraska viii
Needhawk, Mr. 45
Neisby, Abraham 16, 71, 109
Nelson, Peter 83
Neolithic period 138, 139, 150, 165, 170, 196, 198, 212
Neri, Y. 240
Nevada viii
New Amsterdam, New York 10, 18
New Brittaine 9, 15
New Hampshire Division of Historical Resources viii
New Hampshire 19, 54, 57, 60, 101, 110
New Harmony, Indiana 84
New Jersey 85
New London, Connecticut 15, 16, 17, 55, 83
New Mexico viii
New Milford, County, Connecticut 83
New Netherland 9
New Paltz, New York 37, 38, 102, 103
New York xiii, xiv, 9–10, 12, 13, 15, 16, 17, 18, 32–41, 48, 65, 74, 78, 83, 85, 87, 88, 90–92, 101–105, 106–107, 109, 110–114, 117, 202, 203
New York Chamber of Commerce 12, 241
New-York Historical Society xiii, 241
New York, Ontario and Western Railroad 38
New York State Museum xiii, 34, 38, 241
New York State Office of Parks, Recreation and Historic Preservation viii
New York State Secretary's Office 241
The New York Times 248
Newland, David H. 240
Newton County, Mississippi 81
Newton County, Missouri 70
Niedermendig, Germany xiv
Niedermendig Millstone Quarries 146, 147, 148, 149, 150
Niles Register 63, 241
Niles Valley 48
Nisyros Island Quarry 169, 197
Norfolk County, Massachusetts 56
Normandy, France 138
Norris, Tennessee viii
North Africa 197
North Alabama 21, 62
The North American Review 248
North Atlantic Ocean 187
North Carolina viii, ix, 15, 19, 20, 41–43, 54, 57–59, 60, 70, 77–78, 79, 86, 88, 90, 94, 101, 106–108, 109, 110–111, 114–115
North Carolina grit 106, 114, 115
North Carolina Land Company 241
North Carolina Millstone Company 58 -59, 107, 241
North Carolina State Geologist 70, 241
North East End of Birk Bank Quarry 161
North Easton, Massachusetts 81
North Elkhorn Creek 68
North Yorks, England xiv
Northampton County, Pennsylvania 98
Northrup, John 47, 108
Northrup, Miles 13
Northumberland, England 159, 161
Norway xi, 4, 124, 125, 164, 172–178, 197, 212
Norwegian xii, 1, 172
Nova Scotia 126
Noye, John T. 91, 92, 105
Noye, Richard 92
Nuttall, Thomas 241

Oakes, R. A. 241
Oakland County, Michigan 16, 87
oats 82, 106, 205; shelling 6
Obermendig, Germany 150
O'Callaghan, E. B. 241
Ogden, George W. 74, 241
Ohio xi, 15, 17, 20, 60–61, 71–76, 81–82, 85, 88, 94–97, 101, 109, 121
Ohio River 46, 73, 85
Ohio State University xi
Ohio University Press, Athens xiii
Ohio Valley Historical Archaeology xii
Ohio Valley Publishing Company 241
oil mills 145
Old Chester, New York 16
Old Mill News viii, xii, xiii
Old Red Sandstone 7, 151, 154, 157
Old Sturbridge Village viii
Olean Conglomerate 48
Olinger, R. L. 105
olive oil 6
Olot, Spain 182
Olot Quarry 182
Olympie Temple de Zeus Quarry 169
O'Malley, Nancy xiv
Oneida conglomerate 35, 37
Oneida County, New York 16, 92, 105
Ontario and Western Railroad 38
Oparno Millstone Quarry 126
Orangeburg, South Carolina 75
Oregon 54, 87
Oregon State Historic Preservation Office viii
Orengo, Yves 228
Ortiga Castillo, M. 224
Ortuno, Joëlle 241
Orvieto, Italy 170
Orvieto Quarry 171, 196
Osage River 69, 81
Östra Utsjö Quarry 193
Otringsneset Quarry 178

outcrops 6, 214
Outhwaite at Roeburndale Quarry 158, 161
Owen, David Dale 241
Owens River 63
Owens Valley 63,
Oxford, England xii
Oxfordshire, England 151, 154, 222
Ozark County, Missouri 69

Pachaug State Forest 55
paint mills 32, 117, 118
paint staff 30, 204, 214
Palaeolithic 166
Palausi, G. 241
The Palladium 16, 68, 248
Pallara, M. 237
Pallara, Mauro 223
Pantelleria Island Quarry 171
Paris, France xiii, xiv, 92, 127, 139, 162
Paris Basin 61, 72, 119, 120
Parke County, Indiana 86
Parker, Alexander 79, 105
Parker, Edward W. 241
Parker, James 15, 78, 79, 105
Parker, Martha Houston 79
Parker, Steve xi
Parkewood, North Carolina 58–59, 106
Parkewood Millstone Quarries 58–59
Parkhouse, Jonathan 242
Parsons, West Virginia xiii
Pascal Mayoral, Pilar xiv, 182–186
Passemard, E. 242
paste 6
Patapsco Land Company 242
Patros, Greece 167
Pawtucket, Rhode Island xiii
Pay, William's Mill 59
Paynter, W. Elmer 13
Peacock, D. P. S. ix, xii, 242, 250, 251
Peacock, T. B. 242
Peak District, Derbyshire 160
Peak millstones 120, 160, 215
pearling barley 6, 82
Pearson, T. 242
Peck, J. M. 242
Peekskill, New York 85
Pekin formation 43
Pelletreau, William S. 231
Pembroke, Virginia viii, ix
Pen Pits Quarry 157, 162
Penallt, Monmouthshire 164
Pen' rallt Quarry 164
Pennsylvania viii, ix, xiv, 13, 19, 20, 37, 43–49, 58, 61, 65, 83, 86, 88, 90, 97, 98, 101, 108, 109, 110–111, 114–115, 117
Pennsylvania Historical and Museum Commission viii
The Pennsylvania Magazine of History and Biography xiii
Pennsylvanian age 3, 22, 109
Perg , Austria 125
Périgord, France 138
Perry County, Ohio 72, 74
Peters, J. T. 242
Pettersson, Täpp John-Erik 236, 242
Phalen, W. C. 242
Phelps and Gorham 13
Philadelphia, Pennsylvania 12, 63, 65, 88, 97, 98
Philip, G. 242, 243
Philips, Judith T. 243
phosphate mills 117, 118
Pickin, John 163
Pictorius, Georges 2
Pieraggi, Bernard 241
Pierson, Isaac 75, 109
Pierson, Sara 75
pigments 118, 204
Pike Foot Quarry 161
Pilot Knob 25, 29, 30
Pilot Knob Millstone Quarry 28
Pine Mountain 22
Piot, Auguste 243
pitching tools 203, 215
Pitt-Rivers, A. 243
Pitts Copse, England 157
Pittsburgh, Pennsylvania 75, 97, 98
Pittsfield, Massachusetts 68, 78
Plamondon, P. 88
Plamondon & Maher 88
Planke, Floyd 106
plaster 6
plaster of Paris 6, 95
plug and feathers 38, 52, 105, 215
Podgrad Quarry 182
Point Pleasant, Louisiana 16, 55
points 32, 36, 45, 203, 215
Poirier, Dave viii
Polak, Jill P. 243
Poland 197
Poliča Quarry 182
Poling, Catherine 47
Polk, John 15
Polk, R. L. Company 243
Polk County, Arkansas 62
Pollack, David vii
Pollocksville, North Carolina 70
Pommepuy, Claudine 223
Pompeii, Italy 170
Poor, N. Peabody 243
Pope County, Illinois 83
porcelain millstones 201
Poros, Greece 17
porphy granites 182
Post-Medieval Archaeology ix
Potowatemy Mills 14, 16
Pottsville Conglomerate 53
Powell, Eric xiii
Powell County, Kentucky 3, 16, 22, 23–25, 28–32, 210
Power, Patrick 243
Poythress, George 65, 109
Pratt, Joseph H. 243
Prevot, J. 243
Price, Beck 106
Price, Benjamin D. 84
Price, Henry Lewis 104
Price, Hugh 105
Price, Israel and Company 50, 105
Price, Jacob 50, 105
Price, Jimmie L. ix, 53, 233
Price, John "Matt" 106
Price, John Michael 105
Price, Leon 106
Price, Leonard 106
Price, Lester 106
Price, Martin 106
Price, Michael 50, 105
Price, Zachariah 105
Prices Fork, Virginia 50–52, 105
pritchel 205
Procopiou, H. 243
proof staff 204, 215
Prov. Álava (Araba)–Barambio-Garrastatxu Quarry 183
Prov. de Granada–Loja–Camino del Calvario Quarry 183
Prov. de Granada–Loja–Cerro de la Fuente Santa Quarry 183
Prov. de Granada–Moclín Quarry 182
Prov. de Soria–Canos–La Cuerda Quarry 183
Prov. de Soria–Fuentelárbol Quarry 183, 184
Prov. de Soria–Fuentelárbol 2 Quarry 183, 184
Prov. De Vizcaya–Arbaitza–Barrio Arbaitzarte Quarry 183
Prov. De Vizcaya–Manzarraga Quarry 183
Prov. De Zaragoza, Cerro Redondo (Pardos Abanto) Quarry 182
Prussia 90
pseudo-burrhstone 70
The Public Advisor 248
The Public Archaeology Laboratory, Pawtucket, Rhode Island xiii
puddingstone 22, 28
puddingstone quern 162
Pulaski County, Arkansas 54
Pulaski County, Missouri 69, 81
Pulaski County, Virginia 53
Pulido, J. L. 227
punches 203
Pung, Olaf 238
Punta de Sa Mioca Quarry 187, 188
Purcell, Asa 102, 103
Purcell, David 102, 103
Putnam Family Quarry 56–57
Putnam, Samuel 56
Py, Michel 243
Pynchon, John 10
Pyne, W. T. 89

Quaix-en-Chatreuse Millstone Quarry 127, 128, 140
quarry 215
quarry excavations 31–32
quartz 6, 21, 37, 112, 118
quartzite 6, 20, 77–78

quartz-shot sandstone 54
Queens, New York 13
Quern Study Group Newsletter xii
Quernmoor 159
querns 3, 4, 5, 6, 7, 86, 124, 126, 138, 139, 150, 151, 154, 157, 159–161, 162, 163, 166, 169, 170, 177, 178, 179, 182, 183, 187, 190, 191, 193, 194, 196–198, 212, 215
Quincy (Norfolk County), Massachusetts 56
Quinnell, Henrietta 243

Raborg, William A. 243
Raccoon Burr millstones xi
Raccoon Creek 15, 71, 72
Raccoon Creek Burr millstones 16, 72–75, 215
Radley, Jeffrey 243
Rahtz, Philip 243
Rainey, Mary Lynne 237
Raistrick, Arthur 243
Raleigh, North Carolina 58
Ramapo Mountain State Forest 85
Rancho Cantine 62
Randall, John 81
Randolph County, Alabama 54
Randolph County, West Virginia 47, 86
Rapp, Frederick 84
Rascarrel Formation 163
Rawlinson, R. 243
Rawnsley, W. F. 243
Rawson, Marion, N. 243
Raymo, Chet 80, 243
Raymond Brothers' Mill 201
Read, M. C. 243
Reading, England ix
Recoaro Terme, Sentiero delle Mole Quarry 171
Red River Millstone Quarries xii, 23–25, 30, 32
Red River millstones 215
Red Sea 197
Reed, Charles 87
regrinding middlings 6
Reichstein, J. 243
Reifsnyder, Samuel 44, 45, 108
Reigniez, Pascal 243, 244
Reinemund, John A. 244
Reliance Iron Works 98–99
Rema, Greece 167, 168
Renouf, J. T. 244
Renzulli, Alberto 244
Reston, Virginia xiii
Revolutionary War 19
Reye, New York 92
Reyes Mesa, José Miguel 244
Rhine stones 121, 148
Rhode Island 19, 54, 66
Rhone-Alps Historical Research Laboratory 124
rhyolite 126
rhyothic stone 171
Rice, David 67
rice, hulling 7
Richland, Ohio 61
Richland Creek 42, 43
Richmond, David 75
Richmond, New Hampshire 79
Richmond County, North Carolina 58
Richter, A. 233
Ridenour, George L. 244
Riedl, Norbert F. 244
Rifnik Millstone Quarry 179, 180
Risk, William 23
Roanoke Times & World News 53
Robertiello, Barbara 244
Roberts, Ellwood 244
Roberts, Joseph K. 244
Roberts, Niall 244
Robinson, Rosemary 244
Robres del Castillo Quarry 187
Robres del Castillo 2 Quarry 187
Robres del Castillo 3 Quarry 187
Rochester, New York 13, 102
Rock County, Minnesota 109
rock emery millstones 199, 200
rock falls 210
Rock Hill 41
Rock Island County, Illinois 88
Rockcastle County, Kentucky 16, 17, 22, 25–27, 109
Rockingham County, North Carolina 58
Rockwell, Peter 244
Rocky Point 22
Röder, J. 226, 234, 244
Rodney, Thomas 85
Rodríguez Rodríguez, Amelia C. 187, 244
Roe, F. 222
Roe, Fiona 244
Rogers, N. 244
Rogers, Nicola S. H. 244
Rogers, Steve viii, 244
Rogers, William Barton 244
roller mills 5, 38, 91, 92, 99, 117, 118, 215
Rolseth, Ingeniør P. O. 244
Roman ix, 7, 125, 126, 135, 138, 139, 140, 151, 157, 159, 160, 162, 164, 165, 166, 169, 170, 171, 179, 182, 183, 194, 196–197, 212
Roman Carmarthen 197
Roman Catterick 154
Romano-British 154, 157, 162
Römisch-Germanisches Zentralmuseum, Mainz, Germany xii, xiii, xiv
Rondout River 38, 39
Rønneseth, Ottar 244, 245
Rønnset Farm 178
Rosakranase, Eleanor S. x
Rose, Arthur 103
Rose, W. H. 102
Rose, William H. 40
Rosen, Steven A. 245
Rosendale Railroad Station 39
rotary querns 138, 161, 162, 165, 166, 196
Rotenizer, David E. viii, 245
Rotten Point knob 29
Roundstone Creek 27
Roupe, George 108
Roups, George 15, 97
Rowan County, North Carolina 58–59, 100, 106, 107
Rowden (Ashpotts) Wood Quarry 161
Royal Opera House, London 161
Ruben, Walter 245
Rucker, B. H. 245
Rucker, James 67
Rudge, E. A. 245
Ruel, Yves 240
Rule, William 245
rum 10, 16, 55
Runnels, Curtis N. xi, 165, 166, 225, 235, 245
runner stone 5, 215
Russell, John 245
Russell County, Kentucky 84
Russell County, Virginia 77
Russia 145
Rutland County, Vermont 54, 109
Ryan, John 245
Ryan, Thomas R. xiii
rye 125
rynd 6

saddle quern 140, 166, 170, 196–198
Safford, James M. 245
Saharan Africa 197
St. Josen, New York 37, 38, 102
St. Emilon 135
St. John, Samuel 245
St. Lawrence County, New York 78
St. Lawrence River 126
Saint-Lebe, Nanou 226
St. Louis, Missouri 91
Saint Paul, Minnesota viii
Saint-Quentin-la-Poterie Quarries 138
Salac, V. 245
Salem, Oregon viii
Salido Creek 22
Saline County, Arkansas 55
Saline County, Illinois 83
Salisbury, North Carolina 107
Salley, A. S. 245
Salt River 67
Saltsville, Preston 77
Samuel, Delwin 245
San Giovanni di Ruoti 197
San Juan, Guy 245
San Vicente de Robres Quarry 183, 186
San Vicente de Robres (Robres del Castillo) Quarry 183, 185, 186, 187
San Vicente de Robres 2 (Robres del Castillo) Quarry 183
Sanchez Navarro, Joaquin xiv, 188–189, 245
Sand Mountain 20
Sanders, Thomas N. vii

sandstone 6, 20, 49, 80–82, 135, 138, 141, 148, 160, 163, 164, 165, 169, 182, 190, 193, 212, 215
Sanford, Samuel 246
Sankalia, H. 245
Santi, Patrizia 244
Saône-et-Loire 135
S'Aranjif, Spain 187, 189
Sardinia, Italy 169, 170
Sarosptaker Quarry 169
Sass, Jon A. 245
Saugus, Massachusetts 56
Saugus Residential Development 56
Sauldubois, Bernard 245
Saunders, Ruth 246
Sauta Creek 61
Savage, James 246
Savannah, Georgia 64
Savary, Xavier 245
Saville, Roy 106
Saville, W. C. ix, 53, 207–210
Saville, Walter C. 106, 202
Savoie, France 139
Saxon period 157
Saxon querns 162
Saxony, Germany 206
Scandinavia xi, 164, 177, 190, 191
Schaaff, Holger 246
Schauble, M. 235
schist 6, 163
Schmandt, J. 246
Schmidt, Robert G. viii
Schneider, J. 245
Schön, Volkmar 246
Schoonhoven, J. 246
Schoonmaker, Benj., Jr. 102
Schoonmaker, C. 103
Schoonmaker, Cyrus 103
Schoonmaker, Ross 103
Schrader, Frank C. 246
Schüller, Hans 246
Schulte, Herman 90
Schweigmatt Quarry 150
Scientific American 61, 63, 64, 102, 246
Scotland 125, 151, 160, 163–164, 212
Screven County, Georgia 83
scroll dress 204
Second Magnesian Limestone 69
Second Sandstone 69, 81
Section 106 of the National Historic Preservation Act of 1966 vii
Seeley, Thaddeus D. 246
Selbu, Norway 172
Selbu Quarry 172, 173, 176
Selkirk, Thomas Douglas 246
Selmeczi Kovács, Attila 169, 246
Serbia 168
Serce Limani shipwreck 197
Serneels, Vincent 218
Serrano García, Angel 182
Servelle, Christian 246
Seven Years War 164
Shacklett, John 46, 108
Shaffrey, Ruth xii, 246
Shakers 57
shapping debris 215
sharpening debris 31
Shawangunk Conglomerate 32
Shawangunk grit 35, 37, 38, 41
Shawangunk Mountains 34–35, 37, 38, 41, 102
Shawnee town 84
Shealor, Arthur 106
Shealor, Byrd 106
Shealor, Guy 106
Shealor, J. Fred 105
Shealor, John 106
Shealor, Olen 106
Shear, Hazel M. 246
Sheffield District, England 161
Sheldon, Hezekiah Spencer 246
Shell, Mary Mason viii
shelling oats 6
Shepard, Charles Upham 246
Sherman, Lucy 11
Sherman & Howdayer 64
Shetland Islands 164
Shipp, Becky xii
Shirley, George S. 26
Short Mountain 48
Shumacher, Fred 82
Shumard, B. F. 222
Sicily 170, 171
sickle dress 204
Siegert, J. 246
sienite 20, 83
Sierra Nevada 62
Sigaut, François xi, xii, 7, 220, 246
silicosis 208–210
Silver Creek 22, 87
Simpson County, Kentucky 84
Sioux City, Iowa 109
Sitte, Josef 247
Skiba, John B. xiii
Skinner, Raymond J. 247
Skor Quarry 177
sledge hammers 39, 45, 215
sleds 17
sloops 17, 34
Slovene Ethnographic Museum xiv
Slovenia xii, xiv, 4, 125, 179–182, 212
Small, Alsstair M. 247
Smerdel, Inja xii, xiv, 179, 181, 247
Smiley, Dan 40
Smith, D. 221, 247
Smith, Dwight L. 247
Smith, Eugene A. 247
Smith, Gayle 106, 207
Smith, H. P. 247
Smith, Harvey 106
Smith, John 103
Smith, Kim Lady ix
Smith, R. F. 243
Smith, Robert 94
Smith, Robert C., II ix
Smith, William, Jr. 247
Smith, Wm. D. 102
Smyth County, Virginia 77
snakes 210
Snider, R. E. 53, 105
Snider, Stanley 106
snuff mills 6
Snyder, Bradley 247
Snyder, Esse 106
Snyder, John 106
Society for Industrial Archaeology viii, 1
Society for Post-Medieval Archaeology ix
Society for the Preservation of Old Mills x, xii, xiii, 41
Sognnes, K 217
Soike, Lowell J. viii
Solomon Branenburg's Landing 46
Somerset, England 162
Somerset, Ohio 72
Somerset County, Pennsylvania 47, 48, 108
Sopko, Joseph 247
South Africa 197
South Alabama 21
South America 42, 90
South Cadbury, Wales 165
South Carolina 60, 76
South Dakota State Historical Society viii, ix
South Mountain 47
South Union, Kentucky 84
Southampton, New York 9, 15
Southern Levant 197
Southern Publishing Company 247
Spafford, Horatio Gates 34, 247
Spain xii, 4, 7, 124, 125, 139, 182–190, 212
Spain, R. J. 247
Spanager, Nicholas 90
Spang, Adam 97, 108
Spang, Jeremiah 97, 108
Spang, Joseph 97
Spanish 7, 124, 125, 182, 190
Spatafora, F. 228
Spence, Sheila 247
spices 6, 39
spindle 6, 95, 98
Spiteri, Danielle 247
split peas, processing 6
Spofford, Jeremiah 247
spoons 202
Sprague, Mary Gabrielle 247
Spratt, D. A. 231
Spring, Steve xii
Spry, Cornelius 109
squares 203
Stalsberg, A. 217
Stanley, Philip R. 247
Stanly County, North Carolina 70
Stansefer, Gabriel 67
Starbuck, Edward 12
Starbuck, Nathaniel 12, 109
Starr, B. F. & Company 91
State Historic Preservation Officers (SHPOs) vii, 3
State Historical Society of Iowa viii
Station Camp Creek 22
Stauffer, Harry 45
staurolite-schist 172
Steffy, Richard 245

Sterling, Virginia xii
Sterrett, Douglas B. 247
Stewart, I. J. 247
Still, Bayrd 247
Stirling, M. 247
Stirling, Scotland 63
Stocker, Sharon R. 229
Stoddard, B. H. 225
Stokes, Julia S. viii
stone boats 39
stonecutters 39, 47, 215
Stonehill College xii, 82, 247
Stoner, Jacob 247
Storck, John 247
Stoyel, A. 247
Stoyel, Alan 247
straight quarter dress 204
Straub Company 96
Straub Mill Company 65
Straub, Isaac 96
Strawhacker, William 248
striking hammers 202, 203, 215
Stringer, Samuel 33
Strohecker, George 108
Strohecker, Gottleib 48, 108
Strohecker, John, Jr. 48, 108
Strohecker, John, Sr. 48, 108
Strohecker, Samuel 48, 108
Sturbridge, Massachusetts viii
subterranean millstone quarries 215
Sudan 197
Suddern Farm 161
Suède Quarry 193
Suesserott, J. L. 239
Suffield, Connecticut 10, 15, 16, 17
Sulfolk County, Massachusetts 91
Sullivan, George M. 248
Summer, William M. 248
Summers, Cornelius 25
Summers, Eli 11
Summit County, Ohio 81
Sumner County, Tennessee 76
Surface, John Snyder 105
Surface, Robert G. 106, 209
Surface, Robert Huston ix, 53, 106, 202, 207–210
Susa Quarry 171
Susquehanna River 84, 109
Sussenbach, Tom 28, 84, 248
Swain, Frank K. 248
Swaine, John 12, 248
Swallow, G. C. 248
Sweden xi, xiv, 4, 124, 125, 164, 190–193, 212
Swedish 124, 190
Sweet, Palmer C. x
Sweringer 11
Swick, Ray 247
Swift, Michael 39, 248
Swisher, Jacob A. 248
Switzerland xii, xiv, 4, 7, 124, 125, 145, 194–196, 212
Symposium on Ohio Valley Urban and Historic Archaeology xii
Syria 198
Syvritos Kephala 197
Taché, J. C. 248
Taconic slates 41
Takaoğlu, Turan 248
Talbert, Charles G. 248
talc 6, 38
Tallahata burstone 62
Tallapoosa County, Alabama 54
tan bark mills 6, 145
Tanner, John 16, 68, 109
Tarbox, Ensine Samuell 11
Tarragona, Spain 182
Tarragona Quarry 182
Taylor, Ed 58, 107
Taylor, George 58, 107
Taylor, J. E. 58
Taylor, Samuel 17, 26, 109
Taylorsville, Tennessee 79
Tchilakadze, M. 248
Teague, Walter D. 247
Tees Valley of York, England 151, 157
Telford, England viii
Telford, Thomas 248
Tennessee 19, 20, 48–49, 54, 60, 76–77, 84, 86, 88, 90, 98, 108
Tennessee Historical Commission viii
Tephrite querns 157
Teter's Creek 47
Teunissen, Jan 10
Texas viii, 20, 82
Thames Exchange Site 157
Third Magnesian Limestone 69, 70
Thomas, Émilie 222
Thomason, Neave & Brothers 46
Thorpe, Richard S. 251
Thue, Johs. B. 248
Tidewater Atlantic Research viii
Tilleda, Hungary 169
Tioga County, Pennsylvania 47, 48, 108
Tioga State Forest 48
Tippah County, Mississippi 86
Tipton, John 13, 14
Tipton County, Tennessee 86
Todd, J. 91
Todd Lake 85
Toler Millstone Quarry 25, 30, 32
Tollon, Francis 241
Tolski, Otto Paul 40
Tomalin, D. J. 248
Tomlinson, Tom D. 248,
tool catalogues 205
tool related injuries 207–208
tools: for millstone cutters 4, 202–203; for sharpening millstones 203–206
Torrecilla en Camerous Quarry 187
Totternhoe Roman villa 157
trachyte 169, 196
Trapp, New York 32
Trask, John B. 248
Treadway, Moses 109
Treadway, Peter 109
Triboulet, C. 248
Triboulot, Bertrand 222
Trillek Quarries, Monmouthshire 159
Trondheim, Norway 172
Trough Brook Quarry 161
Trousdale County, Tennessee 48
Truax, J. W. 248
Tubb, J. 238
Tucci, P. 248
Tucker, D. Gordon viii, 249
Tucker, Gertrude E. 224, 225
Tuomey, M. 249
Tuomey, Michael 20
Turkey 4, 125, 196, 197, 212
Turkey Creek 58
Turkey Hill Millstone Quarry x, 43–45, 48, 108, 215
Turkish 167, 199
Turland, Michel 237
Turner, Hugh 46, 47, 108
Tuttle, Charles R. 249
Tydal, Norway 172
Tylas, Mr. 13

Ulster Community College 40
Ulster County, New York x, 16, 20, 30–32, 38, 39, 41, 94, 102, 203
Ulster County Historical Society x
underground mines 6, 136–138, 140, 142–149
Union County, Illinois 66
Union County, Tennessee 49
United Kingdom xii
United States viii, ix, 2, 3, 6, 9–118, 212
U.S. Geological Survey xiii
U.S. millstone producers 101–111
U.S. Bureau of Mines and Minerals 100, 101, 118, 119, 215
University of Birmingham, England viii
University of Delaware Library 59
University of Grenoble II xiii, xiv, 2
University of Kentucky xiv, 3
University of Michigan 4
University of North Carolina Library 59
University of Southampton ix
Uppsala area, Sweden 191, 193
urban millstone factories 4, 88–99
Utica, New York 16, 91–94, 105
Utica French Burr Mill-Stone Manufactory 93
Utter, William T. 249

Valentin, Frédérque 231
Valero, J. F. 249
Valette, Bernard 225
Valley of Virginia 53
Valmeriana Quarry 170
Van Cleave, Samuel 86
Van Cleave Mills 16
Vandike and Keller 67
van Doorminck, Frederick H. 245
Van Etten, James S. 102
Vargiolu, R. 243

Venturelli, G. 223
Venturo, D. 237
Vermont 19, 20, 54, 109, 110, 114
Vernon, Kentucky 26
Versailles, Kentucky 67
Viking era 173, 174, 177, 191, 191, 193, 212
Villaroya Quarry 187
Villet, Damien 218
Vincent, Francis 249
Vindolando Fort 154
Viner, L. 239
Vinton County, Ohio xi, 15, 16, 17, 74–76, 109
Virginia ix, xiii, 19, 20, 39, 49–53, 54, 60, 65, 77, 86, 90, 105–106, 107, 109, 110–111, 114–115, 207–210
Virginia Abrasive Company 105
Virginia Gazette 120–121
Virginia Millstone Company 105
Vitali, Daniele 244
volcanoes 140, 150
Volterra, V. 249
Vrettou-Souli, Margarita xii, xiii, 7, 167–168, 249

Waage, Astrid xii, xiv, 172–178
wagons 23, 44
Wailes, B. L. C. 249
Wait, G. 222
Waldshut, Germany 148, 150
Waldshut-Tienge Quarries 145
Wales xii, 151, 164–165, 212
Walker, George 15, 97, 108
Wallace, W. W. 98
Wallcut, Thomas 13, 249
Wallis, F. S. 249
Walton, James 249
Wamsutta Formation 81
War of 1812 19, 65, 120
Ward, Owen H. viii, xi, xii, 54, 127, 233, 249, 250
Ward's, W. & E., French Burr Mill-Stone Manufactory 96
Ware, J. E. 62
Ware Millstone Quarry 26, 29, 30–32
Ware's Mill 62
Warren County, Kentucky 84
Warrior River 20
Warush, Lewis x, 46
Washburn, C. C. 99
Washington, North Carolina viii
Washington County, Missouri 69, 85
Waterford, Connecticut 55
Watertown, New Hampshire 79
Watkins, Richard 71, 73
Watson, J. S. 251
Watson, Thomas 250
Watts, Martin xii, 250
Watts, Sue xii, 250
Wawarsing, New York 38, 102
Waynesboro, Pennsylvania x
Weaver, Valerie 250
Web, John 10
Webb, Mr. 94
Webb, P. C. 251
Webb, William S. 250
Webb & Batchelder 56
Webster County, Missouri 69
wedges 32; and feathers 45, 202; and shims 202; and slips 202
Wefers, Stefanie xii, 250
Weidman, S. P. A. 42, 108
Weidmann, D. 250
Weinhold, Annie 44
Weinhold, William 43, 108
Weise, Arthur James 250
Weisener Quartzite 21, 71
Welch millstones viii, 120, 164, 165, 215
Welfare, A. F. 250
Wesler, Kit xii
Wessex, England 161, 197
West, Karl 81
West Blatchington, England 154
West Fork of Howards Creek 24
West Germany 140, 141, 148
West Indies 17, 55
West Sussex, England 151
West Virginia 19, 86, 101, 106, 107, 110
Westerly, Rhode Island 60
Wetmore, Alphonso 250
Weybridge, England 151
Wharncliffe Quern Workings 159, 161, 162
wheat 6, 12, 38, 67, 79–80, 117, 205; chaffing 6
whiskey 17, 26, 75, 80
White, D. 250
White Clay Creek State Park 83
white flint buhrs 59
White River 22
Whitley County, Kentucky 22, 27
Whittle Hills 159
Whortleberry Hill, Connecticut 78
Whyte, Samuel G. 84
Wilkerson, Joseph 25
Wilkes & Dillingham 89
Wilkesboro, North Carolina 59
Wilkie, Aitken 250
Wilkshire, England 154
Will County, Illinois 87
Willcox, G. 250
Williams, Albert, Jr. 250
Williams, David xii, 250
Williams, Edward T. 234
Williams, Joseph S. 251
Williams, Zebediah 10
Williams-Thorpe, Olwen 226, 242, 243, 251
Williams & Company 251
Williams County, Tennessee 76–77
Wilmington, Delaware x
Wilmington, North Carolina 42
Wilson, Catherine 251
Wilson, John 27
Wilson, Judy xiii
Wilson Patent Stone Dressing Machine 64
Winchester, James 84
Winchester, Kentucky 23, 24
Windy Clough Quarry 161
Winnall Down, Winchester 161
Winston County, Alabama 21, 61
Winthrop, John 9
Winwood, H. H. 251
Wisconsin 61, 88, 98–99
Wisdom, L. 251
Wise, Annie 44
Wise, Mrs. William K. 44
Wissler, Benjamin 42, 108
Witt, C. Wells 239
Wolf, Carol L. 229
Wolfe Creek 13, 18
Wolfram, P. J. 237
wood and emery millstones 200–201
Woodbury, Connecticut 11, 17
Woodford County, Kentucky 16, 17, 66, 67, 68, 109
Woodruff, Elias 10
Wooldridge, J. 245
Worcester, Massachusetts 56
Worcester Vocational High School 56
Worm's Heath 157
Worsham, Gibson 251
Worth, Henry B. 251
Worth, William 12, 109
Wright, J. C. 89
Wright, M. Elizabeth 251
Wright, Nathaniel 9
Wright County, Missouri 69
Würenlos Quarry 194–195
Wyoming viii

Xenophontos, C. 228, 251

Yarncliff, England 159, 162
Yorkshire, England 120, 151, 159, 161, 162
Young, Alexander 251
Young, Andrew W. 251
Youngs, Benjamin S. 84
Yugoslavia 197, 198, 212

Zahouani, H. 243
Zaleski, Ohio 75
Zanca, J. 217
Zanca, Jean 221
Zanesville, Ohio 71
Zanesville, Ohio, Board of Trade 251
Zanettin, E. 237
Zanettin, Eleconora 223
Zerfass, Samuel G. 251
Zirkl, E. J. 251
Zupanc, Stanko 180
Župančič, M. 234